INTRODUCTION TO

THERMODYNAMICS

AND

HEAT TRANSFER

INTRODUCTION TO
THERMODYNAMICS
AND
HEAT TRANSFER

DAVID A. MOONEY

Mechanical Engineer
Jackson and Moreland, Engineers
Formerly Assist. Prof. Mechanical Engineering
Massachusetts Institute of Technology

Englewood Cliffs
PRENTICE-HALL, INC.
1955

Library of Congress Card Catalog Number: 55-9769

49968

PREFACE

This is an introductory text on the principles and applications of engineering thermodynamics and elementary heat transfer. The reader is presumed to have a knowledge of college physics and mathematics, including elementary calculus. With the exception of the chapter on heat transmission, the material in this book is derived directly from my earlier one, *Mechanical Engineering Thermodynamics*, mainly by omitting sections that appeared to be unnecessary for a shorter course. Most of the omissions are from the chapters on applications.

In this book, as in the longer one, thermodynamics is treated as a logical formulation of facts known from experience, with emphasis on the generality of the subject. The purpose of the book is to help the reader toward understanding the subject, rather than to confer facility in solving type problems. Nevertheless, since experience shows that understanding must be built up from a foundation of familiarity with small details, the derivations and basic applications are explained in considerable detail, and many worked examples are given. As a further aid to understanding, the following types of concepts, often confused in a reader's mind, are carefully identified as they appear: facts from experience; principles accepted because they are confirmed by experience; arbitrary definitions, chosen on the basis of their utility for specific purposes; approximations like the gas laws, used because physical data do not conform to simple mathematical formulations; conventions, devices for facilitating thought and communication.

The first part of the book deals with work, heat, and the First Law as applied to general systems. The steady flow equations are then developed to provide material for application of the basic concepts.

The Second Law is presented as a formal statement of the existence of irreversibility, and the thermodynamic temperature scale is defined as a corollary of the Second Law. Entropy is defined, and stated to be a property, but no proof is given.

The properties of substances are presented first from a general viewpoint, using plots to show the relationships among gaseous, liquid, and solid states. Then the special methods of handling property data are explained, with reference to steam table data, gas law data, properties of gaseous mixtures, and gas table data.

In the chapters on applications there is less detail than in the earlier book, but within the narrower scope the object is the same: to show the application of thermodynamics in heat engineering, without going into details of plant engineering, but taking care to point out that thermodynamics shows only one aspect of the total problem.

In the chapter on heat transmission I have tried to combine enough explanation to give a reasonable physical picture, with enough formulas and data to permit some approach to realistic problems. A single chapter, written from this approach, must be sharply limited in scope, but there is sufficient material for the usual short treatment. Because most students now encounter dimensional analysis in fluid mechanics, the development of the dimensionless groups has been omitted to save space.

While it is desirable in any case for an instructor to discuss in more detail, as opportunity arises, various points that are not covered in detail in the text, this is especially desirable with a condensed book. My longer book would provide some material for this purpose, and the references given at the ends of the chapters should also be consulted.

During my years of study and teaching at Massachusetts Institute of Technology, which led to the writing of this book, many instructors and colleagues have helped me toward knowledge and understanding of thermodynamics and its exposition. Of these, I am especially indebted to Joseph H. Keenan for fundamental training and inspiration, and to Warren M. Rohsenow and Carl L. Svenson. In the preparation of the original manuscript for the basic chapters, I was greatly aided by the penetrating criticism of David H. Fax, who acted as a consultant for the publisher. Acknowledgment is given in the text pages to the authors, publishers, and manufacturers from whose publications data or illustrations have been used. Most of the line drawings were made by Donald E. Keyt.

David A. Mooney

Dedham, Mass.

CONTENTS

BACK-COVER CHARTS AND GRAPHS

MATTER AND ENERGY

All that we know about matter relates to the series of phenomena in which energy is transferred from one portion of matter to another, till in some part of the series our bodies are affected, and we become conscious of a sensation. . . .

Hence we are acquainted with matter only as that which may have energy communicated to it from other matter, and which may, in its turn, communicate energy to other matter.

Energy, on the other hand, we know only as that which in all natural phenomena is continually passing from one portion of matter to another.

James Clerk Maxwell
Matter and Motion, 1877

A PRECEPT FOR ENGINEERS

Economy of fuel is only one of the conditions a heat engine must satisfy; in many cases it is only secondary, and must often give way to considerations of safety, strength and wearing qualities of the machine, of smallness of space occupied, or of expense in erecting. To know how to appreciate justly in each case the considerations of convenience and economy, to be able to distinguish the essential from the accessory, to balance all fairly, and finally to arrive at the best result by the simplest means, such must be the principal talent of the man called on to direct and co-ordinate the work of his fellows for the attainment of a useful object of any kind.

S. Carnot
Reflexions sur la Puissance Motrice du Feu, 1824
Translation by H. L. Callendar

Chapter 1

INTRODUCTION

Of all the technical achievements of man the most important is the capacity to control large quantities of energy. Only through the wide availability of power, or controlled energy, can the results of science be put to effective use for the general population. Industry, transportation, agriculture, even scientific research itself, if denied the use of power-driven machinery, would still be struggling along the same roads they followed centuries ago.

The purpose of this book is to explain a method of analysis, based on the science of thermodynamics, for processes involving the control or use of energy.

1-1 Energy. What is energy? It is a symbol or concept which is useful in discussing the interactions of physical bodies. Experience shows that a physical effect such as the hoisting of a weight can be accomplished by maintaining a flow of electric current through a motor, or by maintaining a flow of steam through a turbine, or by maintaining a fire under a hot air engine; furthermore there are quantitative relations between the amounts of electricity, steam, or fuel used, and the amount of weight raised through a given elevation. Something must exist, common to the different physical systems, by which they can all accomplish the same effect. It is therefore assumed that *energy* is transmitted from the electric conductors to the hoisting mechanism, or from the steam to the hoisting mechanism, or from the fire to the hoisting mechanism. The justification for this assumption is found in the agreement between experimental facts and the results of reasoning based upon the assumption. The existence of energy is inferred from the physical effects of *energy transfer*. Experience shows that all cases of energy transfer may be classified as either heat transfer or work transfer. A fundamental difference exists between these two forms of energy as will appear in later chapters.

Although energy can be directly identified only through its transfer from one body or system of bodies to another, yet the possibility of storage of energy within bodies must be considered, because something which can be quantitatively accounted for during its transfer from one place to another must either be stored or created and destroyed before and after its transfer. The fact that energy can be stored in bodies is the substance of the First Law of Thermodynamics. Work, heat, and the First Law will be the subjects of the first part of this book.

The engineering processes which form the subject of the greater part of this book, involve the conversion of energy from one form to another, the transfer of energy from place to place, and the storage of energy in various forms. These operations are accomplished by utilizing a *working substance* which can absorb, hold, or release energy depending upon the conditions imposed upon it. The working substance is almost invariably a fluid (liquid or gas) because fluids are easy to transport from place to place in pipe systems and because fluids are conveniently made to absorb and release energy. Examples of working fluids are: fuel, air, and combustion products in internal combustion power plants; liquid water and steam in steam power plants; ammonia in refrigeration plants; air and water in air-conditioning plants.

1-2 Thermodynamics. Thermodynamics is the science of energy transfer in its relation to the physical properties of substances. It is concerned with the general laws of energy applying to all types of systems, mechanical, electrical, and chemical. It is also concerned with the overall effects of energy transfer rather than with the physical mechanisms by which these effects are accomplished.* Engineering thermodynamics is the special development of thermodynamics as applied to mechanical engineering operations such as power generation from fuels, refrigeration, and air-conditioning.

1-3 Thermodynamic systems. A thermodynamic system is a body or group of bodies upon which attention is to be concentrated in the analysis of a problem. The whole structure of thermodynamic analysis is based upon an accounting scheme in which energy quan-

* Thermodynamic analysis is not applied to the characteristics of individual molecules but only to the statistical characteristics of large aggregations of molecules. Hence it is sometimes said that thermodynamics deals with *macro*scopic phenomena rather than *micro*scopic phenomena. See Zemansky, M. W., *Heat and Thermodynamics*. New York: McGraw-Hill, 1943, chap. 1.

tities are checked into or out of the system. Hence it is necessary to choose a definite system and to keep it under close scrutiny throughout the course of the operation being analyzed in order that all energy quantities involved shall be properly entered in the account. As the term is used in this book, a system consists of the same bodies throughout an operation; no material may be added or taken away. It is usually convenient to visualize a system as enclosed within an imaginary envelope which may change its size, shape, or location but which must always contain the system and nothing but the system. The selection of a suitable system is the most important step in a thermodynamic analysis; in the examples appearing in the later chapters of this book some of the factors which determine the choice of a system will become evident.

1-4 Properties. Every system has certain characteristics by which its physical condition may be described; examples are mass, volume, pressure, temperature, and electrical resistivity. Such characteristics are called properties of the system.

1-5 State changes, processes, and cycles. When all the properties of a system have definite values, the system is said to be in a definite *state*. When a different value exists for any one or more of the properties, the system is in a different state. Any operation in which one or more of the properties of a system changes is called a *change of state*. The *path* of a change of state is the succession of states passed through during the change of state. When the path is completely specified, the change of state is called a *process*. A process is often specified by stating that some particular property remains constant; an example is a constant-pressure process. Another way of specifying a process is by some algebraic relation between properties; an example is a process in which the product of pressure and volume remains constant. A thermodynamic *cycle* is a series of state changes such that the final state is identical with the initial state. Observation of a system before and after the execution of a cycle would reveal no difference in any property of the system, although all the properties might have had different values at some time during the execution of the cycle.

1-6 Units and definitions. The system of units used in this book is the pound-second-foot system. The unit of mass is the pound mass, which is determined by reference to a carefully preserved standard. The unit of force is the pound force, which is the force required to accelerate a pound mass at the standard rate of 32.17 ft/sec². These units are related by Newton's Second Law of Motion, which is expressed by

$$F = kma \tag{a}$$

where F = force in pounds force (lbf)

m = mass in pounds mass (lbm)

a = acceleration in ft/sec²

and k = a proportionality constant $\left(\frac{\text{lbf}}{\text{lbm}} \times \frac{\text{sec}^2}{\text{ft}}\right)$

In the system used here the units of force and mass are arbitrarily chosen so that the weight of a body at the earth's surface will have the same numerical value as its mass. At the earth's surface the acceleration of gravity is 32.17 ft/sec². So, if F is numerically equal to m, then $k = 1/32.17$. The statement of Newton's law then becomes

$$F = \frac{ma}{32.17} \tag{b}$$

which contrasts with the usage of physicists who arbitrarily assign the value unity to the constant k. Note that the 32.17 in Eq. (b) is not an acceleration, nor does it change for different locations.* The weight w of a body of mass m, at a location where the acceleration of gravity has any value g will be

$$w = \frac{mg}{32.17} \tag{c}$$

However, at any elevation normally encountered, even in high-flying airplanes, the value of g differs by only a fraction of a percent from 32.17; hence for practical purposes $w = m$ (numerically). (The symbol g_0 will be used for the dimensional constant 32.17 ft lbm/lbf sec².)

The unit of time for all dimensionally correct equations is the *second.*

The unit of length is the *foot.*

The unit of volume is the *cubic foot.* However it may be convenient to work in terms of *specific volume,* which is the volume of a unit mass (dimensions, cu ft/lb).

Density is the reciprocal of specific volume (dimensions, lb/cu ft).

* See Hawkins, L. A. and S. A. Moss, *Mechanical Engineering,* Vol. 68, No. 2, 1946; pp. 143–144, 660–662.

Pressure is the normal force exerted by a system against a unit area of its bounding surface. Pressures acting outward against the confining walls are taken as positive. Thus, the pressure of a body in compression is positive, but the pressure of a body in tension is negative. In this book only positive pressures are considered, since fluids do not normally sustain appreciable tension.

The basic units of pressure in this book are lb/sq ft. Unfortunately, standard practice in engineering calls for the use of a variety of units. Pressures are variously stated in lbf/sq in. (psi), inches of mercury, and feet, or inches, of water. Since the density of liquids varies with temperature, pressures in terms of liquid head imply a liquid of standard density. Taking the density of water as 62.4 lb/cu ft (corresponding to ordinary room temperatures), we find that

$$1 \text{ ft water} = 62.4 \text{ lb/sq ft} = 0.433 \text{ psi}$$

Since the specific gravity of mercury at room temperature may be taken as 13.6, a pressure of

$$\begin{aligned} 1 \text{ in. mercury} &= 62.4 \times 13.6 \times \tfrac{1}{12} \\ &= 70.7 \text{ lb/sq ft} = 0.491 \text{ psi} \end{aligned}$$

When pressures are given in atmospheres, a standard atmosphere of 760 mm of mercury (29.92 in. of mercury, or 14.696 psi) is implied. Most instruments indicate pressure relative to the atmospheric pressure, whereas the pressure of a system is its pressure above zero, or

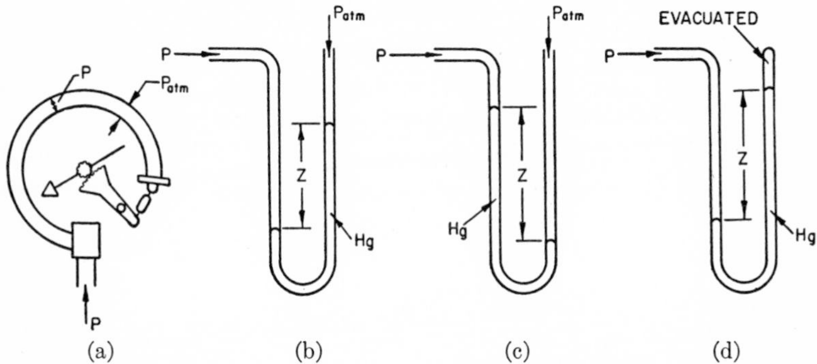

Fig. 1-1. Pressure gages. (a) Bourdon gage measures difference between system pressure inside tube and atmospheric pressure. (b) Open U-tube indicating gage pressure. (c) Open U-tube indicating vacuum. (d) Closed U-tube indicating absolute pressure. If P is atmospheric pressure this is a barometer

relative to a perfect vacuum. The pressure relative to the atmosphere is called *gage pressure;* the pressure relative to a perfect vacuum is called *absolute pressure.*

$$\text{absolute pressure} = \text{gage pressure} + \text{atmospheric pressure}$$

When the pressure in a system is less than atmospheric pressure, the gage pressure becomes negative but is frequently designated by a positive number and called *vacuum.* Figures 1-1 and 1-2 show relationships among the various pressure scales.

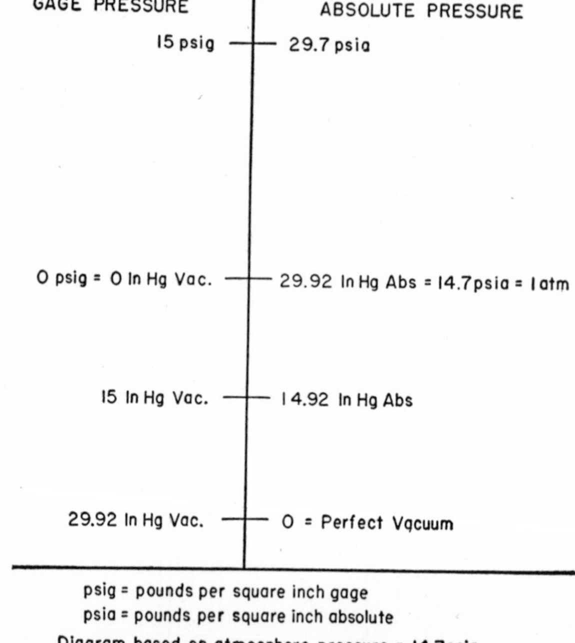

Fig. 1-2. Absolute pressure and gage pressure.

PROBLEMS*

1-1. Find the weight in lbf of a 5 kilogram mass in a location where the acceleration of gravity is 32.00 ft/sec². How much force, in lbf, would be necessary to accelerate the mass at the rate of 32.17 ft/sec²?

* Conversion factors and physical data for these and problems of succeeding chapters may be found in the Appendix.

1-2. The acceleration of gravity is given as a function of elevation above sea level by $g = 980.6 - 3.086 \times 10^{-6}H$, where g is in cm/sec^2, and H is in centimeters. If an airplane weighs 10,000 lb at sea level, what is the gravity force upon it at 20,000 ft elevation? What is the percentage difference from the sea-level weight?

1-3. Convert the following readings of pressure to psia, assuming the barometer reads 29.92 in. mercury (Hg): 76 in. Hg gage; 32 psig; 16 in. Hg vacuum; 3.3 ft H_2O gage.

1-4. A certain pressure is measured by a water-column gage in which the water is at 85°F. If the density of water at 39°F is taken as an arbitrary standard, what will be the percentage correction factor to reduce the reading to a standard density reading? At what water temperature would the correction be 1 percent? Obtain the specific volume of water from the column headed "Sat. Liquid" in the steam table in the Appendix.

1-5. The volume coefficient of expansion of mercury at ordinary temperatures is 101×10^{-6} cu ft/cu ft deg F. If pressures in standard inches of mercury are understood to be in terms of mercury density at 32°F, at what temperature would a mercury column be 1 percent higher than a standard mercury column exerting the same pressure?

1-6. Assume that the pressure p and the specific volume v of the atmosphere are related according to the equation $pv^{1.4} = 75{,}500$, where p is in lb/sq ft, and v is in cu ft/lb. If the acceleration of gravity is constant at 32.17 ft/sec^2, what is the depth of the atmosphere necessary to produce a pressure of 14.7 psia at the earth's surface? (Consider the atmosphere as a fluid column.)

1-7. A common fallacy regarding the engineering system of units is that the engineering form of Newton's Second Law is

$$F = \frac{w}{g}a$$

where w is weight and g is the local acceleration of gravity. Explain the distinction between this equation and equation (b) on page 4. Failure to understand the engineering system of units is often the result of ignorance of this distinction.

Chapter 2

WORK

2-1 Work—definition. Work, one of the basic forms in which energy is transferred, is encountered in the science of mechanics as the result of the action of a force on a moving body. A force is a means of transmitting an effect from one body to another, but a force is not a form of energy, because it never accomplishes a physical effect except when combined with motion. An effect such as the raising of a certain weight through a certain distance can be accomplished by using a small force through a large distance, or a large force through a small distance. Similarly the velocity of a body may be increased a certain amount by using a small force through a large distance, or a large force through a small distance. In every case however, the *product* of force and distance is the same if the same effect is produced. The following definition of work is therefore given in mechanics:

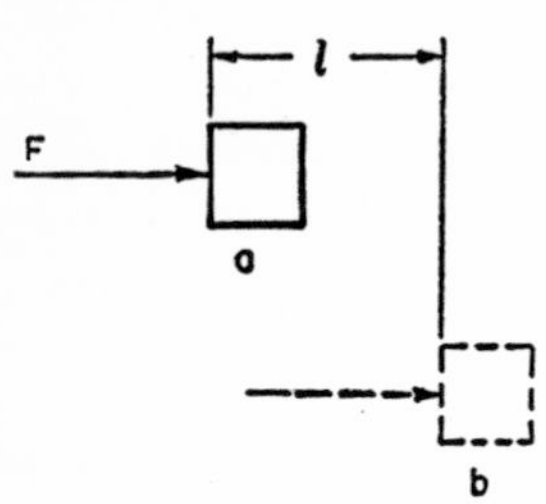

Fig. 2-1.

Work is done by a force when the force acts upon a body moving in the direction of the force.

The magnitude of the work is given by the product of the force and the distance moved parallel to the force.

In Fig. 2-1 the body shown moves from a to b while acted on by the horizontal force F. The horizontal component of the distance moved is l. The work, W, is given by

$$W = Fl$$

The positive directions of force and motion are taken to be the same. Therefore, the work will be positive if the force and displacement have the same direction.

The action of a force through a distance (and the equivalent action of a torque through an angle) is called *mechanical work*, because other forms of work can be recognized in thermodynamics. Energy transfer by an electric current is one example. Most work in the scope of this book is mechanical work. Nevertheless, a definition of work in thermodynamics should allow for forms other than mechanical work, and this definition should be stated in terms of a system. It is also desirable to state the definition in terms of a simple physical picture such as the raising of a weight. The following definition of work is therefore given:

*Work is transferred from a system by a given action of the system if the total external effect of the given action can be reduced to the raising of a weight.**

The work is measured by the amount of weight that would be raised through a specified height if the total external effect were the raising of a weight. Work transferred from a system is taken as positive; conversely work transferred to a system is negative with respect to that system. In using the definition of work, it is assumed that friction and resistance can be eliminated from a hypothetical apparatus upon which the system acts. If, using hypothetical frictionless mechanisms, the total effect of an action can be converted to the raising of a weight, then that action constitutes a work transfer. The reason for requiring that the *total* effect of the action must be reducible to the raising of a weight is that another form of energy transfer from a system, namely heat, may cause the raising of a weight as *part* of its external effect; but even when frictionless mechanisms are permitted, the *total* effect of heat transfer can never be reduced to the raising of a weight.

2-2 Application of the definition of work. Work exists only as an effect transferred across the boundary of a system. Taking a locomotive as a system, it does work upon the train it draws because the force transmitted through the draw-bar as the train moves could have been used for the sole effect of raising a weight. The work is positive because it is done *by* the system. If the train were taken as the system, the work would be unchanged in magnitude but would be negative because it would be done *on* the system. If the locomotive and train together were taken as the system, the work would be zero because the force involved would not act across the boundary of the system. These examples illustrate the importance of working with a definite system.

* An external effect is an effect on things outside the system.

Consider a cylinder and piston machine, Fig. 2-2. Let the gas in the cylinder be a system having initially the pressure p_1, volume V_1, and length l_1. Now allow the piston to move out to a new position and let the new pressure, volume, and length of the system be respectively p_2, V_2, and l_2. At any point in the travel of the piston let the pressure be p, the volume V, and the length l. It is desired to

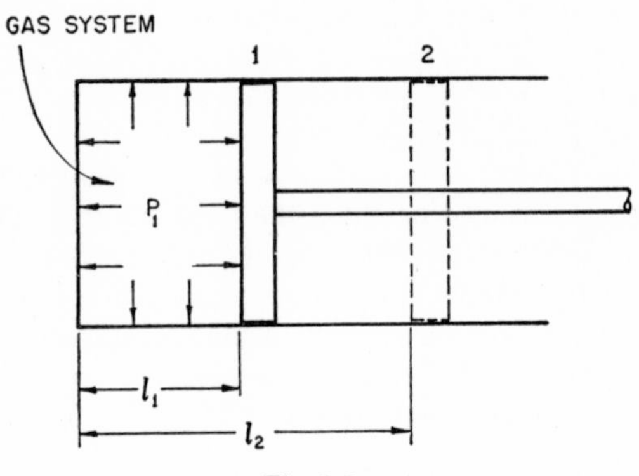

Fig. 2-2.

find the work done by the system as the piston moves from position 1 to position 2. Examining the boundaries of the system, it is seen that the piston is the only boundary that moves. Therefore no work can be done by the pressure force against the other walls, but work will be done upon the piston when it moves. The force against the piston will probably vary as the piston moves, but it may be written, for any position of the piston, as

$$F = pa$$

where F is the pressure force on the piston and a is the area of the piston. For an element of piston travel the element of work may be written

$$dW = Fdl$$

because the motion of the piston dl is parallel to the force upon it. Then

$$dW = pa\,dl$$

But $a\,dl$ may be written as dV, an element of volume; then

$$dW = p\,dV$$

and

$$W = \int_{V_1}^{V_2} p\,dV$$

If p is known as a function of V, W can be evaluated from this equation.

The work done by a system *as a result of a change in volume* is given in general by $\int_{V_1}^{V_2} p\,dV$ for any system in which the pressure p is uniform over the whole surface of the system at all times (though it may vary with time). The pressure will not be uniform over the surface of a system if gravity or acceleration forces are present.* However, with gaseous systems in particular, there are many cases in which the effects of gravity and acceleration forces are small; in such cases the work due to a change in volume may be calculated with good accuracy by the integral of $p\,dV$.

2-3 Work in a stationary system. A system in which motion is negligible is called a *stationary system*. This definition permits relative motion of parts of the system, incidental to state changes, but excludes any motion, in translation or rotation, of the system as a whole.† Work may cross the boundary of a stationary system as the result of a volume change, and such work is measured by the integral of $p\,dV$ if pressure equilibrium prevails in the system. The integral of $p\,dV$, however, does not necessarily represent all the work transferred across the boundary of a stationary system in a given change of state. Consider the stationary fluid system shown in Fig. 2-3. If the system is a real fluid, rotation of drum B in the fluid can be maintained only by the continuous action of a torque (from the weight w) upon the rotating shaft. The work thus done upon the drum is transferred into the fluid system through the friction forces at the surface of the drum.

* Surface tension, magnetism, and electricity may sometimes affect the pressure distribution in fluid systems, but such effects are outside the scope of this development. In solid systems uniform pressure is the exceptional case.

† The pressure of a gas on a moving wall depends on the velocity of the gas molecules relative to the wall. The average velocity of a gas molecule at a temperature of 80°F is about 1600 fps, and at higher temperatures is greater. Since the speeds of pistons seldom exceed 50 fps, pistons may usually be considered stationary relative to the gas molecules, and the gas confined by a piston may be considered stationary, insofar as pressure is concerned.

If a paddle wheel is substituted for the drum, a torque is still necessary to maintain rotation of the paddle wheel, and the work involved is transferred to the fluid. Work introduced into a system by such means as the friction drum or the paddle wheel may be called *paddle-wheel work* or *stirring work*. Such work cannot, in general, be measured by a function of the properties of the system but must be evaluated from some other knowledge of the circumstances.

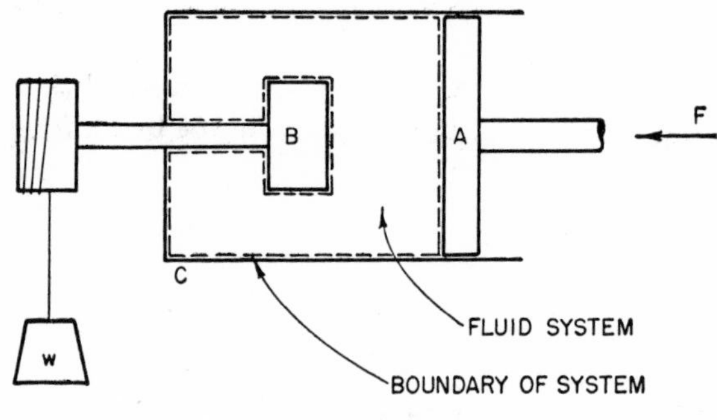

Fig. 2-3.

It is of interest to consider the effect of using an ideal frictionless fluid for the system of Fig. 2-3. The work due to a volume change would still be given by the integral of $p\,dV$ but there would be no paddle-wheel work. Experience shows that as the frictional effects in a fluid approach zero the torque required to maintain steady rotation of a paddle wheel approaches zero. Hence with a frictionless fluid in a stationary system the paddle-wheel work is zero. The only work transfer that will occur, to or from such a system, is the work transfer due to normal forces on moving boundaries. If pressure equilibrium prevails, this work will be measured by the integral of $p\,dV$.*

2-4 Work in various processes—graphical representation. Since many problems involve work given by $\int p\,dV$, it is convenient to make a graphical presentation on the coordinates of pressure and volume. The integral will appear as an area upon such a plot. Figure 2-4 represents a general process in which a stationary system,

* Work transfer by means of surface tension, magnetism, and electricity is possible but is excluded here.

having initially the pressure and volume indicated by point 1, changes, by the path shown, to a state in which the pressure and volume are indicated by point 2. If the shape of the path is known as a function of p and V, the $\int p\,dV$ can be computed by calculus; and if the path is an experimental plot for which no mathematical equation is known, the $\int p\,dV$ can still be computed by graphical integration.

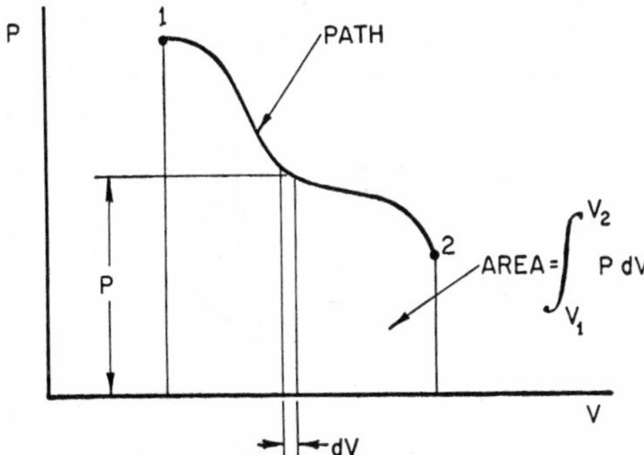

Fig. 2-4. Graphical representation of a process.

Several examples are given below of the functional and graphical representation of $\int p\,dV$ for various processes in a stationary system.

(a) *Constant-Pressure Process.*

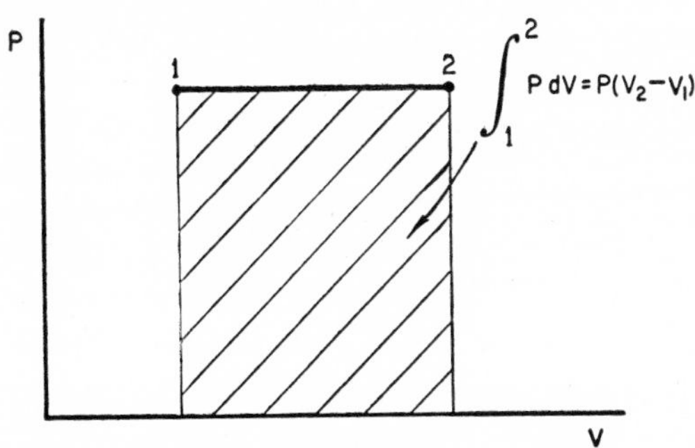

Fig. 2-5. Constant-pressure process.

For this case

$$\int_1^2 p\,dV = p(V_2 - V_1)$$

(b) *Constant-Volume Process.*

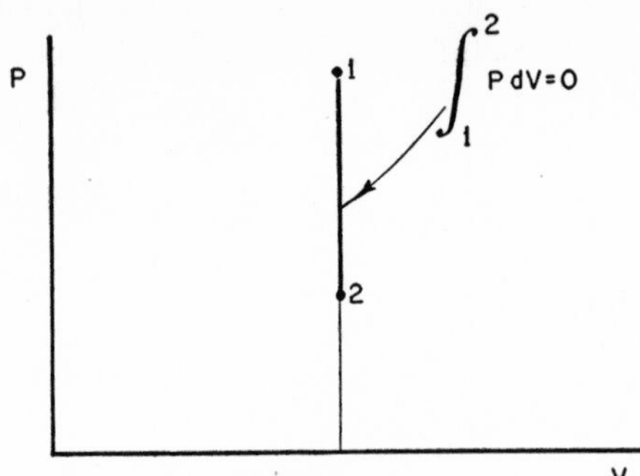

Fig. 2-6. Constant-volume process.

For this case
$$\int_1^2 p\,dV = 0$$

(c) *Process in Which pV = Constant.* (This and the following relation between p and V are functions which approximate the relations observed by experiment in certain types of processes with certain fluid systems.)

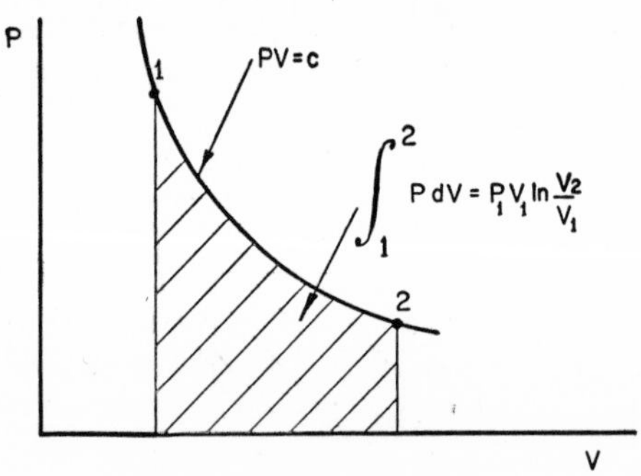

Fig. 2-7.

For this case
$$\int_1^2 p\,dV = p_1V_1(\ln V_2 - \ln V_1)$$
$$= pV \ln \frac{V_2}{V_1} = pV \ln \frac{P_1}{P_2}$$

(d) *Process in Which pV^n = Constant, Where n Is Any Constant.* Figure 2-8 shows the shapes of the pV plots of the function $pV^n = C$

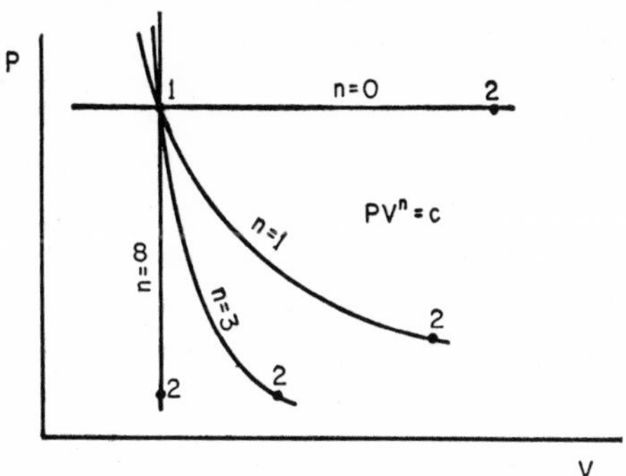

Fig. 2-8.

for positive values of n between zero and infinity. (For negative n the curves occupy the other two quadrants.)

For this case

$$\int_1^2 p\,dV = \frac{p_2V_2 - p_1V_1}{1-n}$$

$$= \frac{p_1V_1}{1-n}\left[\frac{p_2V_2}{p_1V_1} - 1\right] = \frac{p_1V_1}{1-n}\left[\left(\frac{p_2}{p_1}\right)^{(n-1)/n} - 1\right]$$

These equations give an indeterminate result for the case of $n = 1$, but that case was considered in (c) above.

Example 1. A stationary fluid system is subjected to a process in which the pressure and volume change according to the relation $pV^{1.4} = C$. The initial pressure and volume are respectively 100 psia and 3 cu ft, the final pressure is 20 psia. (a) Find the magnitude and direction of the $p\,dV$ work for this process. (b) Is this the total work for the process? Why?

Solution: (a)

$$\int_1^2 p\,dV = \frac{p_2V_2 - p_1V_1}{1-n}$$

All factors are known except V_2.

$$V_2 = V_1\left(\frac{p_1}{p_2}\right)^{1/n} = 3\left(\frac{100}{20}\right)^{1/1.4} = 9.48 \text{ cu ft}$$

$$\int_1^2 p\,dV = \frac{(20)(144)(9.48) - (100)(144)(3)}{1 - 1.4} = +39{,}745 \text{ ft lb}$$

Note the conversion of pressures to lb/sq ft.

(b) It is impossible to say whether this is the total work because the problem does not say whether the process was frictionless or not. If the fluid were frictionless, or if no shearing forces and no paddle-wheel work were involved, the answer to (a) would be the total work.

Example 2. In a stationary fluid system, Fig. 2-3, a paddle wheel supplies work at the rate of 1 horsepower. During a certain period of 1 minute the system expands in volume from 1 cubic foot to 3 cubic feet while the pressure remains constant at 69.4 psia. Find the total system work during the 1-minute period.

Solution:

Paddle-wheel work, W_B:

$$1 \text{ hp for } 1 \text{ min} = 33{,}000 \text{ ft lb}$$

$$W_B = -33{,}000 \text{ ft lb; negative because flowing into system}$$

Piston work, W_A:

$$W_A = p(V_2 - V_1) = (69.4)(144)(3 - 1) = +20{,}000 \text{ ft lb}$$

Total or net work, W:

$$\begin{aligned} W &= W_A + W_B = 20{,}000 - 33{,}000 \\ &= -13{,}000 \text{ ft lb; net work } \textit{to} \text{ the system because negative} \end{aligned}$$

2-5 Indicator diagrams. The plots discussed in the preceding section were all state diagrams; the pressure and volume coordinates of any point represented the corresponding properties of a fluid system at a particular state. Another kind of plot which may represent work is the indicator diagram. This is the trace made by a recording pressure gage, called an engine indicator, attached to a chamber in which the pressure varies. The coordinates of the record in the conventional case are pressure vs. piston travel. An arrangement for making such a record is shown in Fig. 2-9. Referring to this figure, the indicator is connected by a short pipe to the engine cylinder so that the pressure upon the indicator piston A will be always the same as the pressure upon the engine piston F. The indicator piston is loaded by a spring so that it moves in direct proportion to a change in pressure. The motion of the indicator piston causes a pencil on the end of the pencil linkage B to move in a vertical line in direct proportion to the pressure change. The pencil at the end of the linkage B writes upon a strip of paper wrapped about the drum C. If the drum C remained stationary, the record of pressure changes would be a vertical line. However, the drum is rotated about its axis by the cord D, which is connected through a reducing motion E to the piston F of the engine. The result is that the surface of the drum C moves

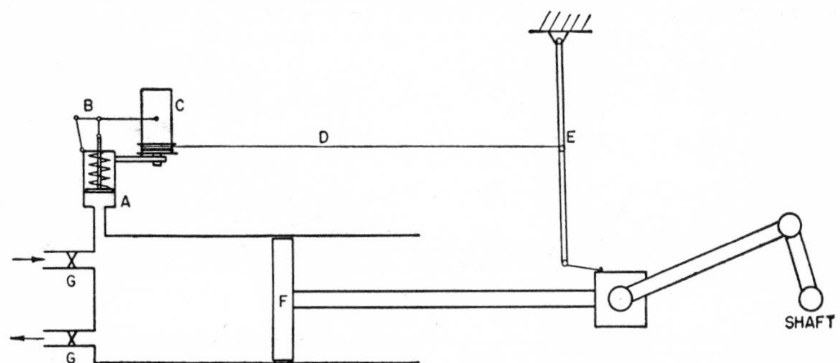

Fig. 2-9. Engine indicator: typical arrangement.

horizontally under the pencil while the pencil moves vertically over the surface, and a plot of pressure upon the piston vs. piston travel is obtained.

Since an engine normally operates in a succession of essentially

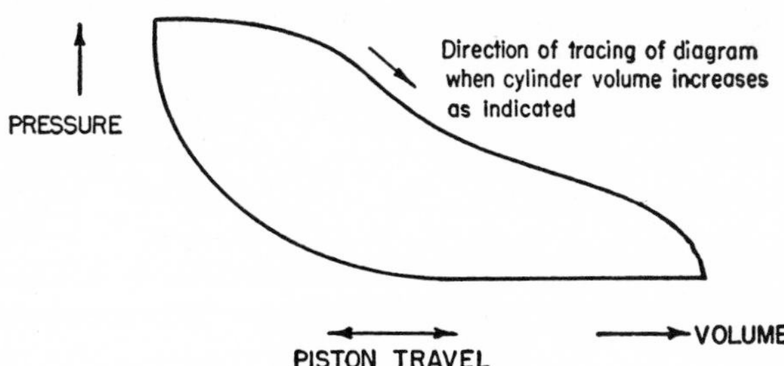

Fig. 2-10. Indicator diagram.

identical machine cycles, the indicator diagram will be a closed curve such as shown in Fig. 2-10. It will be observed that the diagram shows only relative pressures and volumes, there being no reference lines or zero axes. In practice it is customary to provide a pressure reference line at 1 atmosphere by connecting the indicator to the atmosphere and drawing a line at constant pressure of 1 atmosphere before taking the diagram. Figure 2-11 shows the diagram of Fig. 2-10 superimposed upon the atmospheric line.

The ordinary use of the indicator diagram is to obtain the net work done upon the piston during a complete cycle of the machine. The work done upon the piston F is given by $W = \int pa\, dl$ where p

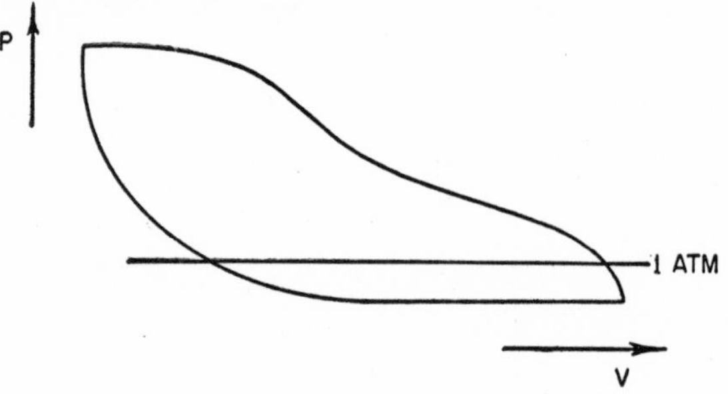

Fig. 2-11.

is the absolute pressure of the fluid, a is the area of the piston, and l is the distance travelled by the piston. The indicator diagram is a plot of p vs. l; a is constant. Therefore the area between the zero pressure line and any line on the diagram represents to a certain scale the work done upon the piston while that line was being drawn. To obtain the work of a complete cycle, however, it is necessary to know only relative pressures, not absolute pressures. For every element of piston travel dl_1 in the positive direction there is a corresponding element of travel dl_2 in the negative direction as shown in

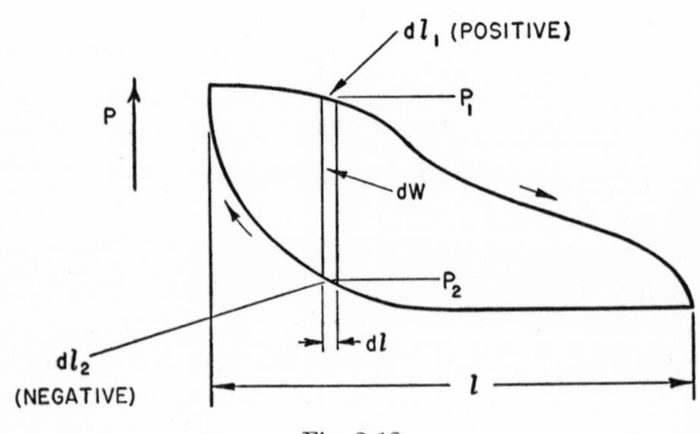

Fig. 2-12.

Fig. 2-12. The net work for the two elements of travel is the element of work dW.

$$dW = p_1 a\, dl_1 + p_2 a\, dl_2$$

but
$$dl_2 = -dl_1 \quad \text{and} \quad dl_1 = dl$$

hence
$$dW = (p_1 - p_2) a\, dl$$

The work for a complete cycle of the machine is then

$$W = \int_0^l (p_1 - p_2) a\, dl$$

which may be written

$$W = P_m a l$$

where P_m (called *mean effective pressure*) is the average value of $(p_1 - p_2)$ over the length of the diagram, and l is the length of stroke of the engine piston. The value of P_m is directly proportional to the average height of the diagram. The average height of the diagram is usually obtained by measuring the area a_d of the diagram with a planimeter and dividing the area by the length l_d of the diagram. Then

$$P_m = \frac{a_d}{l_d} C$$

where C, the *indicator spring constant,* is a constant of the engine indicator, giving the scale relation between diagram height and pressure. The dimensions of C are psi per in.

Example 3. An engine has a bore and stroke of 11 by 15 inches. An indicator diagram taken from this engine has an area of 1.60 sq in., and length of 2.40 in. The indicator spring constant is 80 psi per in. How much net work was done by the fluid in the cylinder upon the engine piston during the engine cycle represented by the diagram?

Solution: Bore means cylinder (and piston) diameter; stroke means piston travel (distance between extreme positions).

Piston area: $a = \dfrac{\pi(11)^2}{4} = 95$ sq in.

Piston travel: $L = \dfrac{15}{12} = 1.25$ ft

Mean effective pressure: $P_m = \dfrac{1.60}{2.40} 80 = 53.33$ psi

Net work: $W = P_m a L = 6{,}360$ ft lb

An indicator diagram is similar to a pressure-volume plot but it does not necessarily represent all the work done by a fluid system, even if the fluid is frictionless. The diagram represents only the work done by the fluid *on the engine piston.* This work will not, in general, be the same as the $\int p\,dV$ for the fluid because in most engines the fluid flows in and out at certain times in the engine cycle, through the valves G in Fig. 2-9. Hence the fluid in the engine does not constitute a stationary system and its work is not given by $\int p\,dV$, nor does the indicator diagram represent a series of states of a fluid system. The indicator diagram, in fact, represents the series of states of the engine piston by giving its surface pressure and its position. Indicator diagram characteristics for various types of apparatus will be discussed in detail in connection with the analysis of the particular apparatus.

2-6 Summary. Definition of Work: *Work is transferred from a system by a given action of the system if the total external effect of the given action can be reduced to the raising of a weight.* From this definition it is clear that work is a name for energy in *transit* from one system to another; energy which does not cross the boundary of a system does not qualify as work with respect to that system. In order to compute a particular work quantity it is necessary to select a system having a boundary across which the particular quantity flows. Work in thermodynamics may include not only mechanical work but also electrical work.

Work transferred *out* of a system is *positive* with respect to that system. Work transferred *in* is *negative.* Since the definition tells only how to identify positive work, it is necessary to identify the negative work of a system by showing it to be positive work of some other system. Any work transfer may appear positive or negative depending upon the system chosen.

In a stationary fluid system (changes of volume being permitted) work may be transferred in or out by pressure force against a moving boundary. If pressure is uniform in the system, this work is given by $W = \int p\,dV$ and is positive or negative as dV is positive or negative. (Only positive pressures are considered.) If the stationary fluid system consists of a real fluid (having friction), it will be possible to transfer work *to* the system (negative work) by boundary friction forces or by paddle-wheel action. Such work cannot be measured in terms of properties of the fluid system; some additional information is necessary to evaluate it. If the stationary fluid system consists of an ideal frictionless fluid, the work of pressure forces is the only work that will be transferred (assuming gravity, electricity, magnetism, and surface tension have negligible influence upon the system).

A plot of a process in a stationary fluid system may be used to represent graphically the work given by $\int p\,dV$.

An indicator diagram is similar to a pressure-volume plot but the indicator diagram is not a state diagram showing the pressure and volume of a fluid system; it is a plot of pressure vs. piston travel and as such may represent work done upon the piston by the fluid. This work, however, will not equal $\int p\,dV$ for the fluid except in the special case that no fluid flows into or out of the engine cylinder during the engine cycle.

PROBLEMS

2-1. A mass of 2000 lb is suspended from a pulley block. Find the magnitude and direction of the work transfer when the mass is hoisted against gravity through 6.5 ft. Show in a sketch the system boundary across which the work is transferred.

2-2. A pump forces 550 gallons per minute of water horizontally from an open well to a closed tank where the pressure is 120 psig. The water temperature is 65°F. How much work must the pump do upon the water in an hour, solely to force the water into the tank against the pressure? Sketch the system upon which the work is done, showing it both before and after the process.

2-3. If the work done in Problem 2-2 upon the water had been used solely to raise the same amount of water vertically against gravity without change of pressure, how many feet would the water have been elevated?

2-4. (a) If the work done in Problem 2-2 upon the water had been used solely to accelerate the water from zero velocity without change of pressure or elevation, what velocity would the water have reached? (b) If the work had been used to accelerate the water from an initial velocity of 30 fps what would have been the final velocity?

2-5. Gas from a bottle of compressed helium is used to inflate a balloon, originally folded completely flat, to a volume of 8.3 cu ft. If the barometer is 30.13 in. Hg how much work is done upon the atmosphere by the balloon? Sketch the system before and after the process.

2-6. The balloon of Problem 2-5 requires no stretching to reach the specified volume: the pressure of the helium in the balloon is negligibly higher than the atmospheric pressure. The helium in the steel bottle is initially at 2000 psia and finally at 1925 psia. Taking the total mass of *helium* as a system sketch the system before and after the process, find the magnitude and sign of the system work. Why can the work be calculated by $\int p\, dV$ when the pressure is not uniform over the whole boundary of the system? Is the p used in calculating the work *the pressure of the system* during the process? Does the system have a definite pressure?

2-7. In a frictionless cylinder and piston machine, Fig. 2-2, the piston is forced against the gas by a spring which exerts a force directly proportional to the volume of the gas. In addition to the spring force the atmospheric pressure of 15 psia acts upon the outer side of the piston. (a) Considering the gas as a system find the work when, from an initial state of 1 cu ft, 30 psia, the gas volume increases to 3 cu ft. Plot the state change on the pV plane. (b) Considering the spring as a system find the work for the same process. (c) Account for the difference in magnitude of the work in (a) and (b).

2-8. A propeller shaft in a ship turns steadily at 200 rpm. The torque applied to the shaft by the driving turbines is 790,000 ft lb. The torque transmitted to the propeller is 780,000 ft lb, the remainder being used to overcome friction at the shaft bearings. (a) At what rate in horsepower does the shaft receive work from the turbines? (b) At what rate does the shaft deliver work to the propeller? (c) Considering the *shaft* as a system, what is the net rate at which work crosses the boundaries of the system? Sketch the system.

2-9. A stationary mass of gas is compressed in a frictionless way from 12 psia, 3 cu ft, to 60 psia, 0.8 cu ft. Assuming that the pressure and volume are related by $pv^n =$ constant, find the work done by the gas system.

2-10. A mass of 1 lb of air is to be compressed from 15 psia to 110 psia in a process for which $pv =$ constant; the initial density of the air is 0.075 lb/cu ft. Find the work done by a piston, to compress the air.

2-11. The indicator diagram for the process in a water pump is a rectangle 3.00 in. long (on the axis of piston travel) and 1.45 in. high; the indicator spring scale is 100 psi/in. The pump process is repeated 60 times per minute. The pump cylinder diameter is 8 in. and the piston stroke is 12 in. Find the rate in horsepower at which the piston does work upon the water.

2-12. A series of indicator diagrams for a diesel engine cylinder yields the following average results: area of diagram 0.232 sq in., length of diagram 2.10 in. The indicator spring constant is 600 psi/in., the engine speed is 295 rpm, the engine cylinder bore and stroke are respectively 10 in. and 18 in., and the engine takes two shaft revolutions to complete each machine cycle. Find the horsepower transferred from the gas to the piston (indicated horsepower).

2-13. A piston and cylinder machine containing a fluid system has a stirring device in the cylinder, Fig. 2-3; the piston is frictionless, and the force F holding it against the fluid system is due only to the atmospheric pressure, 14.7 psia. The stirring device is turned 10,000 revolutions with an average torque against the fluid of 10 in. lb. Meanwhile the piston 2.0 ft in diameter moves out 2.0 ft. Find the net work for the system.

Chapter 3

TEMPERATURE AND HEAT

3-1 Temperature. The concept of temperature comes originally from the reactions of the senses to "hot" and "cold" objects; a body which feels *hotter* than another is considered to have *higher temperature.* Such physiological reactions are too crude and inconsistent for quantitative measurement. Experience shows, however, that certain measurable properties of substances are related to the temperature. For example, a bar of metal is usually found to increase in length when it changes from cold to hot to the touch; a quantity of mercury enclosed in a glass tube may expand in volume faster than the glass and fill a greater length of the tube when it changes from cold to hot; the electrical resistivity of a metallic conductor usually increases as the conductor becomes hotter to the senses. A measurement of length, volume, or electrical resistivity might therefore serve as a measurement of the temperature of a body, but such arbitrary temperatures would apply only to the particular body whose length, volume, or resistivity had been measured.

3-2 Temperature equilibrium. It is a matter of common experience that when a hot body is brought into close contact with a cold body the difference in temperature between the two bodies diminishes with time. If the two bodies are separated from all else so that each is affected only by the other, they will finally reach a state in which no further change occurs. The two bodies are then said to be in *temperature equilibrium.* The manner of approaching temperature equilibrium depends upon the natures of the bodies involved. Usually the hot body becomes cooler and the cold body warmer, as when a piece of hot iron is dropped into a bucket of water to cool off. If, however, the iron were dropped into the middle of a lake, the change would *appear* to be confined to the iron; it could be said that for all practical purposes the iron came to the temperature of the lake and the lake was unaffected. This situation suggests the possibility of comparing the temperatures of different bodies by using a comparison body (or thermometer) small enough to have a negligible effect upon the bodies with which it is brought into contact.

3-3 Thermometers. Suppose a mass of mercury is placed in a glass bulb with a long narrow extension so that the mercury fills the bulb and part of the extension or stem, Fig. 3-1. The level of the mercury in the stem then becomes a sensitive indication of the mercury volume relative to the glass volume. If an arbitrary scale is marked on the stem, the result is a device which can indicate its own temperature in terms of volume. This is the mercury-in-glass thermometer, familiar to everyone. Now if this thermometer is placed in close contact with a body and isolated from everything else, the thermometer will eventually reach temperature equilibrium with the body and the reading of the thermometer can be taken as an indication of the temperature of the body. Three practical points should be noted here: (1) The problem of isolating the thermometer from everything but one body cannot always be satisfactorily solved, but the general method is to immerse the thermometer in a hole in a solid body, or directly in a fluid body. (2) The thermometer will give the temperature of both itself and the body *after temperature equilibrium is reached,* but this may not be the same temperature as the body had before the immersion of the thermometer. The thermometer should be small relative to the body so that the thermometer may have only a small effect upon the body. (3) The thermometer must not be subject to effects such as pressure changes, which might change the volume independently of temperature.

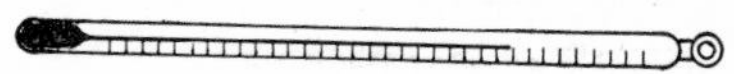

Fig. 3-1. Mercury-in-glass thermometer.

Assume that a thermometer is brought to equilibrium with each of two bodies in succession and that the thermometer gives the same reading with each body. Now if the two bodies are placed together, experience shows that no change due to temperature difference will occur in them. That is to say: *Two bodies each in temperature equilibrium with a third body will be in temperature equilibrium with each other.* This means that an arbitrarily marked thermometer could be used to compare the temperatures of bodies not themselves in contact. Also, any thermometer could be calibrated by comparison with an arbitrary standard and would then show the same temperature for a given body as would the standard thermometer.

3-4 Temperature scales. Among the physical characteristics which are found to be related to the temperature of a substance are the changes from solid to liquid, and from liquid to gas. For example, if an arbitrarily calibrated thermometer is used as a measuring device it is found that pure-water ice always melts into liquid water at a definite temperature, and that pure water always boils into steam at a definite temperature, provided the pressure is kept at a given value. Therefore two points on a thermometer scale can be defined in terms of the melting point and boiling point of pure water at one atmosphere pressure, and these points will be reproducible in any laboratory. On the Centigrade scale the melting point, or ice point, is marked 0 degrees and the boiling point, or steam point, is marked 100 degrees. On the Fahrenheit scale the ice point is marked 32 degrees and the steam point is marked 212 degrees. On both scales the intermediate points are obtained by dividing the distance between the ice point and the steam point into equal subdivisions of scale length. With a mercury thermometer the temperature that would cause the mercury to reach a point midway between the marks at 32°F and 212°F would be called 122°F. With a metal bar thermometer the temperature that would bring the bar to a length midway between its length at 32°F and its length at 212°F would be called 122°F. With an electrical resistance thermometer the temperature that would bring the resistance of the conductor to a value midway between the resistance corresponding to 32°F and the resistance corresponding to 212°F would be called 122°F. Now all these types of thermometers, of whatever materials they may be constructed, must agree at 32°F and at 212°F by the definition of their scales, but if two thermometers of different types or different materials, calibrated as described above, are brought to equilibrium in a place where one of them reads 122°F, the other will, in general, read something different from 122°F. It is an experimental fact that different materials have different relations between expansion and temperature change, such that a temperature scale based upon the expansion of one material will not agree at all points with a scale based upon the expansion of another material. Similar discrepancies exist with respect to electrical thermometers. Furthermore, at certain temperature levels one scale may even show a reversal of direction as compared with some other scale. Such characteristics limit the choice of a scale for a basic standard of temperature.

3-5 Gas thermometers. There exists one group of substances which exhibit excellent agreement among themselves, over wide ranges of temperature, when used as thermometric substances. These are the gases such as hydrogen, nitrogen, oxygen, helium, which are difficult to condense to liquids. (Such gases were once called "permanent gases" in the belief that they had no liquid or solid states.)

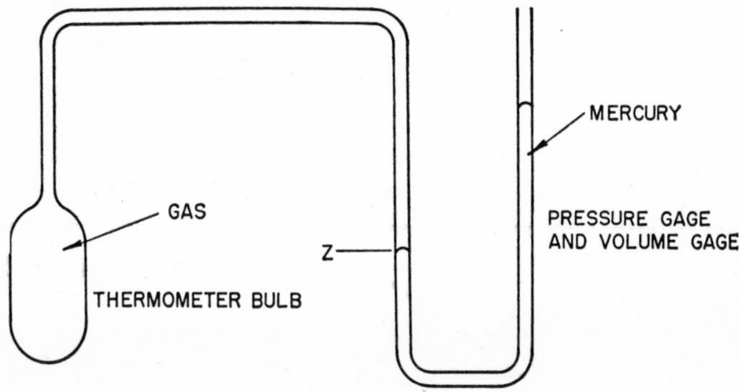

Fig. 3-2. Gas thermometer.

A gas thermometer is a mass of gas in a bulb fitted with means for measuring pressure and volume as shown schematically in Fig. 3-2. The device may be used in either of two ways. In the constant-pressure gas thermometer the mercury is adjusted to maintain the pressure constant in the gas, and the level of the mercury at z is a measure of the gas volume which is taken as the indication of temperature. In the constant-volume gas thermometer the mercury is adjusted to keep the level z always at a fixed reference point, and the pressure required on the gas to do this is taken as the indication of temperature.

3-6 Absolute temperature. The gas thermometer has the following important characteristics:

(a) Both constant-volume and constant-pressure thermometers using permanent gases, when individually calibrated at the ice point and the steam point, show good agreement among themselves at all other temperatures not close to the condensation or dissociation temperatures of the gases used. The agreement among different gases becomes better as the pressure of the gases is reduced. By operating a constant-pressure thermometer at several pressures and extrapolating to find the result that would be obtained at zero pressure, the same results are obtained for all gases. Temperatures so found may be called ideal gas temperatures, being based upon an ideal pressureless gas scale.

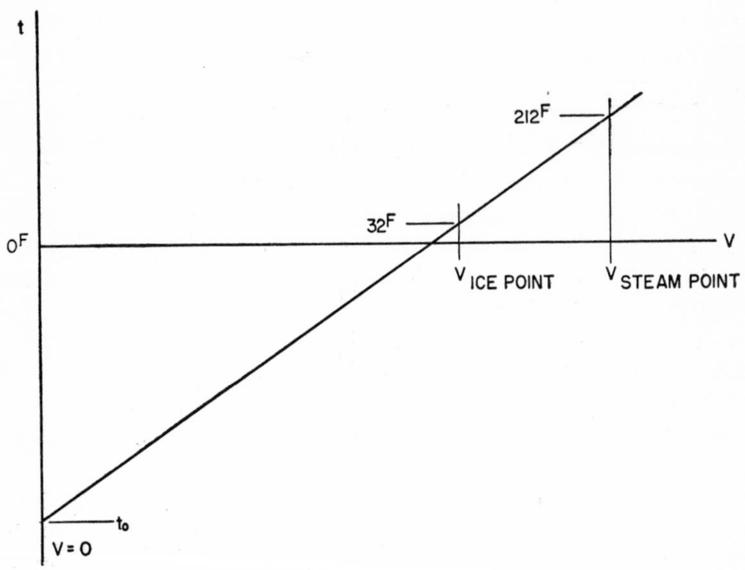

Fig. 3-3.

(b) If the scale of a constant-pressure gas thermometer is extrapolated beyond the ice point to lower temperatures, there must eventually be a point reached at which the extrapolated gas volume becomes zero. This point is found to correspond closely to −460°F (−273°C) for all "permanent" gases. The extrapolation is indicated graphically in Fig. 3-3. It may be expressed algebraically as follows (for the Fahrenheit scale):

$$32 - t_0 = (212 - 32)\,\frac{1}{\dfrac{V_{212}}{V_{32}} - 1} \tag{a}$$

The volume ratio has been found experimentally to be 1.366 for the ideal gas thermometer. This makes the temperature corresponding to zero volume, t_0, equal $-459.7°F$.

(c) If the scale of any constant-volume gas thermometer is similarly extrapolated to zero pressure, the same temperature ($-459.7°F$) is closely reached.

These facts in themselves would be sufficient to justify the acceptance of the ideal gas thermometer scale as an arbitrary standard. But it is also a fact that this ideal scale, approached as a limit by the real gas scales, is the same as a thermodynamic temperature scale which will be introduced as a consequence of the Second Law of Thermodynamics. Until the latter scale is introduced in Chap. 8, the ideal gas scale will be accepted as an arbitrary standard of temperature, called *absolute temperature*, and defined as follows: (1) The ratio of steam-point temperature to ice-point temperature is the same as the ratio of the corresponding volumes of gas in a constant-pressure ideal gas thermometer. (2) The difference between the steam-point temperature and the ice-point temperature is 180 degrees for the Fahrenheit absolute scale or 100 degrees for the Centigrade absolute scale.*

The relations between the absolute temperature scales and the conventional scales are given with sufficient accuracy for general purposes by

$$T^{F_{abs}} = t^{F} + 460 \tag{b}$$

and

$$T^{C_{abs}} = t^{C} + 273 \tag{c}$$

where $T^{F_{abs}}$ = Fahrenheit absolute temperature;
t^{F} = Fahrenheit temperature;
$T^{C_{abs}}$ = Centigrade absolute temperature;
t^{C} = Centigrade temperature.

It is customary to use capital T as a symbol for absolute temperatures and small t as a symbol for other temperatures. The Fahrenheit absolute scale is frequently called the Rankine scale (symbol R), and the Centigrade absolute scale is called the Kelvin scale (symbol K). Conversion equations are as follows:

* The International scale of temperature is a working scale which gives absolute temperatures in terms of more practicable devices than the gas thermometer. See the references at the end of this chapter.

$$t^F = 1.8t^C + 32 \tag{d}$$

$$T^R = 1.8T^K \tag{e}$$

3-7 Heat. It was pointed out in Sec. 3-2 that whenever two bodies at different temperatures are brought into close contact with each other and are isolated from other bodies, the temperature difference between the two bodies will diminish with time. If this happens it follows that at least one of the bodies must change its temperature. It is not necessarily true that both bodies must change temperature. For example, a piece of hot iron dropped into a bucket of ice and water at 32°F may be finally cooled to 32°F while the mixture of ice and water is still found to be at 32°F. This does not mean, however, that no change occurred in the water-ice mixture; in fact it will be found that some of the ice has melted to liquid. A corresponding situation exists in connection with the vaporization of a liquid. Hot iron dropped into a bucket of water at 212°F will not raise the temperature of the water finally, but will cause some of the water to evaporate at 212°F while the iron cools to 212°F.

It is known from experiment that the temperature changes or other effects of bringing two bodies to temperature equilibrium are quantitatively reproducible. For example, 1 lb of iron at 70°F dropped into 1 lb of water at 60°F will come to equilibrium at 60.9°F; the water will rise in temperature 0.9 degrees while the iron will fall 9.1 degrees. One lb of iron at 70°F dropped into a mixture of ice and water at 32°F will come to equilibrium at 32°F, but 0.0268 lb of ice will melt, assuming at least that much ice was present initially. Identical effects in the water and ice can be produced by materials other than iron; identical effects in the iron can be produced by materials other than water or ice. These facts may be summarized as follows: When two bodies at different temperatures are placed in contact with each other and isolated from other bodies, there will always be a change in temperature or physical structure in *both* bodies. These changes are always quantitatively related and, in the case of two temperature changes, are in opposite directions. The magnitudes of the effects produced depend upon the material and the mass of each body and upon the magnitudes of their initial temperatures.

The conclusion drawn from these facts is that energy transfer occurs when two bodies at different temperatures are placed in contact. This energy transfer is not a work transfer because experience shows that it cannot, even with the aid of frictionless mechanisms, have the sole effect of raising a weight. Hence a new term, *heat*, is defined as follows: *Heat is energy transferred from one system to another solely by reason of a temperature difference between the systems.**

The transfer of heat between two bodies in direct contact is called *conduction*. Heat may also be transferred between bodies separated by empty space through the mechanism of *radiation*. Radiation may travel not only through empty space but also through gases and some liquids and solids. A third method of heat transfer between two bodies is fluid *convection*, which is a combination of three operations: (1) conduction from one body to the fluid; (2) motion of the fluid from one body to the other; (3) conduction from the fluid to the second body. Regardless of the mechanism of heat transfer involved, the effect produced is always the result of a *temperature difference between two systems.*

The direction of heat transfer is taken to be from the high-temperature system to the low-temperature system. The algebraic sign relative to a given system is *positive* for heat flow *to* that system. The quantity of heat transferred in a given operation is measured by the mass of water that could be raised through a specified temperature interval, if the water were substituted for the low-temperature system and the identical operation were executed in the high-temperature system. (The water might also be substituted for the high-temperature system while the given operation is executed in the low-temperature system. The water would then fall through the specified temperature interval.)

To show that heat is transferred in a given process, it is necessary to show that a physical effect is the consequence of a temperature difference between two systems. This is necessary because a particular effect could be obtained perhaps by other means than heat transfer; thus a rise in temperature of a body of water may be obtained by transferring paddle-wheel work to the water. Consequently the fact that the temperature rises is no guarantee that heat

* Some people feel that heat, like work, should be defined independently of the concept of energy, because energy has not been defined in a rigorous way. A definition of heat which does not use the word energy follows: *When an effect in a system occurs solely as a result of a temperature difference between the system and some other system, the process in which the effect occurs shall be called a transfer of* heat *from the system at the higher temperature to the system at the lower temperature.*

transfer has occurred. On the other hand it is true that when bodies at different temperatures are placed in contact, or are placed so that radiation can travel between them, heat will always be transferred. The essential relation between heat and temperature is that a temperature *difference* between two systems is necessary for heat to be transferred. The fact that the quantity of heat is measured in terms of the temperature *rise* of water is absolutely arbitrary. It could just as well have been said that the quantity of heat is measured by the mass of ice that could be melted. In that case there would have been no temperature *change* involved in the definition of the quantity of heat.

A process in which no heat crosses the system boundary in either direction is called an *adiabatic* process.

3-8 Units of heat. The unit of heat was formerly defined as the amount of heat required to cause a unit rise in temperature of a unit mass of water at atmospheric pressure. One *British thermal unit* (Btu) would raise the temperature of 1 lb of water 1 deg F; one *calorie* (cal) would raise the temperature of 1 gm of water 1 deg C. Experience showed, however, that these units, being functions of the properties of water, varied in size depending upon the initial temperature of the water. A new definition of the unit of heat, made possible through the First Law of Thermodynamics, is given in terms of basic electrical energy units (see Sec. 5-3). The magnitude of the unit is not changed by the new definition.

3-9 Specific heats. For small temperature changes, experiment shows that the temperature rise of liquid water due to heat transfer to the water is given by

$$Q = mc(T_2 - T_1) \tag{1}$$

where Q is the heat transferred to the water, Btu;
m is the mass of water, lb;
$T_2 - T_1$ is the temperature rise of the water, deg F;
c is an experimental factor, Btu/lb F, called *specific heat.*

The specific heat is found to be a function of temperature so Eq. (1) must be rewritten for generality as

$$dQ = mc\,dT \tag{2}$$

or

$$Q = \int_{T_1}^{T_2} mc\,dT \tag{3}$$

Equation (2) is not limited to use with water, but may be written for any homogeneous system in which a temperature change results from heat transfer *in the absence of work other than $p\,dv$ work.** In dealing with a general system, rather than a mass of a particular substance, the *heat capacity* C of the system would be used in place of mc.

In general the value of c will depend upon the substance in the system, the type of state change involved, and the particular state of the system at the time of transferring the element dQ of heat. For liquids and solids the value of c does not differ much for different processes but does change appreciably with temperature. For gases the value of c differs greatly with different processes and also varies appreciably with temperature; variation with pressure is generally small. Because of the above facts the specific heats for solids and liquids are often quoted without reference to the type of process in which the heat transfer occurs, whereas for gaseous substances it is customary to quote specific heats for two particular types of process, the constant-pressure process and the constant-volume process. The symbols for these two quantities are respectively c_p and c_v. In all cases the specific heat may be given either as a single numerical value, appropriate to a certain range of temperature, or as an equation in terms of temperature.†

From the old definitions of the Btu and the calorie (Sec. 3-8) it is evident that the specific heat of water would, by definition, be unity in both British and metric systems of units. Consequently the specific heats of any substance would be identical, numerically, in both systems of units. In making the present definitions of heat units, the relative magnitudes of the units were preserved so that specific heats have the same numerical values whether in Btu/lb F or cal/gm C. Moreover, the specific heat of water remains unity (with a small variation) at temperatures between 32°F and 212°F.

The use of specific heat data in heat transfer problems is illustrated in the following examples.

Example 1. It is desired to cool iron parts from 500°F to 100°F by dropping them into water initially at 75°F. The specific heat of the iron

* This restriction is necessary because in some cases the effects of heat and work are indistinguishable, and the temperature change is therefore not necessarily determined by the heat transferred.

† For values of specific heats of various substances see the Appendix.

is 0.120 Btu/lb F and the specific heat of water may be assumed to be 1.00 Btu/lb F. Assuming that all the heat from the iron goes to the water and that none of the water evaporates, how many pounds of water are needed per pound of iron?

Solution: The iron and water come to temperature equilibrium at 100°F. Therefore the heat flow to the water is

$$\begin{aligned} Q_w &= m_w c_w\,(T_{2w} - T_{1w}) \\ &= m_w\,1.00[100 + 460 - (75 + 460)] \\ &= m_w\,1.00(100 - 75) \end{aligned}$$

The heat flow from 1 lb of iron is

$$\begin{aligned} -Q_i &= -m_i c_i (T_{2i} - T_{1i}) \\ &= -(1.0)(0.120)[100 + 460 - (500 + 460)] \\ &= -(1.0)(0.120)(100 - 500) \end{aligned}$$

Note the negative sign for heat flow *from* the iron. Note also that when temperature differences are involved it is possible to use either Fahrenheit temperatures or absolute temperatures because the 460 subtracts out. The heat flow to the water is equal to the heat flow from the iron:

$$\begin{aligned} Q_w &= -Q_i = -(1.0)(0.120)(100 - 500) \\ Q_w &= +48.0 \text{ Btu/lb iron cooled} \end{aligned}$$

The pounds of water required are

$$m_w = \frac{Q_w}{1.0(100 - 75)} = \frac{48}{25} = 1.92 \text{ lb water/lb iron.}$$

Example 2. The specific heat of a certain gas at constant pressure is given in Btu/lb F by

$$c_p = 0.248 + 0.448 \times 10^{-8}\,T^2$$

where T is in Fahrenheit absolute degrees. How much heat must be transferred from a system consisting of 5 lb of this gas to cool it at constant pressure from 1540°F to 540°F?

Solution: The heat transferred *to* the system is given by

$$Q = m \int_{T_1}^{T_2} c_p\, dT$$

where T_1 and T_2 are respectively the initial and final temperatures of the process. Note that in this problem the limits of integration, T_1 and T_2, must be in absolute temperature units because c_p is given as a function of absolute temperature.

$$Q = m \int_{T_1}^{T_2} (0.248 + 0.448 \times 10^{-8}\, T^2)\, dT$$

$$= m \left[0.248T + \frac{0.448 \times 10^{-8}}{3} T^3 \right]_{T_1}^{T_2}$$

$$T_1 = 1540 + 460 = 2000^R$$

$$T_2 = 540 + 460 = 1000^R$$

$$Q = 5 \left[0.248(1000 - 2000) + \frac{0.448 \times 10^{-8}}{3} \left(\overline{1000^3} - \overline{2000^3} \right) \right]$$

$$Q = 5[-248 + (0.448/3)(-70)] = -1290 \text{ Btu}$$

Q is the heat transferred *to* the system. The heat transferred *from* the system is then 1290 Btu. The numerical work is shown in detail so that the signs can be kept right. This is very important in more complicated problems since the sign of the result is not always self-evident.

3-10 Latent heats. When the result of heat transfer is a change in physical structure of a substance, instead of a change in temperature, the quantity of heat *in the absence of work other than p dv work** is a function of the quantity of substance changed from one form to the other. For example, when, by heat transfer, ice is melted to water at 1 atm pressure, approximately 143 Btu will flow for each pound of ice melted; the same quantity of heat will flow out for each pound frozen at 1 atm pressure. When, by heat transfer, liquid water boils into vapor at 1 atm pressure, approximately 970 Btu will flow in for each pound of water vaporized; the same quantity of heat will flow out for each pound condensed at 1 atm pressure. Such heat quantities are called *latent heats.* Latent heat related to the melting of a solid is called *heat of fusion;* latent heat related to the vaporization of a liquid is called *heat of vaporization.* Other latent heats exist related to sublimation (vaporization of a solid) and to changes in crystalline structure of solids.

The latent heat of fusion differs for various substances but is little affected by pressure. The latent heat of vaporization is different for various substances and is greatly affected by pressure. Latent heats are always given for constant-pressure state changes (temperature is also constant).

Latent heats and specific heats have other significance than that of a measure of heat transferred in a certain type of process; more general definitions will be given in subsequent chapters, but for the

* See first footnote, p. 33.

present these quantities may be used in the computation of heat transferred *in the absence of work other than $p\,dv$ work.*

Example 3. Ten pounds of solid sulfur at 70°F are to be heated at constant pressure until it is a vapor at 1 atm pressure. How much heat will be required?

Data for sulfur at 1 *atmosphere*

Melting point	235°F
Boiling point	832°F
Specific heat of solid	0.180 Btu/lb F
Specific heat of liquid	0.235 Btu/lb F
Latent heat of fusion	15.8 Btu/lb
Latent heat of vaporization	120 Btu/lb

Solution: Four separate heat quantities must be calculated: (a) to raise solid temperature to 235°F; (b) to melt solid at 235°F; (c) to raise liquid temperature to 832°F; (d) to vaporize liquid at 832°F. (Change of solid crystal forms is ignored.)

$$Q_a = mc_p(T_2 - T_1) = (10)(0.180)(235 - 70) = 297 \text{ Btu}$$
$$Q_b = mL = (10)(15.8) = 158 \text{ Btu}$$
$$Q_c = mc_p(T_3 - T_2) = (10)(0.235)(832 - 235) = 1403 \text{ Btu}$$
$$Q_d = mL = (10)(120) = 1200 \text{ Btu}$$
$$Q = Q_a + Q_b + Q_c + Q_d = 3058 \text{ Btu}$$

Example 4. A piece of ice having an initial temperature of 22°F is dropped into an insulated tank, which contains 40 lb of water at 70°F. If the temperature of the water, after equilibrium is reached, is 40°F, how many pounds did the ice weigh? Assume no heat transfer with other bodies has occurred.

Data

Specific heat of ice	0.501 Btu/lb F
Specific heat of water	1.00 Btu/lb F
Latent heat of fusion	143.3 Btu/lb
Melting point of ice	32°F

Solution: Four separate heat quantities must be calculated: (a) to raise ice from 22°F to 32°F; (b) to melt ice at 32°F; (c) to raise melted ice from 32°F to 40°F; (d) to lower water from 70°F to 40°F. Let the mass of ice be m_i lb.

$$Q_a = mc_p(T_2 - T_1) = m_i(0.501)(32 - 22) = 5.01m_i \text{ Btu}$$
$$Q_b = mL = m_i 143.3 = 143.3m_i \text{ Btu}$$

$$Q_c = mc_p(T_3 - T_2) = m_i(1.00)(40 - 32) = 8.00m_i \text{ Btu}$$
$$Q_d = mc_p(T_3 - T_4) = (40)(1.00)(40 - 70) = -1200 \text{ Btu}$$
$$Q_a + Q_b + Q_c = -Q_d$$
$$(5.01 + 143.3 + 8.00)m_i = 1200$$
$$m_i = \frac{1200}{156.3} = 7.68 \text{ lb of ice}$$

Suppose the water had been contained in a tank of copper, weighing 10 lb, which was in equilibrium with the water at the beginning and end of the process. How much additional ice would have been required? Specific heat of copper is 0.092 Btu/lb F.

Solution: An additional heat quantity must be calculated for the copper:

$$Q_e = mc_p(T_3 - T_4) = (10)(0.092)(40 - 70) = -27.6 \text{ Btu}$$
$$(5.01 + 143.3 + 8.00)m_i' = 27.6$$
$$m_i' = \frac{27.6}{156.3} = 0.176 \text{ lb ice additional}$$

3-11 Summary. Temperature can be sensed directly but an objective measurement of temperature can be made only indirectly through measurements of other properties such as volume, pressure, or electrical resistivity.

Any group of bodies at different temperatures, if they could be brought together and isolated from all other bodies, would eventually reach a common temperature at which they would be in temperature equilibrium.

A thermometer is a device calibrated to read its own temperature in terms of some property such as volume and arranged to be conveniently brought to temperature equilibrium with other bodies, while exerting a minimum effect upon the temperature of the other bodies.

Ordinary thermometer scales are arbitrary. The melting and boiling temperatures of water at 1 atm pressure are assigned definite values, and the scale is filled in by assuming a linear variation of temperature with the reading of some arbitrarily selected standard thermometer. The most suitable standard has been found to be the gas thermometer, which is used in physical laboratories to determine the reference points at which other thermometers are calibrated.

An absolute temperature scale may be constructed by taking the temperature as directly proportional to the volume of the gas in a constant pressure ideal gas thermometer. This scale, based upon a gas at zero pressure, is the physical realization of a logically formulated absolute thermodynamic temperature scale, which will be introduced in connection with the Second Law of Thermodynamics. For ordinary purposes absolute temperatures may be found from Fahrenheit and Centigrade temperatures by the equations

$$T^{\mathrm{F_{abs}}} = t^{\mathrm{F}} + 460, \qquad T^{\mathrm{C_{abs}}} = t^{\mathrm{C}} + 273$$

Heat is energy transferred from a system at a given temperature to a system at a lower temperature solely as a consequence of the existence of a temperature difference between the two systems. The quantity of heat is measured by the number of pounds of water that could be raised through a given temperature interval by the given heat flow. Heat flowing to a system is positive with respect to that system.

The unit of heat is the British thermal unit (Btu), which was formerly defined as the heat required to raise the temperature of a pound of water 1 degree F at a specified temperature level. The Btu is now defined in terms of electrical energy units but its magnitude is essentially the same as before.

Specific heat is the rate of heat flow per unit temperature rise per unit mass, when the temperature of a mass is changed by heat flow in the absence of work other than $p\,dv$ work. Specific heats have a broader significance which will be brought out in Chap. 5. Specific heats for solids and liquids are not appreciably affected by the type of process but do vary with temperature. Specific heats for gases and vapors are affected by the type of process and are therefore usually given for two particular types of process, the constant-pressure process and the constant-volume process. Both of these specific heats vary with temperature and, to a much smaller extent, with pressure.

Latent heat is the heat flow, in the absence of work other than $p\,dv$ work, associated with changes such as melting or vaporization in the physical structure of substances. It is measured in Btu per unit mass of substance changed from one structure to the other, while at constant pressure and temperature. Latent heat of fusion is the heat required to melt a unit mass of solid substance at constant pressure and temperature. Latent heat of vaporization is the heat required to vaporize a unit mass of liquid at constant pressure and temperature. Heats of fusion are little affected by pressure change. Heats of vaporization are greatly affected by pressure change.

PROBLEMS

3-1. A thermometer is constructed by arranging an aluminum rod so that its length can be measured accurately by a scale made of a nickel alloy, *invar*. The thermometer is calibrated at the melting point and boiling point of water, and the distance between these two points is divided uniformly into 100 degrees of equal length on the scale. Find the difference between the reading of this thermometer and the gas-thermometer temperature at 0, 25, 50, and 100°C on the gas thermometer scale. The linear expansion of aluminum may be represented by the equation

$$l_t = l_0(1 + 0.2221 \times 10^{-4}\, t + 0.114 \times 10^{-7}\, t^2)$$

where l_t is the length at t on the Centigrade gas thermometer scale, and l_0 is the length at 0°C. The coefficient of expansion of invar between 0°C and 100°C is 0.9×10^{-6} cm/cm deg C.

3-2. Two mercury-in-glass thermometers are made of identical materials and are accurately calibrated at 0°C and 100°C. One has a tube of constant diameter, while the other has a tube of conical bore, ten percent greater in diameter at 100°C than at 0°C. Both thermometers have the length between 0 and 100 subdivided uniformly. What will the straight bore thermometer read in a place where the conical bore thermometer reads 50°C?

3-3. In a constant-volume gas thermometer the following pairs of pressure readings were taken at the boiling point of water and at the boiling point of sulfur, respectively:

Water bp:	50.0	100	200	300
Sulfur bp:	96.4	193	387	582

The numbers are the gas pressures, mm Hg, each pair being taken with the same amount of gas in the thermometer, but the successive pairs being taken with different amounts of gas in the thermometer. Plot the ratio of $S_{bp}:H_2O_{bp}$ against the reading at the water boiling point, and extrapolate the plot to zero pressure at the water boiling point. This gives the ratio of $S_{bp}:H_2O_{bp}$ on a gas thermometer operating at zero gas pressure, i.e. an ideal gas thermometer. What is the boiling point of sulfur on the gas scale, from your plot?

3-4. In testing electrical apparatus the average temperature of a coil of wire is often obtained by measuring the electrical resistance of the wire. The resistance of a certain magnet coil at room temperature (86°F) was 1239 ohms; after operation under test the resistance was 1433 ohms. The resistance of a copper wire at temperature t°C is $R = R_{20}[1 + 0.00393(t - 20)]$ where R_{20} is the resistance at 20°C. Find the average temperature of the coil after test, in degrees F.

3-5. Experience shows that the time rate of heat transfer by *conduction* between two systems is closely proportional to the temperature difference between the systems; the rate of heat transfer by *radiation* is proportional to the difference between the fourth powers of the absolute temperatures of the systems. For both conduction and radiation, what value does the rate of heat transfer approach as the system temperatures approach equality? (In later sections of this book hypothetical heat transfer processes will be discussed in which the temperature difference between the two systems is infinitesimal.) How long would it take to transfer a finite quantity of heat with infinitesimal temperature difference?

3-6. (a) Explain why the temperature change of a system in a process is not necessarily a measure of the heat transferred in the process. (b) Devise a process in which there is heat transfer without temperature change in either of the systems involved. (c) Devise a process in which, for one system, the heat transfer is of opposite sign from the temperature change.

3-7. The average specific heat of green vegetables is about 0.9 Btu/lb deg F. A ton of vegetables at 75°F is placed in a cold room in which the air is maintained constantly at 40°F. The refrigeration plant can remove heat at a maximum rate of 400 Btu/min, but 150 Btu/min of this capacity is required to compensate for heat leakage from the surroundings to the cold room. What will be the minimum time in which the vegetables can be cooled to 40°F? (The actual time will be longer, and will depend upon the heat transfer rates that can be obtained; thermodynamics can determine only the limiting time, with perfect heat transfer.)

3-8. Fifteen hundred pounds of fish at 40°F are to be frozen and stored at 10°F. The specific heat of the fish above freezing is 0.76 Btu/lb F, and below freezing is 0.41 Btu/lb F. The freezing point is 28°F and the latent heat of fusion is 101 Btu/lb. How much heat must be removed to cool the fish, and what percent of this is latent heat?

3-9. The heat of fusion of aluminum at 657°C is 94 cal/gm. If the average specific heat of solid aluminum between 0°C and 657°C is 0.247 cal/gm deg C, and of liquid aluminum is 0.256 cal/gm deg C, how much heat will be required to raise the temperature of 25 pounds of aluminum from 80°F to 1350°F?

REFERENCES

American Institute of Physics, *Temperature, Its Measurement and Control in Science and Industry*. New York: Reinhold, 1941.

Keenan, J. H., *Thermodynamics*. New York: Wiley, 1941, chaps. 1, 21.

Messersmith, C. W. and C. F. Warner, *Mechanical Engineering Laboratory*. New York: Wiley, 1950, chaps. 3, 12.

Zemansky, M. W., *Heat and Thermodynamics*. New York: McGraw-Hill, 1943, chaps. 1–4, 9.

Sears, F. W., *Thermodynamics, Kinetic Theory and Statistical Mechanics*. Cambridge: Addison-Wesley, 1950, chap. 1, chap. 4, secs. 4-3 and 4-4.

Chapter 4

PROPERTIES OF SYSTEMS

4-1 Properties—physical concept and general concept. A *property* of a system was defined in Sec. 1-4 as a characteristic by which the state of a system may be described. The simple physical concept of a property involves observable characteristics such as pressure, temperature, mass, velocity, and location relative to the earth. When all the observable characteristics have definite values, the state of the system is definite; hence it is possible to define a property in a more general way as *any quantity which has a definite value for each definite state of the system.* From this more general viewpoint observable properties are not the only properties of a system.

4-2 Properties mathematically derived. Given the observable properties of a system, it is possible to form any number of mathematical functions of these properties such that the value of the function will be fixed if the values of the observable properties are fixed. Such functions then become properties of the system under the general definition of a property. Consider, for example, the ratio of the mass to the volume of a system. When the mass and the volume are fixed, the ratio of mass to volume will have a definite value; it is therefore a property of the system (it will be recognized as the density). As another example consider the product of pressure and temperature. This quantity, chosen at random, has no physical meaning, yet it is a property of a system. A number of property functions which appear frequently in thermodynamic analysis, and which are themselves properties, have been given names and symbols as a matter of convenience. Some of these properties will be encountered in subsequent chapters.

4-3 Properties derived from general laws. In addition to observable properties and mathematically derived properties, there exist two properties which are derived from the two general laws of thermodynamics. These properties, *internal energy* derived from the First

Law, and *entropy* derived from the Second Law, are not observable properties nor can they be arrived at by mathematical manipulations of other properties. They can be shown to be properties only through the use of the general laws of thermodynamics. It is not intended to discuss these two properties at this point but only to indicate the existence of this third class of properties.

4-4 Point functions and path functions. The general requirement for a property, that it have a definite value for any state of the

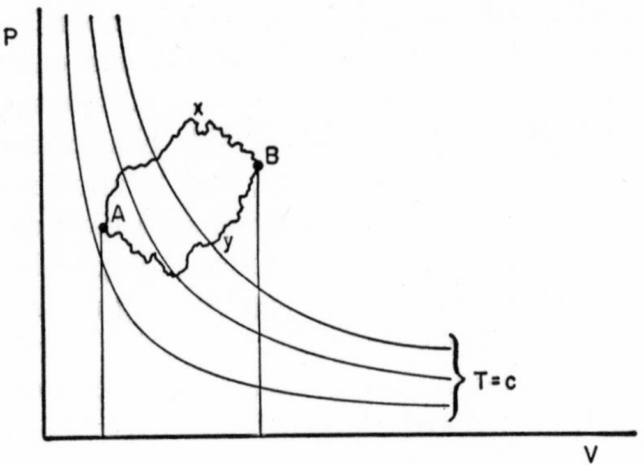

Fig. 4-1. State points and paths.

system, may be expressed by saying that a property is a *point function*. This term refers to the fact that in a graphical representation of properties a point on the plot will correspond to a definite value for each property. If a plot on the coordinates of pressure and volume is constructed to show the relation between pressure, volume, and temperature for a unit mass of a gas, the result will look like Fig. 4-1. At any point on the plot, such as A, there can be read off definite values of pressure, volume, and temperature. It is only necessary to mark the point A and the properties at A are fixed. If another point, such as B, is marked, the properties at B are fixed. Hence the name, point function, for a property.

It is often more important to know how much a property changes when a change of state occurs than to know its value at either end state. The change in any property during a change of state from A to B would be given by the difference between the value at B and the value at A. For example

$$\Delta T_{AB} = T_B - T_A$$

where ΔT_{AB} signifies the change in temperature during a change of state from A to B, and T_A to T_B are, respectively, the temperatures at states A and B. Similarly

$$\Delta V_{AB} = V_B - V_A$$

and

$$\Delta P_{AB} = P_B - P_A$$

These relations would hold whether the change of state followed the path x or the path y in Fig. 4-1. *The change in the value of a property during a change of state is independent of the path of the change of state and depends only upon the end points.*

Work is not a property. It will be recalled that in Chap. 2 computations were made of the work done during certain processes. The work done by a system through pressure against a moving boundary was given by

$$W_{AB} = \int_A^B p\,dV$$

Now, referring to Fig. 4-1, it will be clear that the $p\,dv$ work for the state change along the path x from A to B would be represented by the area between the line AxB and the axis of volumes, whereas the $p\,dv$ work for the state change along the path y from A to B would be represented by the area between the line AyB and the axis of volumes. It is obvious that these areas are not equal. It is then obvious that the $p\,dv$ work is entirely dependent upon the *path followed* between A and B rather than upon the location of points A and B. Hence the work is called a *path function* to distinguish it from a property.

It was emphasized in Chap. 2 that work is not always related to the properties of a system by *any* function. Now it is shown that even when the work *is* related to the properties of the system there is a fundamental difference between work and a property. Work is determined by the *path* between two states and not by the end states; property changes are determined by the end states and not by the path.

Although it cannot be readily demonstrated at this point, it is an experimental fact that heat, like work, is a path function. Heat is therefore not a property of a system.

4-5 Heat and temperature. It is well to point out here some possible sources of confusion in connection with heat, which is not a property, and temperature, which is a property.

The word heat has often been used by physicists to denote two entirely different things. (1) The energy transferred between two systems due to temperature difference; this is not a property. (2) The kinetic and potential energy of the molecules and atoms of a substance; this is a property. This dual usage of the word is bad enough in itself but the difficulty has been compounded in some cases by defining heat as (2) and using it as (1). It has become generally accepted in the field of thermodynamics that the word heat shall be used solely to denote energy *transfer* due to a temperature difference. This usage will be followed strictly in this book.

Temperature is frequently defined in physics as a measure of the kinetic energy of molecules and atoms. This definition, though perfectly proper in its place, cannot be used in thermodynamics because the broad approach of thermodynamics does not take account of the detailed mechanisms of energy storage in molecules and atoms. That subject belongs to other branches of science which are beyond the scope of this book. We may recognize that energy is stored as potential and kinetic energy in molecules and atoms but we are not in a position to study these factors in detail. Instead, we utilize empirical information in regard to the energy stored in substances, and avoid the problems of molecular and atomic structure. The science of molecular and atomic structure is, in fact, of great aid to thermodynamics in supplying useful data on the properties of substances. These data are determined on the basis of a certain physical picture or mechanism for the internal structure of a substance, but they are used in thermodynamics simply as empirical facts without reference to the physical picture by which they were computed. Similarly temperature is accepted in thermodynamics as a fact, without attempting to explain the fact.

The use of a physical picture or mechanism may often be a valuable aid in the formulation of a problem, but it may also lead to false results if the picture is not well chosen. The great value of the thermodynamic method (and its great limitation) lies in its independence of any particular picture of the mechanism of energy transfer and energy storage in systems.

Chapter 5

FIRST LAW OF THERMODYNAMICS

5-1 Background of the first law. Work and heat, which have already been discussed separately, are related by the First Law of Thermodynamics. This law is a statement of the general principle of the *conservation of energy*. The hypothesis of energy conservation was developed by a number of scientists during the first half of the nineteenth century but direct experimental tests of the hypothesis were first made by J. P. Joule in the eighteen-forties.

The early theories of heat were based on the assumption that heat was a fluid which could be stored in substances and transferred from one substance to another. Some phenomena can be explained on this basis. For example, a certain amount of heat transferred into a system consisting of a mass of water at constant pressure will raise the temperature of the water from t_1 to t_2; if then the same amount of heat is transferred out of the system, the temperature will fall from t_2 to t_1. This might lead to the theory that the heat supplied is stored in the system at t_2 and released when the temperature returns to t_1. However, when this idea is put to the test of general experience it is found that all the heat supplied to a system during a change of state need not always reappear when the system is returned to its initial state.

It is also a matter of experience that the work done on a system during a change of state need not always reappear when the system is returned to its initial state; in fact it often happens that *none* of the work done in a state change reappears when the system is returned to its initial state.

In all cases, however, when a system is caused to execute a cycle, it is found that a failure to recover all the *heat* supplied is accompanied by the transfer of more *work from* the system than was put in during the cycle. Also, a failure to recover all the *work* supplied is accompanied by the transfer of more *heat from* the system than was put in during the cycle. This indicates that neither heat nor work is conserved as such but that some means exists whereby a system working in a cycle can take in one of these entities and give out the other.

The facts stated above lead to the following hypothesis: *Heat and work are different forms of a single entity which is conserved.* The name *energy* is applied to this entity. On the basis of this hypothesis, energy which enters a system as heat may leave the system as work, or energy which enters the system as work may leave as heat; consequently the energy in storage in the system is neither work nor heat but requires another name. The name *internal energy* is used for energy in storage in a system.

5-2 Equality of heat and work in a cycle. An experiment for checking the hypothesis of energy conservation is shown schematically in Fig. 5-1. Work input to the paddle-wheel is measured by the fall of the weight, while the corresponding temperature rise of the

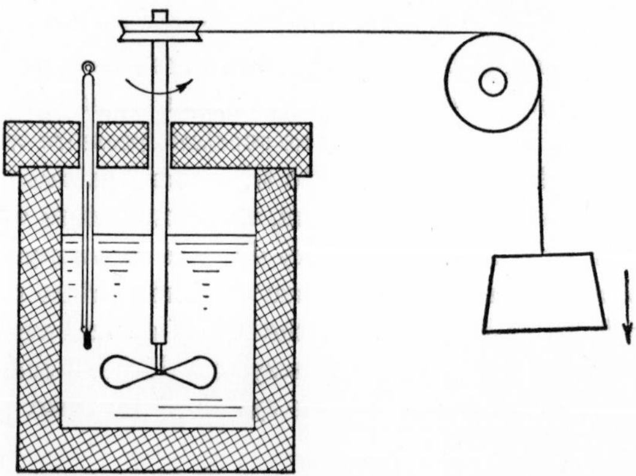

Fig. 5-1.

liquid in the insulated container is measured by the thermometer. It is already known from experiments on heat transfer that temperature rise can also be produced by heat transfer. Experiments show: (1) that a definite quantity of work (778 ft lb) is always required to accomplish the same temperature rise obtained with 1 Btu of heat; (2) that regardless of whether the temperature of the liquid is raised by work transfer or by heat transfer, the liquid can be returned by negative heat transfer to the identical state from which it started. Such results agree with the hypothesis that work and heat are different forms of something more general, which is called *energy*.

Going beyond the simple paddle-wheel experiment, it can be stated as an invariable experience that whenever a physical system passes through a complete cycle the algebraic sum of the work transfers during the cycle $(\Sigma W)_{\text{cycle}}$ bears a definite ratio to the algebraic sum of the heat transfers during the cycle $(\Sigma Q)_{\text{cycle}}$. This is expressed by the equation

$$(\Sigma W)_{\text{cycle}} = J(\Sigma Q)_{\text{cycle}} \tag{1}$$

where J, the experimental constant of proportionality, has the value 778.16 ft lb/Btu.

Experience has shown that Eq. (1) holds for any cycle and any system whatever. The cycle may involve chemical reactions, or physical state changes, the only restriction being that no material crosses the system boundary.

Equation (1) is the First Law of Thermodynamics. It is accepted as a general law of nature because no violation of it has ever been demonstrated. No independent general law of nature can ever be proved except in this negative way because there are always new experiments which might disprove the law; but when years of experience produce many experiments which confirm the law, and yet fail to produce the single contrary experiment necessary to disprove it, the law becomes as firmly established as any human knowledge can be.*

5-3 Units of energy. Up to this point work and heat have been measured in different units, each in terms of the physical picture by which it was defined. But now the quantitative relation between work and heat and the acceptance of the hypothesis that they are both forms of energy, permit the use of a single unit for both. Heat transfer quantities can be expressed in foot pounds by substituting 778 foot pounds for each Btu; work can be expressed in Btu by substituting 1/778 Btu for each foot pound. If either of these were done,

* No concern need be felt about the effect upon this law of the recent discoveries in the conversion of mass to energy; mass simply becomes another form in which energy may appear. In any event, within the scope of Engineering Thermodynamics, transitions of mass to energy are not encountered.

the constant of proportionality J would not be needed in Eq. (1). From this point on it will be assumed that equations involving work and heat are written with the same units for both work and heat. Equation (1) will now be written as

$$(\Sigma W)_{\text{cycle}} = (\Sigma Q)_{\text{cycle}} \tag{2}$$

where the same units are used for both W and Q.

Since work *to* a system is negative and heat *to* a system is positive, Eq. (2) states that, for a cycle, the algebraic sum of all energy transfers to a system is zero.

The fact that work and heat may be expressed in the same units does not mean that work and heat are interchangeable in all operations, for they are not. Nevertheless, in all operations in which work and heat *are* interchangeable, 778 ft lb of work will accomplish the same effect as 1 Btu of heat. Therefore 1 Btu can represent a perfectly definite amount of work, or 1 ft lb can represent a definite amount of heat. It must always be remembered, however, that changing the units of heat to foot pounds does not give heat the characteristics of work. The basic definitions of work and heat are unchanged, whatever the units of measurement.

It may be noted at this point that the definition of the Btu is no longer based upon the properties of water but is arbitrarily set at a certain ratio to the foot pound. Actually the calorie (Sec. 3-8) is defined as 1/860 watt hour, and when these units are converted to Btu and ft lb, the relation obtained is 1 Btu = 778.16 ft lb.

5-4 The first law and internal energy. Equation (1) is a statement of the First Law of Thermodynamics but it is in an inconvenient form for general use, since it applies only to cycles. If a system executes a change of state which is not a cycle, the summation of all energy transfers across the boundary of the system is not necessarily zero. In accordance with the hypothesis of energy conservation, a net energy transfer *to* a system results in an equal *increase* of internal energy stored in the system. This may be written as follows:

$$Q = \Delta E + W \tag{3}$$

where Q is the heat transferred to the system during the process;
W is the work transferred from the system during the process;
ΔE is the change in the internal energy of the system during the process, and all these terms are expressed in the same units.

For convenience the equation is written in terms of a single heat quantity and a single work quantity; if several heat quantities are involved in the process, Q symbolizes the algebraic sum of these quantities, and similarly W is the algebraic sum of the work quantities.

Equation (3) is the usual statement of the First Law. It says that in any change of state the heat supplied to a system is equal to the increase of internal energy in the system plus the work done by the system. This means that energy is conserved in the operation. From this viewpoint the First Law is a particular formulation of the principle of the conservation of energy. Equation (3) may also be considered the definition of internal energy. The equation does not contain the internal energy E explicitly, but contains the change of internal energy ΔE for the process. Hence the definition does not give an absolute value for internal energy. Although no absolute value can be obtained for the internal energy of a system, it is possible to show that the internal energy has a definite value at every state of a system and is, therefore, a property of the system.

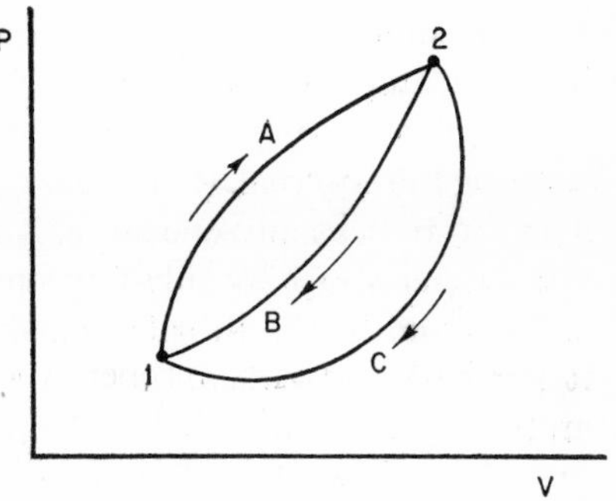

Fig. 5-2.

Consider a system to be initially in a state 1 which is changed to state 2, as represented on the pressure-volume plane by the points 1 and 2 in Fig. 5-2. Let the change of state from 1 to 2 be represented by the path A, for which the following relation will hold:

$$Q_A = \Delta E_A + W_A \tag{a}$$

Now complete a cycle by returning from 2 to 1 via a different path B. Then for path B

$$Q_B = \Delta E_B + W_B \tag{b}$$

The combined processes A and B constitute a cycle for which the sum of the work transfers must equal the sum of the heat transfers according to the First Law, Eq. (2). Hence

$$Q_A + Q_B = W_A + W_B$$

or

$$Q_A - W_A = W_B - Q_B$$

Substituting from (a) and (b)

$$\Delta E_A = -\Delta E_B \tag{c}$$

Now suppose the system had returned from 2 to 1 by any other path C. By the same reasoning,

$$\Delta E_A = -\Delta E_C$$

then

$$\Delta E_C = \Delta E_B \tag{d}$$

Hence the change of internal energy ΔE_{2-1} is the same for *any* path between 2 and 1. Assuming an arbitrary value assigned to the internal energy at 1, the value at any other point, 2, will depend solely upon the location of the point 2 and not upon the path traversed in going from 1 to 2. This shows that the internal energy E is a point function and therefore is a property of the system. (See Chap. 4 for the characteristics of a property.)

Since the internal energy is a property, if an arbitrary value is assigned to it at any one state, it is possible to determine the values at all other states by substitution in Eq. (3) of measured quantities of heat and work. In subsequent discussion of the internal energy it is to be understood that no absolute value is known but some arbitrary base value may have been assigned. In many cases it is unnecessary to establish a scale of internal energy since the change of internal energy ΔE satisfies the needs of the problem.

5-5 Internal energy—general and special concepts. The change of internal energy ΔE defined by Eq. (3) may appear in several different physical forms. The work done on the system may have increased the velocity of the system, in which case part of ΔE is kinetic energy; if the elevation of the system above the earth has changed, part of ΔE will be gravity potential energy; if the temperature or pressure of the system has changed, part of ΔE may be in the potential and kinetic energies of the molecular and atomic structure of the system. Then the internal energy of a system may be written in detail as

$$E = E_k + E_p + E_u \tag{4}$$

where E_k is the kinetic energy associated with observable velocities,
E_p is the gravity potential energy of the system,

E_u is the remaining internal energy of the system as defined by (4); that is, the internal energy which is not a function of motion and gravity.*

The quantity E_u is the portion of the internal energy which is stored in the molecular and atomic structure of the system. For a particular system this portion of the internal energy can be evaluated, without going into details of atoms and molecules, by measurements of work and heat transfer. But a more general approach is available through the properties of pure substances.

A pure substance is a substance of uniform constant chemical composition throughout its mass. This includes mixtures having constant composition. In a system composed of a pure substance, experience shows that in general the quantity E_u is fixed if the mass and any two other independent properties (excluding velocity and elevation) are fixed. Furthermore at any given state of such a system, E_u is directly proportional to the mass of the system. Hence *for a unit of mass of a pure substance there exists at each state a value of E_u which can be determined once and for all by measurements of heat and work.* These values can then be formulated as a function of two observable properties such as temperature and pressure. The value of E_u for a unit mass of a pure substance is called the *specific internal energy of the substance.*† The specific internal energy of a pure substance will be designated by a small u. The corresponding property for a system consisting of m lb of a pure substance will be designated by capital U.‡ Then $U = mu$. The First Law for a system composed of a *pure substance* in the absence of internal energy changes due to motion and gravity may then be written

$$Q = \Delta U + W \tag{5}$$

The property u is a function of two independent properties of a substance and may be formulated either as an equation, $u = f(p,t)$

* E will also depend upon surface tension, electrical, and magnetic effects all of which are assumed absent in this book. If they were present, additional terms would appear for them in Eq. (4).

† Specific internal energy may be evaluated not only by measurements of work and heat, but in part by other methods based upon theories of atomic and molecular structure. The manner in which the data are obtained has no bearing upon their use in thermodynamics.

‡ The distinction between E_u and U is that E_u applies to *any* system, where U applies only to a *pure substance.*

for example, or as a tabulation of the same relationship.* This special concept of internal energy is of great importance, first because many systems encountered in engineering thermodynamics may for purposes of analysis be considered as pure substances, and second because it provides a basis for the computation of the internal energy of more complicated systems.

To summarize: The general symbol E refers to the internal energy of any system, and includes all forms in which energy may be stored within the system. The special symbol U refers to the internal energy of a system composed of a pure substance and does not include internal energy associated with motion, gravity, surface tension, electricity, or magnetism. E is a function of motion, position in the gravity field, mass, composition, and the states of the system's components. U is a function of the mass and state of a system composed of a pure substance and is different for different substances. U is not a function of motion or position in the gravity field.

5-6 Specific heat at constant volume. In Chap. 3 specific heats were presented as measures of heat transferred in processes involving no work other than $p\,dv$ work. More general definitions can now be given. The specific heat of a substance at constant volume is defined as the rate of change of specific internal energy with respect to temperature when the volume is held constant. In the notation of the calculus this is written

$$c_v = \left(\frac{\partial u}{\partial T}\right)_v \tag{6}$$

From (6) it follows that for any *constant-volume* process,

$$(\Delta u)_v = \int_{T_1}^{T_2} c_v\,dT \tag{7}$$

The specific heat at constant volume as defined in Chap. 3 is a special case of the specific heat defined in Eq. (6); this is shown as follows. The First Law may be written for a stationary system composed of a unit mass of a pure substance,

$$Q = \Delta u + W \tag{8}$$

or in differential form,

$$dQ = du + dW \tag{9}$$

* See, for example, Keenan, J. H. and F. G. Keyes, *Thermodynamic Properties of Steam*. New York: Wiley, 1936; Keenan, J. H. and J. Kaye, *Gas Tables; Thermodynamic Properties of Air*. New York: Wiley, 1948. See also the tables in the Appendix.

Now for a process in the absence of work other than $p\,dv$ work

$$dW = p\,dv$$

and

$$dQ = du + p\,dv \tag{10}$$

If the volume is held constant, $dv = 0$ and $(dQ)_v = (du)_v$. Hence, in the absence of work other than $p\,dv$ work, Eq. (7) may be written for a constant-volume process as

$$(Q)_v = \int_{T_1}^{T_2} c_v\,dT$$

which is Eq. (3) of Chap. 3 written for a unit mass in a constant-volume process.

The specific heat at constant volume as defined in this section is a property of a pure substance and may be expressed as a function of any two observable properties such as temperature and pressure. The specific heat at constant volume may be used to calculate the internal energy change in *any* constant-volume process, but it may be used to calculate heat transferred *only* when all work is $p\,dv$ work (in which case work will be zero).

5-7 Enthalpy. The specific heat at constant pressure may also be defined in terms of properties of a substance. For this purpose a property is needed which bears the same relation to heat in a constant-pressure process as internal energy does in a constant-volume process. Internal energy change is equal to the heat transferred in a constant-volume process involving no work other than $p\,dv$ work. From Eq. (10) it is possible to derive an expression for the heat in a constant-pressure process involving no work other than $p\,dv$ work. In such a process in a stationary system composed of unit mass of a pure substance

$$dQ = du + p\,dv \tag{10}$$

At constant pressure $\quad p\,dv = d(pv)$; then

$$(dQ)_p = du + d(pv)$$

or

$$(dQ)_p = d(u + pv) \tag{11}$$

The quantity $(u + pv)$ will obviously be a property of a substance because for each set of definite values assigned to u, p, and v, the quantity $(u + pv)$ will have a definite value. A new property called *enthalpy* (en-thal′-py) is therefore arbitrarily defined as the sum of the internal energy and the pressure-volume product. Using the symbol h for specific enthalpy,

$$h = u + pv \tag{12}$$

Observe that Eq. (12) is an *arbitrary definition*. The reasons for choosing this definition involved heat, and a particular type of process at constant pressure, but the definition itself contains no restrictions. The enthalpy of a system of mass m is given by $H = mh$. No special symbols are used to distinguish between the enthalpy of a pure substance and the enthalpy of a general system. For a general system

$$H = E_u + pV \tag{13}$$

and for a pure substance

$$H = U + pV \tag{14}$$

In any given problem only one of these concepts will ordinarily be involved so no confusion need arise from lack of another symbol.

Since no absolute value for internal energy is known there can be no absolute value for enthalpy. However, since pressure and volume have absolute values, if any arbitrary base is set for internal energy, the values of enthalpy at all states will be fixed. In the steam tables of Keenan and Keyes an arbitrary base was set for enthalpy, and the values of internal energy thereby became fixed. Care must be exercised in substituting numerical values in the equation $h = u + pv$ since u and h are frequently in Btu/lb while the pv product will be in ft lbf/lbm when the usual units are used. The equation as written demands consistent units.

5-8 Specific heat at constant pressure. The specific heat of a substance at constant pressure is defined as the rate of change of specific enthalpy with respect to temperature when the pressure is held constant. This is written

$$c_p = \left(\frac{\partial h}{\partial T}\right)_p \tag{15}$$

From this it follows that for any *constant-pressure* process

$$(\Delta h)_p = \int_{T_1}^{T_2} c_p \, dT \tag{16}$$

The specific heat at constant pressure defined in Chap. 3 is a special case of the specific heat defined in Eq. (15). It is evident from

Eqs. (11) and (12) that for a constant-pressure process in the absence of work other than $p\,dv$ work, the heat transferred could be substituted for $(\Delta h)_p$ in Eq. (16), giving

$$(Q)_p = \int_{T_1}^{T_2} c_p\,dT$$

which is Eq. (3) of Chap. 3 written for a unit mass in a constant-pressure process.

The specific heat at constant pressure is a property of a pure substance. It may be used to calculate the enthalpy change in *any* constant-pressure process, but it may be used to calculate the heat transferred *only* when no work other than $p\,dv$ work is transferred.

The reason for defining specific heats in terms of properties rather than in terms of heat quantities is that the definition automatically makes the specific heats properties. This permits the development of many useful relations among the specific heats and other properties. The specific heats still serve a useful purpose in heat transfer computations as before but they are now useful in many more ways.

5-9 Latent heats, which were discussed in Sec. 3-10, may be defined as enthalpy changes. Hereafter in this book, when the term *latent heat* is used it will mean *change of enthalpy at constant pressure* for a given phase change.

5-10 Applications of the first law to stationary systems. In this section some examples will be given of the use of the First Law in problems involving stationary systems.

Example 1. A stationary mass of gas is compressed without friction from an initial state of 10 cu ft and 15 psia to a final state of 5 cu ft and 15 psia, the pressure remaining constant during the process. There is a transfer of 34.7 Btu of heat from the gas during the process. How much does the internal energy of the gas change?

Solution: Let the mass of gas be the system. By the First Law $Q = \Delta E + W$. Q is given, ΔE is to be found; hence seek a means of determining W. In Sec. 2-3 it was pointed out that when a stationary fluid system of uniform pressure is frictionless the only work associated with it is that given by $\int p\,dV$. Hence $W = \int p\,dV$. For constant pressure

$$W = p(V_2 - V_1)$$
$$W = (15)(144)(5 - 10)$$
$$W = -10{,}800 \text{ ft lb}$$

the negative sign indicating work done *on* the system. Now substituting in the First Law equation and using units of Btu,

$$-34.7 = \Delta E - \frac{10{,}800}{778}$$

Note that the heat is negative since it is transferred *from* the system. Then

$$\Delta E = -34.7 + 13.9 = -20.8 \text{ Btu}$$

The internal energy of the gas decreases 20.8 Btu in the process. Observe the necessity for adherence to consistent units and sign conventions.

Example 2. A stationary system of constant volume experiences a temperature rise of 35°F when a certain process occurs. The heat transferred in the process is 34 Btu. The specific heat at constant volume for the pure substance in the system is 1.2 Btu/lb F, and the system contains 2 lb of substance. Determine the internal energy change and the work done.

Solution: The First Law applies in the form

$$Q = \Delta U + W$$

Q is given; ΔU can be found in a constant-volume process from

$$\Delta U = m \int_{T_1}^{T_2} c_v \, dT$$

Then

$$\Delta U = 2 \int_{T_1}^{T_2} 1.2 \, dT = 2.4(T_2 - T_1)$$

$$\Delta U = 2.4\,(35) = 84.0 \text{ Btu}$$

Then

$$W = Q - \Delta U = 34 - 84 = -50 \text{ Btu}$$

Observe that even though the volume was constant the work was not zero.

Example 3. In the steam tables of Keenan and Keyes it is set forth that during the evaporation of 1 lb of water at 500 psia and 467.01°F the specific volume increases from 0.0197 cu ft/lb to 0.9278 cu ft/lb, while the enthalpy increases from 449.4 to 1204.4 Btu/lb. How much work is done by a stationary system consisting of 1 lb of water when, because of an inflow of heat, the system changes from liquid to vapor at 500 psia? How much does the internal energy change?

Solution: The only work that can be done *by* a stationary system of uniform pressure is given by $\int p \, dV$ (in the absence of surface tension etc.). Therefore the work of the system in this constant-pressure process is

$$W = p(V_2 - V_1)$$

$$W = (500)(144)(0.9278 - 0.0197)$$

$$W = 65{,}380 \text{ ft lb} = 84.0 \text{ Btu}$$

Since enthalpy, pressure, and volume are given for both end states of the process, the internal energy change may be found from the definition of enthalpy,

$$h = u + pv$$

Then $$\Delta h = \Delta(u + pv) = \Delta u + \Delta(pv)$$

or $$\Delta u = \Delta h - \Delta(pv)$$

$$\Delta u = 1204.4 - 449.4 - \frac{(144)(500)}{778}(0.9278 - 0.0197)$$

$$\Delta u = 755.0 - 84.0 = 671.0 \text{ Btu}$$

Observe the conversion of psia to psfa (lb/sq ft abs) and the conversion of the ft lb units of the pv product to Btu. Another method may be used to find the internal energy change. In a constant-pressure process in which the only work is given by $\int p\, dv$, the heat transferred is given by

$$Q = \Delta h$$

as shown in Sec. 5-7, Eqs. (11) and (12). But for *any* process in a stationary system consisting of a pure substance the First Law may be written

$$Q = \Delta U + W$$

Therefore in this problem

$$\Delta h = \Delta u + W$$

or $$\Delta u = \Delta h - W$$

Then $$\Delta u = 755.0 - 84.0 = 671.0 \text{ Btu}$$

which is the result obtained by the first method.

Example 4. The internal energy of a certain substance is given by the following equation:

$$u = 0.480pv + 35$$

where u is given in Btu/lb, p is in psia, and v is in cu ft/lb. A system composed of 3 lb of this substance expands from an initial pressure of 75 psia and volume of 6 cu ft to a final pressure of 15 psia in a process in which pressure and volume are related by $pv^{1.2}$ = constant.

(a) If the expansion is frictionless, determine Q, ΔU and W for the process.

(b) In another process the same system again expands according to the same pressure-volume relationship as in part (a), and from the same initial state to the same final state as in part (a), but the heat in this case is 30 Btu. Find the work for this process.

(c) Explain the difference in work in parts (a) and (b).

Solution: (a) ΔU can be determined from the equation given.

$$\Delta U = U_2 - U_1 = m(u_2 - u_1)$$
$$\Delta U = 3(0.480p_2v_2 + 35 - 0.480p_1v_1 - 35)$$
$$\Delta U = 3(0.480)(p_2v_2 - p_1v_1)$$

but
$$v = V/m = V/3$$

so
$$\Delta U = (0.480)(p_2V_2 - p_1V_1)$$

p_1, V_1 and p_2 are given, and $p_1V_1^{1.2} = p_2V_2^{1.2}$

Then
$$V_2 = V_1\left(\frac{p_1}{p_2}\right)^{1/1.2} = 6\left(\frac{75}{15}\right)^{1/1.2}$$
$$V_2 = 6(3.83) = 22.98 \text{ cu ft}$$
$$\Delta U = (0.480)[(15)(22.98) - (75)(6)]$$
$$\Delta U = (0.480)(275.8 - 450) = -83.6 \text{ Btu}$$

Since the process is frictionless, $W = \int p\,dV$. For the relation $pV^{1.2} =$ constant,

$$W = \frac{p_2V_2 - p_1V_1}{1 - 1.2}$$
$$W = \frac{(275.8 - 450)(144)}{-0.2}$$
$$W = 125{,}400 \text{ ft lb} = 161 \text{ Btu}$$

The heat can now be found from the First Law.

$$Q = \Delta U + W$$
$$Q = -83.6 + 161 = 77.4 \text{ Btu}$$

(b) The system changes along a path identical to that for part (a). Therefore $\int p\,dV$ is the same as for part (a). But it is also true that the system changes between the same end states as in part (a). Therefore ΔU is the same as for part (a).

$$\Delta U = -83.6 \text{ Btu}$$
$$Q \text{ is given} = 30 \text{ Btu}$$

By the First Law,

$$W = Q - \Delta U = 30 - (-83.6) = 113.6 \text{ Btu.}$$

(c) The work in (b) is not equal to $\int p\,dV$. This simply means that process (b) is not a frictionless process. The point in this example is that W is not always equal to $\int p\,dV$ but depends upon the type of process as well as the path. The change of internal energy, ΔU, on the other hand, depends only upon the end states and not at all upon the process. The First Law also applies regardless of the nature of the process. Hence the work in (b) can be determined from the known ΔU and the given Q, by use of the First Law.

5-11 Summary. The First Law of Thermodynamics is a special statement of the law of conservation of energy. It may be written for a cycle as

$$(\Sigma Q)_{\text{cycle}} = (\Sigma W)_{\text{cycle}}$$

or for a process as

$$Q = \Delta E + W$$

The latter equation may be taken as a definition of the quantity *internal energy, E*. Internal energy has no absolute value but its change between any two states of a system does have a definite value. Hence internal energy is a property of a system.

The general concept of the internal energy of a system includes energy in storage as kinetic energy associated with observable velocities, as gravity potential energy of the system as a whole, and as the potential and kinetic energy of the atoms and molecules of the system.* The first two may often be computed separately by the laws of mechanics but the third, from the viewpoint of thermodynamics, is obtained from measurements of heat and work relative to the system, or to the substances of which the system is composed.

A special concept of internal energy is the specific internal energy of a pure substance. This is the internal energy of a unit mass of a pure substance in the absence of gravity and motion (and surface tension, electricity and magnetism). Although the specific internal energy change Δu can be determined originally only by experimental measurements, it can be formulated as a function of any two independent properties of a substance for subsequent use. The symbol U is used to represent mu, the internal energy of m pounds of a substance.

The specific heat at constant volume of a pure substance is defined in terms of the internal energy of the substance by

$$c_v = \left(\frac{\partial u}{\partial T}\right)_v$$

This definition includes as a special case the definition given in Chap. 3. By the use of the more general definition the specific heat at constant volume provides a means of calculating internal energy changes in *any* constant-volume process while still providing the means of calculating heat transfer in constant-volume processes involving no work.

* See first footnote on p. 51.

The specific enthalpy of h of a substance is a property arbitrarily defined by $h = u + pv$. All terms in this equation must have the same units, for example ft lb. The symbol H is used to represent mh, the enthalpy of m pounds of a substance.

The specific heat at constant pressure of a pure substance is defined in terms of the specific enthalpy of the substance by

$$c_p = \left(\frac{\partial h}{\partial T}\right)_p$$

This definition includes as a special case the definition given in Chap. 3. By the use of the more general definition the specific heat at constant pressure provides a means of calculating enthalpy changes in *any* constant-pressure process while still providing the means of calculating heat transfer in constant-pressure processes involving no work other than $p\,dv$ work.

In a frictionless constant-volume process in the absence of motion the heat transferred is equal to the change of internal energy. In a frictionless constant-pressure process, in the absence of motion, the heat transferred is equal to the change of enthalpy. These two statements presume no work from electrical, magnetic or surface-tension effects.

PROBLEMS*

5-1. (a) A system consists of the air in a tire pump and the connected tire. The tire-pump plunger is pushed in and the temperature of the pump cylinder is observed to rise. Give the signs of Q and W for the process. (b) In the same apparatus the fluid system is a frictionless, incompressible liquid; are W and ΔE positive or negative for the fluid system when the pump plunger is pushed in?

5-2. A certain combustion experiment is made by burning a mixture of fuel and oxygen in a constant-volume "bomb" surrounded by a water bath. During the experiment the temperature of the water bath is observed to rise. For the system comprising the fuel and oxygen give the signs of Q, ΔE and W for the process.

5-3. (a) A truck powered by an electric storage battery climbs a hill at constant speed; the storage battery and the motor are at higher temperatures than the surroundings. Considering the entire truck as a system, indicate on a sketch the streams of heat and work crossing the boundary of the system;

* In these problems, when the sign of a quantity is asked for, if the quantity is zero write zero as the answer.

would ΔE be positive or negative? (b) A trolley car climbs a hill at constant speed; the motor is at a higher temperature than the surroundings. Considering the entire car as a system, indicate on a sketch the streams of heat and work crossing the boundary of the system; would ΔE be positive or negative? (c) A truck powered by a gasoline engine climbs a hill at constant speed; the engine takes in air from the atmosphere and discharges hot exhaust gases. Why cannot a simple system be chosen to analyze this process, similarly to cases (a) and (b), above?

5-4. A slow chemical reaction takes place in a fluid at the constant pressure of 15 psia. The fluid is surrounded by a perfect heat insulator during the reaction which begins at state 1 and ends at state 2. The insulation is then removed and 100 Btu of heat flows to the surroundings as the fluid goes to state 3. The following data are observed for the fluid at states 1, 2, and 3:

State	V (cu ft)	t (F)
1	0.1	70
2	10	700
3	2	70

For the fluid system determine the internal energy at states 2 and 3 if the internal energy at state 1 is zero.

5-5. A piston and cylinder machine contains a fluid system which passes through a cycle composed of four processes a-b, b-c, c-d, and d-a. During the cycle, the total negative heat transfer is 150 Btu. The system completes 100 cycles per minute. (a) Complete the following table; show your method for each item.

Process	Q (Btu/min)	W (ft lb/min)	ΔE (Btu/min)
a-b	0	-1.6×10^6	———
b-c	20,000	0	———
c-d	−2,000	———	−34,960
d-a	———	———	———

(b) Find the net rate of work output in horsepower.

5-6. A household refrigerator is loaded with fresh food and closed. Consider the whole refrigerator and contents as a system. The machine uses 1 kwh of electric energy (work) in cooling the food, and the internal energy of the system decreases 5000 Btu as the temperature drops. Find the magnitude and sign of the heat transfer for the process.

5-7. One pound of liquid having a constant specific heat of 0.6 Btu/lb F is stirred in a well insulated chamber, Fig. 5-1, causing the temperature to rise 25°F. Find the change of internal energy of the liquid; find the work for the process; explain your solution on the basis of the First Law.

5-8. The same liquid as in Problem 5-7 is stirred in a conducting chamber.

During the process 15 Btu of heat are transferred from the liquid to the surroundings, while the temperature of the liquid is rising 25°F. Find the change of internal energy of the liquid, and the work for the process; compare with the previous problem, and explain the solution on the basis of the First Law.

5-9. The properties of a certain substance are related as follows:

$$u = 673 + 0.320t, \qquad pv = 0.600(t + 460)$$

where u is specific internal energy, Btu/lb; t is temperature, F; p is pressure, psia; v is specific volume, cu ft/lb.

(a) Find the specific heat at constant volume. (b) Find the specific heat at constant pressure.

5-10. A system composed of 2 lb of the substance of Problem 5-9 expands in a cylinder and piston machine from an initial state of 150 psia, 200°F to a final temperature of 60°F. If there is no heat transfer find the net work for the process.

5-11. If all the work in the expansion of Problem 5-10 is done on the moving piston, show that the equation representing the path of the expansion in the pv plane is $pv^{1.35}$ = constant.

5-12. (a) A mixture of gases expands at constant pressure from 100 psia, 1 cu ft to 100 psia, 2 cu ft, with 80 Btu positive heat transfer; there is no work other than that done on a piston. Find the change of internal energy of the gaseous mixture. (b) The same mixture expands through the same state path while a stirring device does 10 Btu of work on the gas. Find the change of internal energy, the work, and the heat transfer for the process.

5-13. A fluid is confined in a cylinder by a spring-loaded frictionless piston so that the pressure in the fluid is a linear function of the volume ($p = a + bv$). The internal energy of the fluid is given by the following equation:

$$U = 32 + 0.004pV$$

where U is in Btu, p in lb/sq ft, and V in cu ft. If the fluid changes from an initial state of 25 psia, 1 cu ft, to a final state of 60 psia, 2 cu ft, with no work other than that done on the piston, find the direction and magnitude of the work and the heat transfer.

5-14. In a certain process in a stationary system, in which all work is $p\,dV$ work, the pressure is given by $p = (300V + 1500)$ and the internal energy is given by $U = 125 + 0.012pV$ where U is in Btu, p in lb/sq ft, and V in cu ft. Find the work and the heat transfer if V changes from 6 to 12 cu ft.

5-15. A well-insulated constant-pressure chamber (1 atm pressure) contains 10 lb of liquid water and 1 lb of chopped ice at a certain instant. A motor-driven stirrer has been operating in the mixture for some time, and continues to operate, with a constant motor output to the stirrer of 25 watts.

The specific heat of liquid water is 1 Btu/lb F, the specific heat of ice is 0.5 Btu/lb F, and the latent heat of fusion of water is 143 Btu/lb. The volume decrease of the water during melting is 0.00145 cu ft/lb; so small that $\Delta(pV)$ is negligible, and $\Delta U = \Delta H$. How much will the temperature change, (a) in one hour? (b) in three hours?

Chapter 6

FLOW PROCESSES: FIRST LAW ANALYSIS

6-1 Flow processes—control surface. The development of the First Law in Chap. 5 resulted in an equation which applies to any system in any process,

$$Q = \Delta E + W \tag{5-(3)}$$

Since E is a function of many variables it was expanded in Eq. 5-(4) as follows:

$$E = E_k + E_p + E_u \tag{5-(4)}$$

the terms on the right side representing, in order, the kinetic energy of the system, the gravity potential energy of the system, and the residual internal energy not otherwise accounted for. From this point Chap. 5 was devoted to the special case of a pure substance in the absence of motion and gravity. Since many practical engineering operations involve flow of fluids it becomes desirable to develop the First Law for convenient application to flow processes.

From Eqs. 5-(3) and 5-(4) a general equation which accounts specifically for the effects of motion and gravity may be written as follows:

$$Q = \Delta E_k + \Delta E_p + \Delta E_u + W \tag{1}$$

As in Chap. 5 the system may now be considered to consist of a pure substance. On this basis, Eq. (1) may be written

$$Q = \Delta E_k + \Delta E_p + \Delta U + W \tag{2}$$

Equation (2) is the First Law for a system composed of a pure substance in the absence of surface tension, electricity, and magnetism. The system is a particular mass of substance, but it is free to move from place to place. In Fig. 6-1 a portion of a power plant is shown diagrammatically; steam flows to the turbine at high pressure, expands in the turbine while doing work on the rotor, and then leaves at low pressure through the exhaust pipe. If a thermodynamic system consisting of a mass of steam is used to analyze this operation by means of Eq. (1) or (2), the system must be followed in its travels through the turbine, account being taken of work and heat transfers all the way through. Although the system approach is quite valid there is another approach which will be presented here. Instead of concentrating attention upon a certain mass of working substance, which constitutes a moving system, attention is concentrated upon a

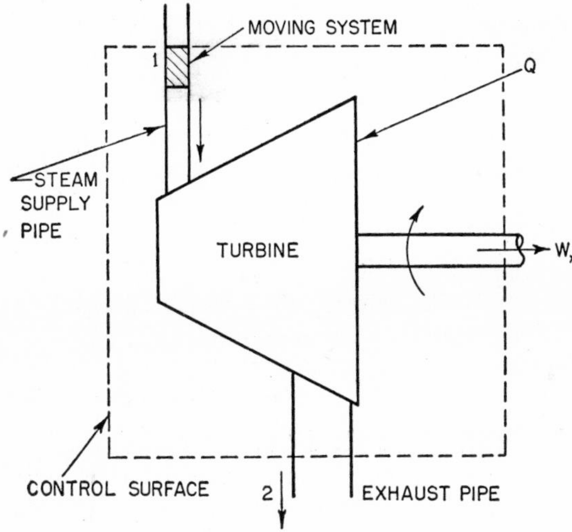

Fig. 6-1. Flow process.

certain *fixed region in space* called a *control volume,* through which the working substance flows. In Fig. 6-1 the broken line represents the surface of a control volume; this surface is known as a *control surface.* The method of analysis is to inspect the control surface and account for all quantities of energy transferred through this surface. By the conservation of energy the net inflow of energy through the control surface in a certain time must be equal to the increase during the same time of the energy stored within the control volume. Since the control volume experiences not only energy transfer, but also mass transfer, and since the energy transfer is a function of the mass transfer, it becomes necessary to make an accounting of mass as well as of energy. By the principle of the conservation of mass the net inflow of mass through the control surface must equal the increase of mass stored within the control volume. In the mechanics of fluids the principle of the conservation of momentum is similarly applied by means of the control surface.

6-2 Steady flow—definition. In the general case, an arrangement such as illustrated in Fig. 6-1 would be subject to changing conditions from time to time so that the rates of flow of mass and energy through the control surface would change with time and the quantities of mass and energy within the control volume would also change with time. For many purposes, however, it is satisfactory to make analyses upon the assumption of steady flow conditions. *A steady flow process is one in which there is no change in conditions within the control volume* during the time under consideration. This means that the velocity, internal energy, specific volume, and all other properties of the working fluid at each point within the control volume do not vary with time. The operation of a machine like a reciprocating engine in which conditions do change from time to time may be treated from the external viewpoint as a steady flow process if the operation consists of the repetition, at a constant rate, of a particular process; for, during a period of time involving a large number of identical operations, the conditions within the control volume have average values which remain constant as time goes on. Hence no overall change is observed from an external viewpoint. The conditions of a steady flow process, from the external viewpoint, are two: (1) the streams of mass crossing the control surface are constant and add to a net value of zero; (2) the streams of energy crossing the control surface are constant and add to a net value of zero.

Considering the process illustrated in Fig. 6-1 as a steady flow process, the *material balance* (or *mass balance*) would have only two terms since there is only one inflow (at 1) and one outflow (at 2). For steady flow, the mass entering at 1 per unit time must equal the mass leaving at 2 per unit time. The *energy balance* would have at least four terms to account for heat flow through the control surface, work flow through the control surface, and energy carried through the control surface as the internal energy of each of the two mass streams. The algebraic sum of all the energy transfer quantities per unit time must be zero for steady flow.

6-3 Material balance and energy balance in simple steady flow processes. In many processes involving flow only two streams of fluid appear, one entering and one leaving the apparatus. Figure 6-2 is a diagram of such a case in which the apparatus under consideration is enclosed by a control surface. The control surface is located so that it cuts the fluid streams entering and leaving the apparatus at sections where flow patterns and properties are steady

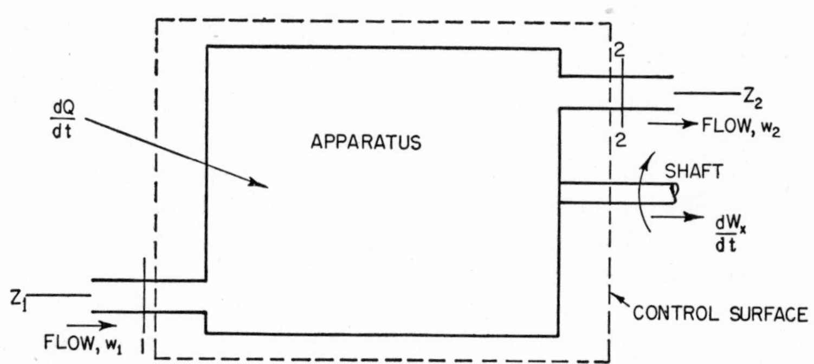

Fig. 6-2. Steady flow process.

with time. The inflowing stream is cut normal to its flow in section 1, and the outflowing stream similarly in section 2. The remainder of the control surface is so located as to facilitate accounting for any work or heat that may be flowing through the control surface. Since the control surface is stationary in space, work can cross it only in connection with the fluid flow or by means of shafts or other mechanisms extending through the control surface. (Work could also cross the control surface by electrical or magnetic effects, and would be classified as "shaft work" in the energy balance.)

The following quantities are defined with reference to Fig. 6-2.

A_1, A_2*	cross-section of stream, sq ft
w_1, w_2	mass flow rate, lb/sec
p_1, p_2	pressure, absolute, lb/sq ft
v_1, v_2	specific volume, cu ft/lb
u_1, u_2	specific internal energy, ft lbf/lbm
$\overline{V}_1$, $\overline{V}_2$	velocity, ft/sec

* Subscripts 1 and 2 refer to the corresponding sections.

z_1, z_2 elevation above an arbitrary datum, ft

dQ/dt net rate of heat transfer through the control surface, ft lb/sec

dW_x/dt "shaft work"—net rate of work transfer through the control surface, *exclusive of work done at sections 1 and 2 in transferring the fluid through the control surface*, ft lb/sec

It is assumed that at sections 1 and 2 all properties (pressure, volume, velocity, elevation etc.) are constant throughout the area of the section. In practice average values are used for properties which vary over the section. Though not strictly correct, this is normally a satisfactory approximation when the flow through the control surface is parallel and steady over the area of the section. It is further assumed that no state change is experienced by the fluid while crossing the control surface.

By the conservation of mass, if no mass change occurs in the control volume, the mass flow rate entering must equal the mass flow rate leaving or

$$w_1 = w_2 \tag{3}$$

Figure 6-3 is an enlarged view of the region near section 1. The cross-hatched region represents the fluid which will cross section 1

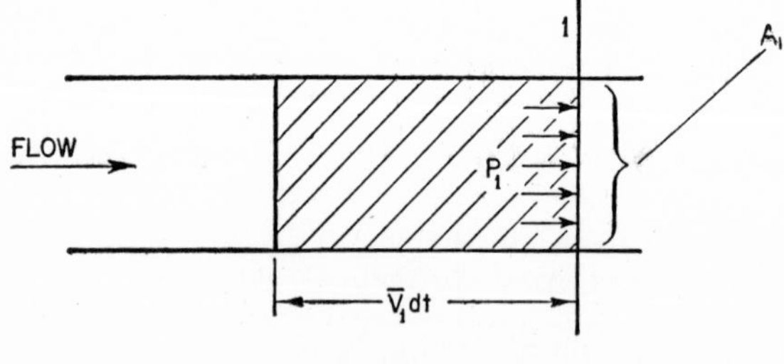

Fig. 6-3.

in the time interval dt; this fluid has a volume $A_1\overline{V}_1\,dt$. The mass of this volume is

$$\frac{A_1\overline{V}_1}{v_1}\,dt = w_1\,dt \tag{4}$$

Therefore the mass crossing section 1 in unit time is

$$w_1 = \frac{A_1\overline{V}_1}{v_1} \tag{5}$$

By similar reasoning at section 2

$$w_2 = \frac{A_2\overline{V}_2}{v_2} \tag{6}$$

and from Eqs. (3), (5) and (6)

$$\frac{A_1\overline{V}_1}{v_1} = \frac{A_2\overline{V}_2}{v_2} \tag{7}$$

This equation, known as the *Equation of Continuity*, expresses the conservation of mass for the process.

By the conservation of energy the total rate of flow of all energy streams entering the control volume must equal the total rate of flow of all energy streams leaving the control volume. This may be expressed in the following equation:

$$w_1e_1 + w_1W_1 + \frac{dQ}{dt} = w_2e_2 + w_2W_2 + \frac{dW_x}{dt} \tag{8}$$

where e_1 is the internal energy carried into the control volume with unit mass of fluid at section 1,

W_1 is the work transferred into the control volume while introducing unit mass of fluid at section 1,

e_2 is the internal energy carried out of the control volume with unit mass of fluid at section 2,

W_2 is the work transferred out of the control volume while extracting unit mass of fluid at section 2, and

w_1, w_2, Q, and W_x are as defined in relation to Fig. 6-2.

The internal energy e_1 can be expressed by writing Eq. 5-(4) for a system of unit mass,

$$e = e_k + e_p + e_u \tag{9}$$

where e_k is kinetic energy per unit mass $= \overline{V}^2/2g_0$; g_0 is the dimensional constant 32.17 ft lbm/lbf sec^2.

e_p is gravity potential energy per unit mass $= z(g/g_0)$. The fraction g/g_0 may be taken as numerically equal to unity.

e_u is internal energy of the fluid substance per unit mass $= u$, the specific internal energy for a pure substance. Then

$$e_1 = \frac{\overline{V}_1^2}{2g_0} + z_1\frac{g}{g_0} + u_1 \tag{10}$$

The work W_1, often called *flow work*, may be derived as follows. Referring to Fig. 6-3, the fluid crossing section 1 exerts a normal pressure p_1 against the area A_1 giving a total force p_1A_1. In time dt this force moves in its own direction a distance $\overline{V}_1\,dt$, thereby doing work. The work in time dt is $p_1A_1\overline{V}_1\,dt$, or the work per unit time is $p_1A_1\overline{V}_1$. But by the definitions of w_1 and W_1 the work per unit time is also w_1W_1. Then

$$w_1W_1 = p_1A_1\overline{V}_1 \tag{11}$$

Substituting from Eq. (5)

$$W_1 = p_1v_1 \tag{12}$$

A similar equation can be written for W_2.

Equation (8) may now be rewritten, using Eqs. (10) and (12) and their counterparts for section 2, giving

$$w_1\left(\frac{\overline{V}_1^2}{2g_0} + z_1\frac{g}{g_0} + u_1\right) + w_1p_1v_1 + \frac{dQ}{dt} = w_2\left(\frac{\overline{V}_2^2}{2g_0} + z_2\frac{g}{g_0} + u_2\right) + w_2p_2v_2 + \frac{dW_x}{dt} \tag{13}$$

Since $w_1 = w_2$, let $w = w_1 = w_2 = dm/dt$, and divide through by dm/dt, giving

$$\frac{\overline{V}_1^2}{2g_0} + z_1\frac{g}{g_0} + u_1 + p_1v_1 + \frac{dQ}{dm} = \frac{\overline{V}_2^2}{2g_0} + z_2\frac{g}{g_0} + u_2 + p_2v_2 + \frac{dW_x}{dm} \tag{14}$$

This is the *steady flow energy equation* for the special case of a single stream of fluid entering and a single stream of fluid leaving the control volume. All terms representing energy flow *into* the control volume are on the left side of the equation and all terms representing energy flow *out* are on the right side. All terms in Eq. (14) represent energy flow per unit mass of fluid flow. Equation (13) is the same equation written in terms of energy flow per unit time. Either basis may be used as may be convenient in any particular problem. It is common practice to write simply Q and W_x in place of the derivatives with respect to time or mass, since in any one problem all terms will be reduced to one basis or the other, and the time rate or mass rate is understood.

The basis of energy flow per unit mass is usually more convenient when only a single stream of fluid entering and leaving is involved. The basis of energy flow per unit time is more convenient when more than one fluid stream is involved. The equations as presented would have units of ft lb for all energy terms, but obviously the units may be changed to Btu by dividing through by 778. In solving problems a definite choice of basis should be made before attempting to write the energy equation. It should be decided at the beginning whether the analysis will be on a time basis or on a mass basis and, before substituting any numerical values, a definite choice of energy units should be made.

Equations (13) and (14) contain the group of properties $(u + pv)$ which has already been given the name enthalpy and the symbol h (Sec. 5-7). Thus Eq. (14) may be, and frequently is, written

$$\frac{\overline{V_1^2}}{2g_0} + z_1\frac{g}{g_0} + h_1 + \frac{dQ}{dm} = \frac{\overline{V_2^2}}{2g_0} + z_2\frac{g}{g_0} + h_2 + \frac{dW_x}{dm} \tag{15}$$

It is largely this fact which accounts for the great practical utility of the enthalpy property, since many engineering processes are steady flow processes, and the use of Eq. (15) reduces computation when the enthalpy is known.

When more than one stream of substance enters or leaves the control volume, the mass equation and the energy equation for steady flow are easily written by adding the terms for the individual streams

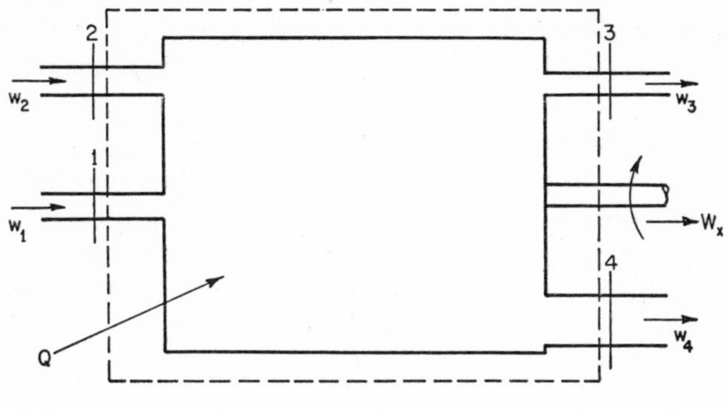

Fig. 6-4.

of fluid. Consider, for example, the control volume shown in Fig. 6-4. Fluid is entering at sections 1 and 2, and leaving at sections 3 and 4. For steady flow the mass balance will give

$$w_1 + w_2 = w_3 + w_4 \tag{16}$$

or
$$\frac{A_1\overline{V}_1}{v_1} + \frac{A_2\overline{V}_2}{v_2} = \frac{A_3\overline{V}_3}{v_3} + \frac{A_4\overline{V}_4}{v_4} \tag{17}$$

The energy balance will give, on a time basis,

$$\left(h_1 + \frac{\overline{V}_1^2}{2g_0} + z_1\frac{g}{g_0}\right)w_1 + \left(h_2 + \frac{\overline{V}_2^2}{2g_0} + z_2\frac{g}{g_0}\right)w_2 + Q$$
$$= \left(h_3 + \frac{\overline{V}_3^2}{2g_0} + z_3\frac{g}{g_0}\right)w_3 + \left(h_4 + \frac{\overline{V}_4^2}{2g_0} + z_4\frac{g}{g_0}\right)w_4 + W_x \tag{18}$$

The symbols Q and W_x are understood to denote time rates of energy flow.

VALIDITY OF STEADY FLOW ENERGY EQUATION

The assumptions upon which the above development of the steady flow energy equation depends are generally satisfied for pipe line flow through the control surface, but should be scrutinized with some care when applied to the internal analysis of machinery. With large variations in velocity across a stream, the square of the average velocity may not give the average kinetic energy. With large temperature differences, and in combustion processes, the assumption of no state change in the substance crossing the control surface may not be valid. The existence of shear stresses in the fluid crossing the control surface may result in a work transfer not accounted for in the equation.

The steady flow energy equation is not at all a "general" energy equation, despite popular use of this misnomer.

6-4 Applications of the steady flow equations. The steady flow energy equation applies to a wide variety of processes including pipe line flow, heat transfer processes, combustion processes, and mechanical power generation in engines and turbines. The relative importance of the various terms in the equation differs in the different applications and some terms are usually zero or negligible in each problem. In solving a problem, however, it is best to write the complete equation first and then eliminate the unnecessary terms. There can be no general rule as to when a given term may be considered negligible; this must be decided in the light of the circumstances and the objectives of the particular problem. Several examples will now be given to show possible orders of magnitude for the different terms and to illustrate the use of the steady flow equations in various types of processes.

Example 1. A turbine operates under steady flow conditions, receiving steam at the following state: pressure 170 psia, temperature 368.4°F, specific volume 2.675 cu ft/lb, internal energy 1111.9 Btu/lb, velocity 6000 ft/min, and elevation 10 ft. The steam leaves the turbine at the following state: pressure 3.0 psia, temperature 141.5°F, specific volume 100.9 cu ft/lb, internal energy 914.6 Btu/lb, velocity 300 ft/sec, and elevation 0 ft. Heat is lost to the surroundings at the rate of 1000 Btu/hr. If the rate of steam flow through the turbine is 2500 lb/hr what is the power output of the turbine in horsepower?

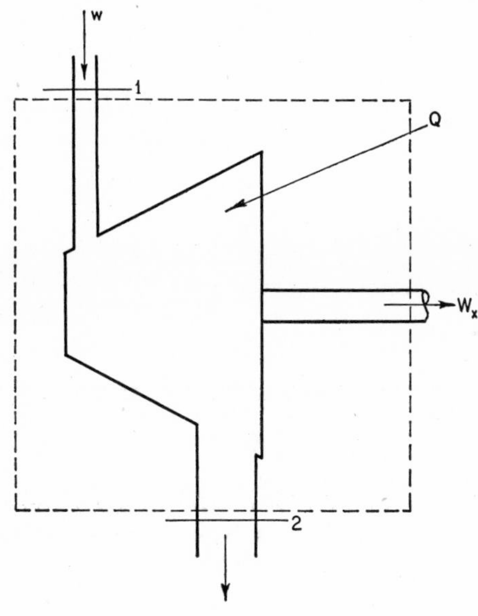

Example 1.

Solution: Sketch the process and locate a control surface. Choose a basis and units; take energy per pound as basis, Btu as energy unit, and lb/hr as flow-rate unit. Write the equations for mass and energy conservation.

$$w_1 = w_2 = w$$

$$u_1 + p_1v_1 + \frac{\overline{V}_1^2}{2g_0} + z_1\frac{g}{g_0} + Q = u_2 + p_2v_2 + \frac{\overline{V}_2^2}{2g_0} + z_2\frac{g}{g_0} + W_x$$

Substitution of numerical values with conversion to units of Btu/lb gives:

$$1111.9 + \frac{(170)(144)(2.675)}{778} + \frac{(6000/60)^2}{2(32.2)(778)} + \frac{10}{778} - \frac{1000}{2500}$$
$$= 914.6 + \frac{(3)(144)(100.9)}{778} + \frac{(300)^2}{2(32.2)(778)} + 0 + \frac{W_x}{2500}$$

$$1111.9 + 84.2 + 0.2 + 0.013 - 0.4 = 914.6 + 56.1 + 1.79 + 0 + \frac{W_x}{2500}$$

$$W_x = 558{,}500 \text{ Btu/hr} = 219 \text{ hp}$$

The values appearing in the above problem are typical for a steam power plant problem. The kinetic energy terms for the steam and exhaust pipes are not very large, and the potential energy terms are exceedingly small. However, the importance of the various terms depends upon the conditions of the particular problem. Observe that it is not necessarily permissible to ignore some terms in the equation simply because they are small compared to some other terms. It is necessary to consider the magnitude of the final answer rather than the magnitudes of individual terms to determine the relative importance of the terms. In the above example, in order to obtain 1 percent precision in the final result it is necessary to have 0.2 percent precision in the internal energy values; however, the *pv* product need be known

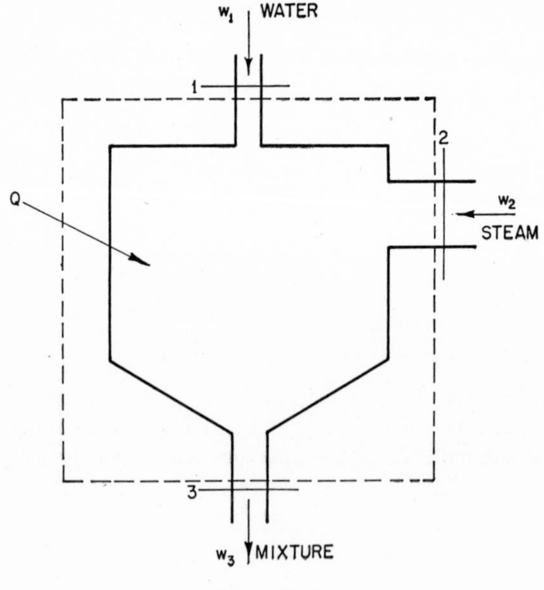

Example 2.

within only 3 percent, and the kinetic energy within only 100 percent to obtain the same overall precision.

Example 2. A certain water heater operates under steady flow conditions receiving 500 lb/min of water at temperature 165°F and enthalpy 132.9 Btu/lb. The water is heated by mixing with steam which is supplied to the heater at temperature 215°F and enthalpy 1150 Btu/lb. The mixture leaves the heater as liquid water at temperature 212°F, enthalpy 180 Btu/lb. How many pounds per hour of steam must be supplied to the heater? State the assumptions necessary in the solution.

Solution: Sketch the process and locate a control surface. By conservation of mass

$$w_1 + w_2 = w_3 \tag{a}$$

Choose a basis for the energy equation. Since more than one stream is entering the control volume a basis of energy per unit time will be best. Use energy units of Btu/min and flow units of lb/min. The energy equation may be written

$$w_1\left(h_1 + \frac{\overline{V}_1^2}{2g_0} + z_1\frac{g}{g_0}\right) + w_2\left(h_2 + \frac{\overline{V}_2^2}{2g_0} + z_2\frac{g}{g_0}\right) + Q = w_3\left(h_3 + \frac{\overline{V}_3^2}{2g_0} + z_3\frac{g}{g_0}\right) + W_x \tag{b}$$

By the nature of the process there is no shaft work. Potential and kinetic energy terms are usually small, and since no information is available they must be assumed to balance to zero. Since no information is given on heat transfer it is necessary either to estimate the heat transfer or to assume it is negligible. It will be assumed that the heater is insulated so that the heat transfer to the surroundings is negligible. Then Eq. (b) reduces to

$$w_1h_1 + w_2h_2 = w_3h_3 \tag{c}$$

Substituting values, and using Eq. (a) for w_3,

$$(500)(132.9) + w_2(1150) = (500 + w_2)(180)$$

$$w_2 = 24.3 \text{ lb/min} = 1458 \text{ lb/hr}$$

Example 3. Steam is flowing steadily through a pipe line 2.00 inches in diameter in which there is a pressure drop due to friction. The pipe is thoroughly insulated so that heat loss is negligible. At a certain section in the pipe the steam pressure is 100 psia, the specific volume is 4.937 cu ft/lb and the enthalpy is 1227.6 Btu/lb. At another section, downstream from the first, the pressure is 90 psia, the specific volume is 5.434 cu ft/lb and the enthalpy is 1223.9 Btu/lb. Find the average velocity at each of the sections mentioned, and the rate of flow in lb/sec.

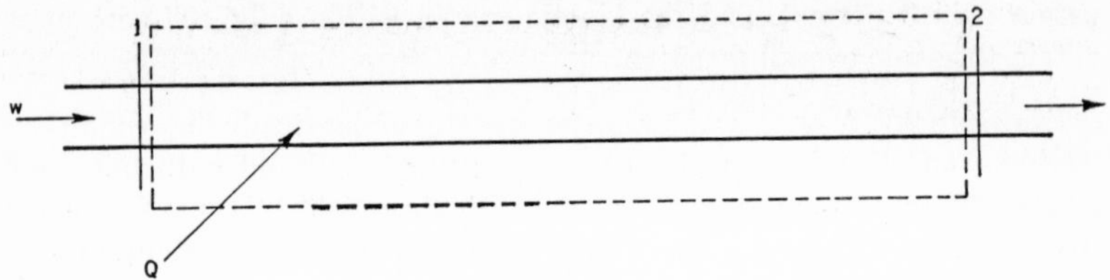

Example 3.

Solution: Sketch the process and locate a control surface. Choose a basis for the computation; use ft lb/lb of steam and lb/sec for units of energy and flow rate respectively. By the conservation of mass

$$w_1 = w_2 = w \tag{a}$$

By the conservation of energy

$$h_1 + \frac{\overline{V}_1^2}{2g_0} + z_1\frac{g}{g_0} + Q = h_2 + \frac{\overline{V}_2^2}{2g_0} + z_2\frac{g}{g_0} + W_x \tag{b}$$

From the nature of the problem $W_x = 0$; Q is given $= 0$; assume the elevation change negligible. Then

$$h_1 + \frac{\overline{V}_1^2}{2g_0} = h_2 + \frac{\overline{V}_2^2}{2g_0} \tag{c}$$

The enthalpies are given but there are two unknown velocities in the equation. However, Eq. (a) may be written in terms of velocities:

$$\frac{A_1\overline{V}_1}{v_1} = \frac{A_2\overline{V}_2}{v_2} = w \tag{d}$$

Also, since the pipe is of constant diameter $A_1 = A_2$, and

$$\frac{\overline{V}_2}{\overline{V}_1} = \frac{v_2}{v_1} = \frac{5.434}{4.937} \tag{e}$$

or

$$\overline{V}_2 = 1.100\ \overline{V}_1 \tag{f}$$

Substituting (f) in (c)

$$h_1 - h_2 = \frac{(1.100\ \overline{V}_1)^2 - \overline{V}_1^2}{2g_0}$$

$$\overline{V}_1 = \sqrt{\frac{(64.4)(1227.6 - 1223.9)(778)}{(1.100)^2 - 1}}$$

$$\overline{V}_1 = 940 \text{ fps}$$

$$\overline{V}_2 = (1.10)(940) = 1034 \text{ fps}$$

From (d)

$$w = \frac{A_1\overline{V}_1}{v_1} = 4.15 \text{ lb/sec}$$

Example 3 illustrates the simultaneous use of the equations of mass and of energy in solving a problem. This is the basis of many flow-metering schemes.

6-5 Summary. The First Law may be applied to flow processes either by the system technique or by the control surface technique.

A control surface is the boundary of a control volume, which is a fixed region in space upon which attention is to be concentrated in analyzing a problem.

A steady flow process is a process in which all conditions within the control volume remain constant with time. The steady flow analysis may be applied to a process which is not a true steady flow process, provided conditions at the control surface satisfy the requirements for steady flow. That is, the mass streams and the energy streams must be constant with time and must respectively add to a net value of zero for the entire surface. If the operation within the control volume is a continuous repetition of a process, at a constant rate, it can satisfy this requirement.

For a simple steady flow process involving one mass stream entering and one mass stream leaving the control volume the following equations will hold: mass balance or *continuity equation*, written on the basis of unit time,

$$w = \frac{A_1\overline{V}_1}{v_1} = \frac{A_2\overline{V}_2}{v_2}$$

Energy balance or *steady flow energy equation*, written on the basis of unit mass of fluid,

$$\frac{\overline{V}_1^2}{2g_0} + z_1\frac{g}{g_0} + u_1 + p_1v_1 + Q = \frac{\overline{V}^2}{2g_0} + z_2\frac{g}{g_0} + u_2 + p_2v_2 + W_x$$

Mass balances and energy balances can easily be made for a steady flow process involving several streams crossing the control surface by summation of the quantities belonging to the individual streams of fluid.

PROBLEMS

6-1. Water at 70°F flows steadily from the bottom of a large open tank through a nozzle to the atmosphere. The tank is kept constantly filled to a depth of 12 ft above the nozzle. (a) If the velocity of the stream leaving the nozzle is uniform, and if no heat transfer or internal energy change occurs for the water passing through the nozzle what will be the water velocity? (b) If the nozzle diameter is 1.5 in. what is the flow rate in lb/hr?

6-2. In a real nozzle there will be some friction, so the velocity will not be as great as found in Problem 6-1. If the measured velocity is 0.985 times the computed velocity how much does the internal energy of the water change while passing through the nozzle? Assume no heat transfer.

6-3. A steam condenser consists of a metal shell or tank through which metal tubes pass; the metal tubes carry a flow of cooling water. Steam fills the space around the tubes and condenses on the tubes, transferring the latent heat of vaporization to the water. The resulting liquid (*condensate*) drips from the tubes to the bottom of the shell where there is a drain hole to remove the condensate. In a certain condenser steam enters at a steady rate and condensate leaves at the same rate. Cooling water flows steadily through the tubes.

Assuming there is no heat transfer with the surroundings, write the steady flow energy equation for (a) the steam-condensate stream; (b) the cooling water stream; (c) the condenser process as a whole. Define each term used, referring to a schematic flow diagram.

6-4. Air flows steadily at the rate of 50 lb/min through an air compressor, entering at 20 fps velocity, 14.5 psia pressure, and 13.5 cu ft/lb volume, and leaving at 15 fps, 100 psia, and 2.6 cu ft/lb; the internal energy of the air leaving is 38 Btu/lb greater than that of the air entering. Cooling water in the compressor jackets absorbs heat from the air at the rate of 3350 Btu/min. (a) Compute the rate of shaft work input to the air in horsepower. (b) Find the ratio of inlet pipe diameter to outlet pipe diameter. State assumptions made.

6-5. A gasoline engine operates steadily with a work output rate of 50 horsepower. Heat transfer from the engine to a stream of cooling water is at the rate of 120,000 Btu/hr. The sum of all other heat transfer to the surrounding atmosphere and the engine foundations is 20,000 Btu/hr. The only stream of fluid other than the cooling water is the stream which enters as fuel-air mixture at 80°F and leaves as combustion products (exhaust gas) at 1500°F. Determine the enthalpy increase of this stream (Btu/hr) in passing through the engine.

6-6. A certain heat exchanger consists of a bundle of metal tubes submerged in a stream of water; a stream of oil passes through the tubes and is heated by the water. Under steady flow conditions the water enters at 200°F and leaves at 110°F while the oil enters at 60°F and leaves at 150°F. There are no net changes in elevation or velocity, and the heat exchanger is perfectly insulated from the surroundings. The enthalpy of water is given as a function of Fahrenheit temperature by $h = t - 32$; the enthalpy of the oil is given by $h = 0.4t + 0.00025t^2$. What is the water flow rate required to heat 10,000 lb/hr of oil?

6-7. An incompressible fluid (specific volume constant) flows steadily in a horizontal insulated converging tube. The pressure, velocity, and tube diameter at the inlet section are fixed. If the flow is frictionless the specific internal energy of the fluid will be constant; for this case derive an expression for the pressure in the fluid as a function of the tube diameter.

6-8. Pressure drop due to friction in a pipe is usually calculated by the equation

$$p_1 - p_2 = f\frac{l}{Dv}\frac{\overline{V}^2}{2g_0}$$

where l is the length of pipe, ft; D is the pipe diameter, ft; $\overline{V}$ is the average velocity, ft/sec; v is the average specific volume, cu ft/lb; and f is the friction factor, dimensionless. (The friction factor depends upon the fluid properties, the pipe geometry and the velocity; values are given in treatises on fluid mechanics.) When water at 70°F flows through a pipe line of 12 in. diameter at a velocity of 10 fps the friction factor has the value 0.014. In a certain case these conditions exist for a pipe 500 feet long.

(a) Calculate the pressure drop due to friction. (b) If the pipe is perfectly insulated against heat transfer how much will the specific internal energy of the fluid change during flow through the pipe? (c) If the same pipe, instead of being insulated, is kept at constant temperature, the internal energy of the water will remain constant. How much heat will be transferred (Btu/lb)?

Chapter 7

SECOND LAW OF THERMODYNAMICS

7-1 Introduction to the second law. The First Law of Thermodynamics states that, when a process is carried out, a certain energy balance will hold. The fact that a given hypothetical process would satisfy the First Law is, however, no guarantee that the process could be carried out. It is a necessary but not a sufficient condition. There are certain directional laws in regard to events in the physical world: bodies at different temperatures, if placed in contact, always approach a common temperature; rivers always run down-hill; people always grow old; the reverse of these processes is never observed. In thermodynamics the fact that a directional law exists results in a limitation on energy transformation other than that imposed by the First Law. For example, when an automobile is stopped by a friction brake the brake gets hot; the internal energy of the brake increases by the absorption of the kinetic energy of the automobile. This satisfies the First Law, but the First Law would be equally well satisfied if the hot brake suddenly cooled off and gave its internal energy back as kinetic energy, causing the automobile to resume its motion. The latter process, however, *never happens*. The action of the brake in stopping the car is an *irreversible* process. Because such situations constantly arise, a general law, called the Second Law of Thermodynamics, has been formulated to express the directional characteristic of processes. The existence of the Second Law has been tacitly assumed from the beginning of this book; it is the existence of this law which necessitates the distinction between work and heat as different forms of energy. For engineering purposes the Second Law is best expressed in terms of the conditions which govern the production of work by a thermodynamic system operating in a cycle. Hence a brief discussion of heat engine cycles is in order.

7-2 Heat engine cycles. A heat engine cycle is a thermodynamic cycle in which there is a net heat flow *to* the system and a net work flow *from* the system. The system which executes a heat engine cycle is a heat engine. A heat engine may be of various forms such as a mass of gas in a cylinder and piston machine, Fig. 7-1a, or it may be

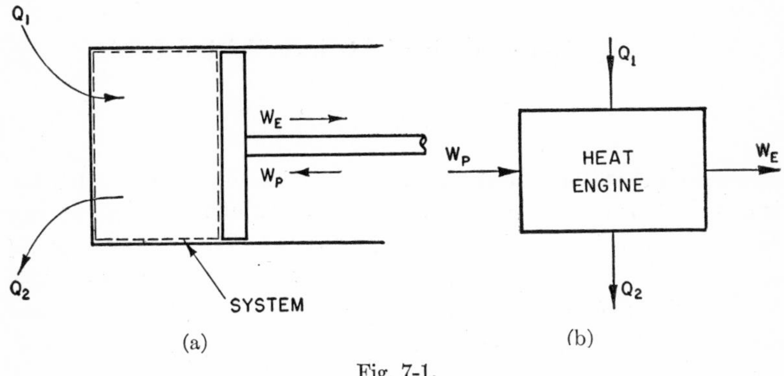

(a) (b)

Fig. 7-1.

a mass of water passing in steady flow through the apparatus of a steam power plant; in any case the heat engine may be represented by the conventional block diagram of Fig. 7-1b.* The heat and work quantities indicated in Fig. 7-1 are defined as follows:

Q_1, *heat supplied;* the heat transferred *to* the system during a cycle,

Q_2, *heat rejected;* the heat transferred *from* the system during a cycle,

W_E, *engine work;* the work done *by* the system during a cycle,

W_P, *pump work;* the work done *on* the system during a cycle. All of these quantities are customarily treated as positive numbers even though two of them represent negative transfers with respect to the system. Hence for a heat engine cycle

$$\Sigma Q = Q_{\text{net}} = Q_1 - Q_2 \tag{1}$$

and

$$\Sigma W = W_{\text{net}} = W_E - W_P \tag{2}$$

Diagrams are drawn consistent with this usage so that Q_2 is associated with an arrow pointing out from the system whereas the general symbol Q is associated with an arrow pointing into the system.

* Note that a heat engine is not at all an *engine* in the sense of a piece of machinery, although it may utilize such apparatus.

The purpose of a heat engine cycle is to produce work; hence the *net* work is of primary interest. When the terms *cycle work* or *work of the cycle* are used they refer to the net work. For many purposes it is unnecessary to break down the work into positive and negative parts. The heat quantities, on the other hand, are always treated separately because the positive heat Q_1 represents a cost or input and is therefore of primary interest.

The efficiency of a heat engine (or of its cycle) is defined by

$$\eta = \frac{\text{net work output}}{\text{gross heat input}} = \frac{W_{\text{net}}}{Q_1} \tag{3}$$

where η (eta) is the symbol for efficiency. The efficiency here defined is often called *thermal efficiency*. By the First Law,

$$\Sigma Q = \Sigma W \tag{4}$$

Substituting from (1) and (2),

$$Q_1 - Q_2 = W_E - W_P = W_{\text{net}} \tag{5}$$

then

$$\eta = \frac{Q_1 - Q_2}{Q_1} = 1 - \frac{Q_2}{Q_1} \tag{6}$$

Note that Eq. (6) is simply a deduced result; Eq. (3) is the *definition* of the efficiency of a cycle.

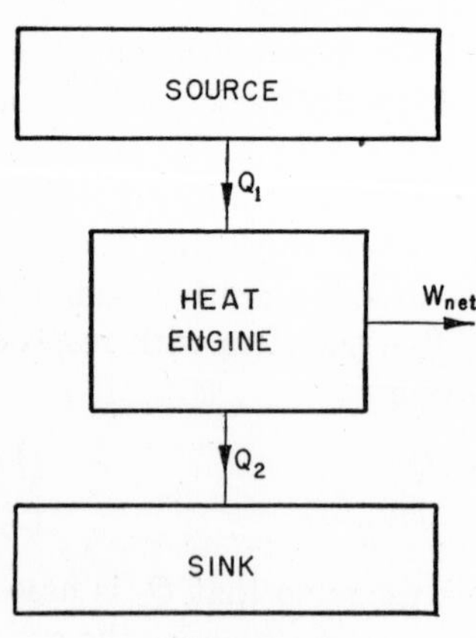

Fig. 7-2.

When heat is transferred to or from a heat engine there must exist other systems with which the heat engine exchanges heat. A more complete diagram than that of Fig. 7-1b would show (Fig. 7-2) not only the heat engine, but a body called a *source*, from which heat Q_1 is transferred to the engine, and a body called a *sink*, to which heat Q_2 may be transferred from the engine. A typical source would be the mass of burning fuel in a furnace; a typical sink would be a body of water such as a lake or a river. Usually, but not necessarily, sources and sinks are assumed to maintain constant temperature.

7-3 The second law. The First Law requires of a heat engine only that the net work be equal to the net heat. Experience shows, however, that other restrictions exist; the net work is always less than the heat supplied Q_1, that is to say there always exists some heat rejected Q_2. Moreover there is no net work in a cycle unless a temperature difference exists between the source and the sink. The formal statement of these empirical facts constitutes the Second Law of Thermodynamics. The earliest recognition of this fundamental principle was by Carnot who said, with reference to a cyclic engine, "Wherever there exists a difference of temperature, motive power can be produced." * This positive assertion, however, is not as precise as the negative assertions of later writers since it may imply, but does not state categorically, that the temperature difference is essential to the production of net work in a cycle. Kelvin, Clausius, and Planck each produced a commonly quoted statement of the Second Law. The following statement, after Kelvin and Planck, is suitable for the purposes of this book: *The Second Law: It is impossible for a heat engine to produce net work in a complete cycle if it exchanges heat only with bodies at a single fixed temperature.*†

This statement appears rather limited in scope but its consequences are extremely broad in their applications. It will be noticed that the form of the law is negative; it states that a certain class of heat engine cycles is impossible of realization. Now the possibility, under the Second Law, of any proposed process may be tested as follows: construct a cycle composed of the proposed process and some other processes known by experience to be possible. If such a cycle can be constructed so that it would produce net work while the system exchanges heat solely with bodies at a single fixed temperature, then the proposed process is impossible. The hypothetical heat engine whose cycle would thus violate the Second Law is frequently called a per-

* *Reflections on the Motive Power of Heat* by N. L. Sadi Carnot, Paris, 1824. Trans. by R. H. Thurston, Amer. Soc. Mech. Engrs., New York, 1943. Carnot's work preceded the establishment of the First Law, hence his reasoning is inconsistent with the present concept of heat. Nevertheless his conclusions are sound because the Second Law is a general law of nature entirely independent of the First Law. Carnot not only presented the first analysis of the effects of what is now known as the Second Law but he also originated the use of cycles in thermodynamic analysis.

† See Section 7-9 for Clausius' statement of the Second Law. For a discussion of the statements of the Second Law see Zemansky, M. W., *Heat and Thermodynamics*, New York: McGraw-Hill, 1943, chap. 7.

petual motion machine of the second kind. (A perpetual motion machine of the first kind is a device which violates the First Law by creating energy.)

The acceptance of the Second Law as a fundamental law of nature rests upon the fact that no one has ever been able to carry out physically a process which, by the above type of reasoning, could be proved impossible under the Second Law. There is no way to prove that experience will always thus confirm the Second Law but since no single violation has ever been demonstrated the validity of the law is unquestioned.

7-4 Reversibility. The one-directional characteristic of many processes is a matter of common experience. A high-temperature body in low-temperature surroundings always goes to lower temperature. A moving body may give up its kinetic energy to increase the internal energy of a brake, but the reverse process never occurs. A formal way of describing such experiences is to say that the processes are *irreversible.*

Although all real processes are probably irreversible, some processes are capable of being carried out almost precisely in reverse. For example a free pendulum starts with potential energy of elevation at one extreme of its swing, exchanges the potential energy for kinetic energy during its fall, and reverses the process during its rise. Experience shows that the reversal is not perfect, for the pendulum gradually loses amplitude and eventually comes to rest. But by placing the pendulum in a vacuum the coming to rest can be delayed, since air friction is eliminated. By care in construction pivot friction may be minimized; thus the process of the downward swing of the pendulum may be made to approach the condition of a *reversible* process. It is thus possible to imagine the approach to a reversible process as a limit, and to treat the limiting case as a hypothetical process. Such limiting cases are of the utmost importance in the development of the applications of the Second Law; consequently the concept of reversibility must be formalized.

DEFINITION: *A process is reversible if, after it has been carried out, it is possible by any means whatsoever to restore both the system and its entire surroundings to exactly the same states they were in before the process. A process that is not reversible is irreversible.*

This definition means that, if a process is reversible, it is possible to undo it in such a way that there will be no trace *anywhere* of the fact that the process occurred. A process that is *irreversible* may be undone in the sense that a specific system is restored to its initial state, but it is not possible to do this without leaving some permanent change in things external to the system.

7-5 Irreversible processes. The definition of a reversible process is general; it specifies no particular physical conditions necessary for reversibility. However, the Second Law requires, and experience confirms, that the presence of certain specific conditions during a process will render the process irreversible; among these conditions are:

(1) heat transfer through a finite temperature difference;

(2) lack of pressure equilibrium between the system and its confining walls;

(3) free expansion; the expansion of a system to a larger volume in the absence of work done by the system;

(4) friction, solid or fluid (and electrical resistance);

(5) transfer of paddle-wheel work to a system.

The first three items involve a lack of equilibrium during the process; the others involve so-called dissipative effects in which work is done without producing finally an equivalent increase in the kinetic or potential energy of any system.

7-6 Mechanical and thermal reversibility—reversible systems. Experience shows that no real process can be strictly reversible. Nevertheless, in many cases, certain effects of real processes differ little from the effects computed by assuming a reversible process. For example the expansion of a real gas in a cylinder and piston machine generally involves so little friction and so little disturbance of pressure equilibrium that the work done is given accurately by computations based on a reversible expansion. It is often said that such a process is considered *mechanically reversible;* a process involving paddle-wheel work would be *mechanically irreversible*. Similarly, if the process involves heat transfer it may be designated *thermally reversible* or *thermally irreversible* depending upon the absence or presence of a temperature difference.

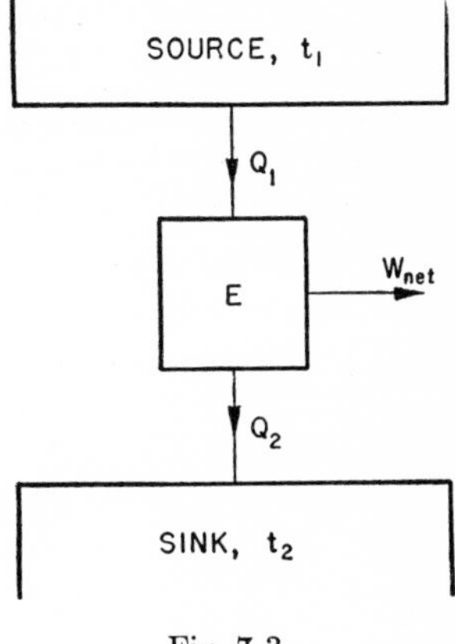

Fig. 7-3.

A thermodynamic system is frequently chosen in such a way as to exclude from the system most causes of irreversibility; the system is then assumed to experience no irreversible effects within its boundaries. Such a system is called an *internally reversible system* or simply a *reversible system.**

7-7 The Carnot cycle. In the application of the Second Law much use is made of reversible cycles, which are hypothetical cycles composed entirely of reversible processes. The classical reversible cycle for the development of theory is the cycle first proposed by Carnot and named for him. The Carnot cycle is a heat engine cycle in which all heat supplied to the engine is transferred reversibly from a constant-temperature source, all heat rejected is transferred re-

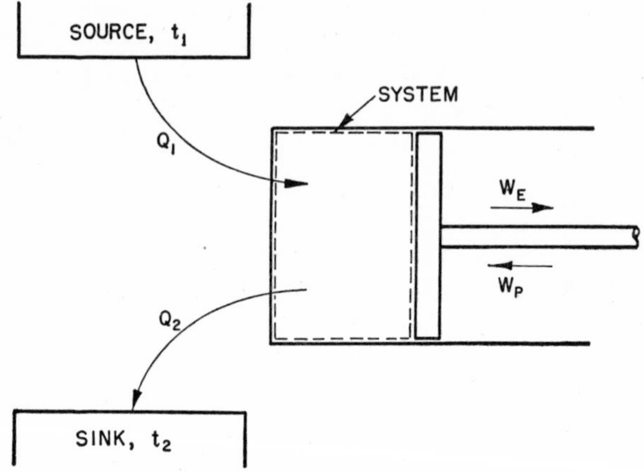

Fig. 7-4. Carnot heat engine: stationary system.

versibly to a constant-temperature sink, and all processes in the cycle are reversible. Referring to Fig. 7-3, if the engine E is to operate in a Carnot cycle, the heat Q_1 must be transferred while the system is at temperature t_1, and the heat Q_2 must be transferred while the system is at temperature t_2. To complete the cycle, the system must go from t_1 to t_2, and back from t_2 to t_1, by adiabatic (zero heat transfer) processes. Hence the cycle consists of four processes: (1) a revers-

* For a discussion of various specific cases of reversible and irreversible processes see Zemansky, M. W., *Heat and Thermodynamics*. New York: McGraw-Hill, 1943; chap. 8.

ible isothermal (constant-temperature) process in which heat Q_1 enters the system; (2) a reversible adiabatic process in which the temperature changes from t_1 to t_2; (3) a reversible isothermal process in which heat Q_2 leaves the system; (4) a reversible adiabatic process in which the temperature changes from t_2 to t_1. The details of the mechanisms by which these processes could be carried out are not important, but if a physical picture is desired the system may be considered to be a mass of gas in a cylinder and piston machine, Fig. 7-4. There are available a source of heat at temperature t_1 and a sink at tempera-

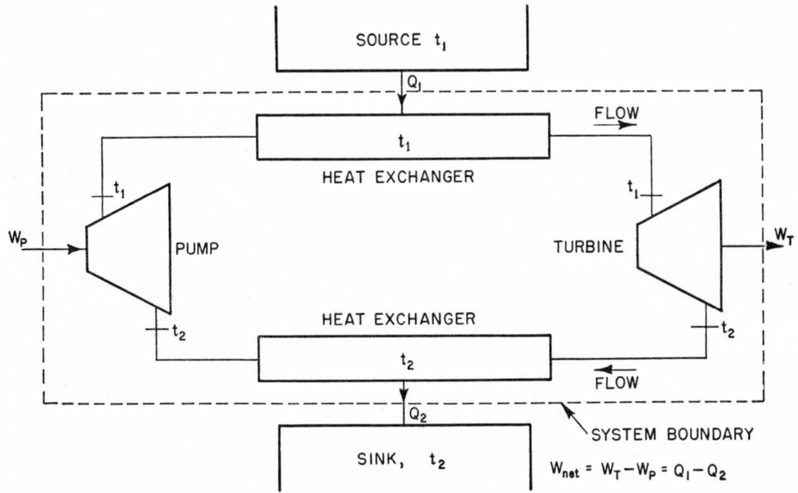

Fig. 7-5. Carnot heat engine: steady flow process.

ture t_2 (which is less than t_1). It is assumed that means are available to bring either the source or the sink into contact with the cylinder head or to cover the cylinder with heat insulation. Then by bringing the source into contact with the cylinder head and allowing the piston to move out slowly a reversible isothermal heat transfer at t_1 would be carried out (hypothetically, of course). By covering the cylinder with insulation and allowing the piston to move out farther the system temperature would be reduced to t_2 in a reversible adiabatic process. By bringing the sink into contact with the cylinder head and forcing the piston slowly toward the cylinder head, a reversible isothermal heat transfer at t_2 would be carried out. By insulating the cylinder and forcing the piston farther toward the cylinder head, the system temperature would be increased to t_1 in a reversible adiabatic process. This is the classical picture of the Carnot cycle, but other pictures are possible, for example, a steady flow system such as shown in Fig. 7-5. In this case the isothermal heat transfer processes take place in heat exchangers while the reversible adiabatic processes take place in the turbine and the pump. To satisfy the conditions for the Carnot cycle there must be neither friction nor heat transfer in the pipe lines through which the working fluid flows.

Carnot, by ingenious reasoning, showed that the cycle described above is capable of producing the maximum work output for cycles working between a source and a sink at given temperatures. In the next chapter Carnot's conclusion will be shown to be a consequence of the Second Law.

7-8 The reversed heat engine. A reversible heat engine cycle such as the Carnot cycle need not be used only for a heat engine. Since all the processes of the Carnot cycle are reversible it is possible to imagine that the processes are individually reversed and carried out in reverse order. When a reversible process is reversed all the energy transfers associated with the process are reversed in direction but remain the same in magnitude. Hence, in a reversed Carnot cycle, heat Q_1 would flow *from* the system at temperature t_1, heat Q_2

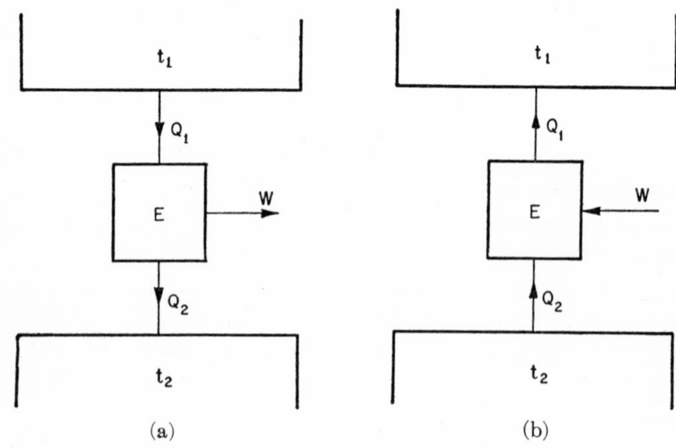

Fig. 7-6.

would flow *to* the system at temperature t_2, and the net work would be done *on* the system instead of by the system.

The reversal of a cycle may be visualized by using the block diagrams of Fig. 7-6 in which (a) shows the normal heat engine and (b) the reversed heat engine. If E is a reversible engine, the quantities Q_1, Q_2, and W will have the same magnitudes, respectively, whether the engine operates normally or reversed. Only the directions of these quantities will change upon reversal of the engine.

The reversed heat engine takes heat *from* a low-temperature body, discharges heat *to* a high-temperature body, and receives an *inward* flow of net work. The names *heat pump* and *refrigerating machine* are applied to the reversed heat engine. For the heat pump the same energy relation holds as for the heat engine,

$$W = Q_1 - Q_2$$

It is possible to operate actual heat pump cycles; they are used in refrigerators and in heating systems of various kinds. However, since an actual cycle cannot be reversible, an actual heat pump cycle is only approximately the reverse of an actual heat engine cycle.

7-9 Clausius' statement of the second law. The Second Law may be stated in terms of the heat pump: Clausius stated that it is impossible for a system working in a complete cycle to accomplish as its sole effect the transfer of heat from a body at a given temperature to a body at a higher temperature. This means that a heat pump must have a net work input. The complete equivalence of the above statement and the statement of the Second Law given in Sec. 7-3 can easily be shown.

7-10 Summary. A heat engine cycle is a thermodynamic cycle in which there is a net flow of heat to the system and a net flow of work from the system. The system which executes a heat engine cycle is a heat engine. The efficiency of a heat engine, or of its cycle, is $\eta = W_{\text{net}}/Q_1$ where Q_1 is the heat transferred to the system in a cycle, and W_{net} is the net work of the cycle. This efficiency is called *thermal efficiency*.

A heat pump or refrigerating machine is a system to which there is a net flow of work and from which there is a net flow of heat in a cycle.

The Second Law is stated as follows (after Kelvin and Planck): *It is impossible for a heat engine to produce net work in a complete cycle if it exchanges heat only with bodies at a single fixed temperature.*

An equivalent statement (after Clausius) follows:

It is impossible for a system working in a complete cycle to accomplish as its sole effect the transfer of heat from a body at a given temperature to a body at a higher temperature.

A process is reversible if, after the process has been carried out, it is possible by any means whatsoever to restore both the system and the entire surroundings to exactly the same states they were in before the process. A process that is not reversible is irreversible. A reversible process is a process that can be undone in such a way that no trace remains *anywhere* of the fact that the process occurred.

Reversible processes exist only as ideal limiting cases which real processes may approximate to greater or lesser degree. The presence of any of the following conditions will make a process irreversible: (1) heat transfer through a finite temperature difference; (2) lack of pressure equilibrium; (3) free expansion; (4) friction of any kind; (5) paddle-wheel work. A truly reversible process would require infinite time since any process at a finite rate involves some disequilibrium.

A reversible cycle is a cycle composed entirely of reversible processes. The classical example is the Carnot cycle which consists of two reversible isothermal processes and two reversible adiabatic processes.

PROBLEMS

7-1. Heat is supplied to a cyclic heat engine at the rate of 1000 Btu/min, and the engine does work at the rate of 10 hp. Find the efficiency of the heat engine and the rate of heat rejection.

7-2. A dry-cell battery can produce work in the form of an electric current while exchanging heat solely with a constant-temperature atmosphere. Does this violate the Second Law? Explain.

7-3. An automobile engine produces net work while exchanging heat only with the atmosphere; why does this not violate the Second Law?

7-4. Would an atomic energy power plant violate either the First or the Second Law of Thermodynamics? Explain.

7-5. A household refrigerator is in thermal communication solely with the atmosphere of a constant-temperature room. The refrigerator works in a complete cycle, receiving work from the electrical lines and exchanging heat only with the atmosphere. Does this violate the Second Law? Explain.

7-6. An electric storage battery which can exchange heat only with a constant-temperature atmosphere goes through a complete cycle of two processes. In process 1; 2800 watt-hours of electrical work flow into the battery, while 700 Btu of heat flow out to the atmosphere; in process 2; 2400 watt-hours of work flow out from the battery. (a) Find the magnitude and direction of the heat transferred in process 2. (b) If process 1 has occurred as described above, does the First Law or the Second Law limit the *maximum possible* work of process 2? How much is the maximum possible work? (c) If the maximum possible work were obtained in process 2 how much heat would be transferred in process 2, and in which direction?

7-7. If a hard steel ball is dropped on a hard steel plate it may rebound almost to the original elevation. If the same ball is dropped on a lead plate it will rebound to a much smaller height. (a) Explain how the First Law is satisfied in each case. (b) Is the process in each case almost reversible or irreversible?

7-8. A fluid flows steadily at constant velocity, without heat transfer, through a pipe in which there is a pressure drop due to friction. Prove, by the Second Law, that this process is irreversible.

7-9. Considering a reversible process to be one which approaches reversibility as a limit (Sec. 7-6), each process described below falls into one of three categories: (1) it is reversible; (2) it *might* be reversible; (3) it *cannot possibly* be reversible. Place each process in its proper category, and identify the sources of irreversibility.

(a) A pound of water is evaporated at constant temperature by an inflow of heat. (b) A pound of water is evaporated at constant temperature by an inflow of work. (c) A mass of air in a constant-volume container is heated slowly from 100°F to 200°F by an inflow of heat. (d) A stationary mass of air is compressed slowly in a frictionless, non-conducting cylinder and piston machine. (e) A gas goes through the isothermal expansion process of a Carnot cycle. (f) A stream of steam at 212°F mixes with liquid water at 50°F in a non-conducting heater. (g) Hot stagnant gases in the cylinder of a non-conducting, frictionless cylinder and piston machine expand as the piston moves out slowly. (h) Hot turbulent gases in the cylinder of a water-cooled automobile engine expand as the piston moves out rapidly.

7-10. A gaseous system in a cylinder and piston machine executes a cycle in which there are three heat transfers as follows: $Q_1 = -180$ Btu/cycle; $Q_2 = 60{,}000$ ft lb/cycle; $Q_3 = 75{,}000$ IT calories/cycle. Is the system acting as a heat engine or a heat pump? Show reasoning.

7-11. A refrigerator utilizing a heat pump cycle has a power input of 0.75 kw, and a heat flow rate from the cold region of 150 Btu/min. How much heat must be transferred per hour to the hot region?

Chapter 8

BASIC APPLICATIONS OF THE SECOND LAW

The Second Law is seldom used directly but several general corollaries or propositions derived from it are of great usefulness in the development of theory, and in the solution of problems. In this chapter these immediate consequences of the Second Law are developed.

8-1 The efficiency of reversible cycles. The first application is to demonstrate the following proposition:

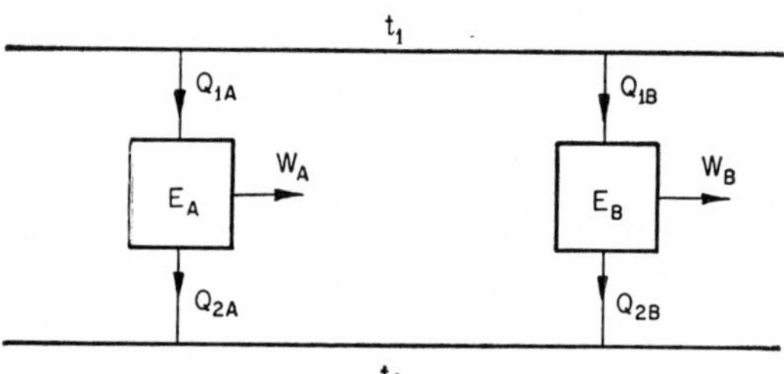

Fig. 8-1.

(1) Of all heat engines operating between a given constant-temperature source and a given constant-temperature sink, none has a higher efficiency than a reversible engine.

Proof: Let *any* heat engine E_A and *any reversible* heat engine E_B be allowed to operate between a source at temperature t_1 and a sink at temperature t_2, as shown in Fig. 8-1. Assume that the theorem to be proved is not true; thus assume E_A more efficient than E_B. Now let the rates of working of the engines be such that

$$Q_{1A} = Q_{1B} = Q_1 \tag{a}$$

By assumption

$$\eta_A > \eta_B$$

or

$$\frac{W_A}{Q_{1A}} > \frac{W_B}{Q_{1B}} \tag{b}$$

From Eq. (a) and inequality (b) it follows that

$$W_A > W_B \tag{c}$$

Now let E_B be reversed. Since E_B is reversible the magnitudes of the heat and work quantities will remain unchanged but their directions will be reversed, giving the result shown in Fig. 8-2. Now take

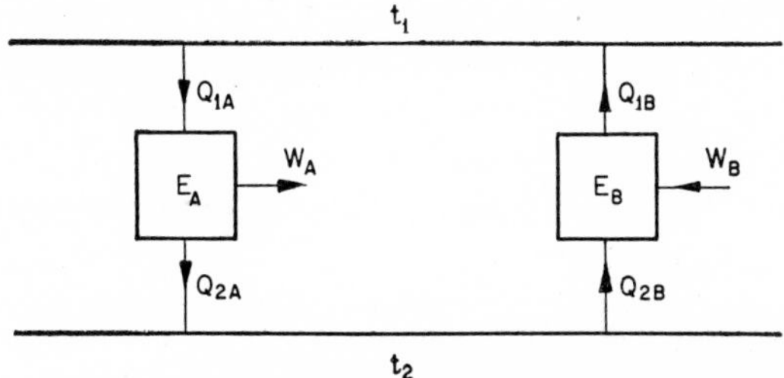

Fig. 8-2.

part of W_A to supply W_B; also, use Q_{1B} to supply Q_{1A}. This will permit the elimination of the source at t_1, giving the result shown in Fig. 8-3. The combined systems E_A and E_B now form a heat engine which is producing net work $(W_A - W_B)$ while exchanging heat only with a single constant-temperature body at t_2. This is a direct violation of the Second Law.* Consequently the assumption that the ef-

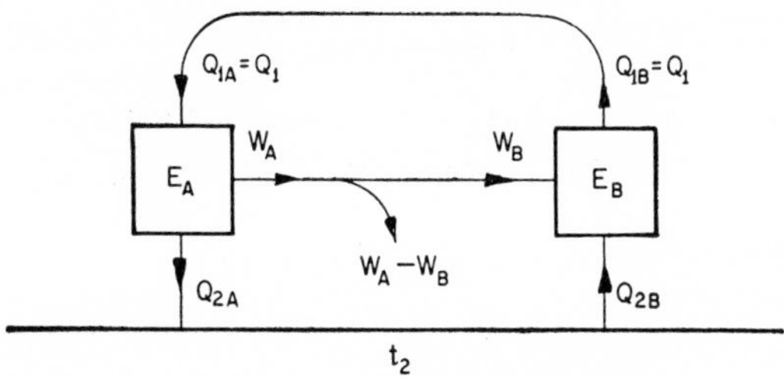

Fig. 8-3.

* Observe that the operation in Fig. 8-3 does not violate the First Law.

ficiency of E_A is greater than the efficiency of E_B must be false, and proposition (1) is proved.

From proposition (1) a second proposition follows: (2) The efficiency of all reversible heat engines operating between the same temperature levels is the same. Proposition (2) is proved by letting both E_A and E_B be reversible heat engines. Then by proposition (1) the efficiency of E_A cannot be greater than the efficiency of E_B, and the efficiency of E_B cannot be greater than the efficiency of E_A; hence their efficiencies must be equal and the proposition is proved.

Observe that in the above reasoning no restrictions were placed upon the type of engine or system to be used or upon any details of operation; thus the results are completely general. No matter what new types of engine or new working substances may be invented or discovered, so long as the Second Law holds, these two propositions will hold. Consequently, if the efficiency of one reversible cycle (for example the Carnot cycle) can be determined as a function of temperature, the limit of efficiency for all cycles will be known. This information is of great practical value in guiding the development of heat power cycles.

8-2 The absolute thermodynamic temperature scale. The efficiency of any heat engine cycle receiving heat Q_1 and rejecting heat Q_2 is given by

$$\eta = \frac{W}{Q_1} = \frac{Q_1 - Q_2}{Q_1} = 1 - \frac{Q_2}{Q_1} \tag{1}$$

By the Second Law it is necessary to have a temperature difference $(t_1 - t_2)$ to obtain work from any cycle. For a reversible cycle operating between fixed temperatures the efficiency depends solely upon the temperatures t_1 and t_2 at which heat is transferred, or

$$\eta = f(t_1, t_2) \tag{2}$$

where f signifies some function of the temperatures. Combining (1) and (2), for a reversible cycle operating between t_1 and t_2

$$1 - \frac{Q_2}{Q_1} = f(t_1, t_2) \tag{3}$$

In terms of a new function F

$$\frac{Q_1}{Q_2} = F(t_1, t_2) \tag{4}$$

If t_1 and t_2 are temperatures on an existing scale, Eq. (4) defines the function F, but if a certain functional relationship is *assigned* between t_1, t_2, and Q_1/Q_2, Eq. (4) becomes the definition of a temperature scale. Such a scale is independent of the properties of particular substances, and is known as an absolute thermodynamic temperature scale.

The absolute thermodynamic temperature scale, as proposed by Kelvin, is *defined* by

$$\frac{T_1}{T_2} = \frac{Q_1}{Q_2} \tag{5}$$

where T_1 and T_2 are temperatures on the absolute thermodynamic temperature scale, and Q_1 and Q_2 are respectively the heat quantities exchanged by a reversible heat engine with a source at T_1 and a sink at T_2.

The definition (5) does not fix the scale completely since it gives only the ratio of any two temperatures; the size of the degree of temperature is arbitrary and is taken, for convenience, the same as the conventional degree. Thus there is a Centigrade absolute scale called the Kelvin scale and a Fahrenheit absolute scale called the Rankine scale.

It is an experimental fact that the ideal gas temperature scale described in Sec. 3-6 is identical with the absolute thermodynamic scale. Therefore the following relations hold:

$$T^{\mathrm{R}} = t^{\mathrm{F}} + 459.7, \qquad T^{\mathrm{K}} = t^{\mathrm{C}} + 273.2, \qquad T^{\mathrm{R}} = 1.8\ T^{\mathrm{K}}$$

where T signifies absolute thermodynamic temperature, t signifies conventional temperature, K signifies Kelvin scale, and R signifies Rankine scale.

Whenever temperature is referred to in thermodynamic analysis the absolute thermodynamic temperature is implied, although data are frequently given in terms of the conventional temperature scales.

By substitution from Eq. (5) in Eq. (1) the efficiency of a *reversible* heat engine cycle in which heat is received solely at T_1 and heat is rejected solely at T_2 is found to be

$$\eta = 1 - \frac{T_2}{T_1} = \frac{T_1 - T_2}{T_1} \tag{6}$$

Equation (6) sets a limit to the possible efficiency of a heat engine operating between T_1 and T_2 because no such engine can exceed in efficiency the reversible engine.

It is not necessary to operate a reversible heat engine in order to establish the relationship of the thermodynamic temperature scales to the conventional scales. It is possible to derive certain relations between the absolute temperature and other properties of substances, by which the absolute temperature may be computed from measurements of other properties.*

8-3 Entropy. Because temperature is defined as a function of the heat quantities transferred in the operation of a reversible cycle it is possible to set up, for a reversible process, a relation between heat transferred and the properties of a system. It was shown in Chap. 2 that under certain special conditions it is possible to express work as a function of pressure and volume; now, for a *reversible process*, it is possible to express heat as a function of temperature and a new property called *entropy* (en′tropy). The existence of this new property can be proved as a consequence of the Second Law. However, we shall simply state without proof that *entropy*, defined as follows, is a property.

DEFINITION. *The entropy s of a system is a quantity which satisfies the equation*

$$ds = \frac{dQ_{\text{rev}}}{T} \tag{7}$$

where Q_{rev} is the heat transferred in a *reversible* process. Integrating,

$$\Delta s = \int_{\text{rev}} \frac{dQ}{T} \tag{7a}$$

In words, the change of entropy in a system is the integral, *for a reversible process*, of dQ/T.

Since entropy is a property, the change in entropy between any two states A and B is independent of the path followed in evaluating it, and the change in entropy for a complete cycle is zero. Like internal energy, entropy has no absolute base value, but if a base value is arbitrarily assigned to the entropy of a system at any one state the entropy at all other states becomes fixed. The entropy of a system is usually given in units of Btu/F_{abs}. The specific entropy of a *substance* is given in units of Btu/lb F_{abs}. In a pure substance the en-

* Planck, Max, *Treatise on Thermodynamics;* tr., Ogg. New York: Dover, 1945; pp. 134–138. Keenan, J. H., *Thermodynamics.* New York: Wiley, 1941; chap. 21. Zemansky, M. W., *Heat and Thermodynamics.* New York: McGraw-Hill, 1943; chap. 9.

tropy, like other properties, is a function of two independent variables such as pressure and volume, pressure and temperature, or internal energy and volume.

It is desirable to dwell upon the significance of Eq. (7) by which entropy is defined. The equation says that in a reversible process the entropy change is measured in terms of heat; it does *not* say that entropy changes only in a reversible process. The entropy change from A to B is identical for *all* processes between A and B. $\int dQ/T$ is equal to the entropy change for all reversible processes between A and B, but has different values for irreversible processes.* These different values of $\int dQ/T$ in irreversible processes are immaterial to the determination of entropy change, since *by definition* the entropy change from A to B is $\int dQ/T$ *for a reversible process*. Hence if it is desired to compute the entropy change for any process, reversible or irreversible, between A and B, all that need be done is to *assume* a convenient *reversible* path between A and B and compute the value of $\int dQ/T$ for this assumed reversible path.

Since the definition of *adiabatic* is zero heat transfer, the integral of dQ/T will be zero in any adiabatic process. Therefore in a *reversible adiabatic* process the change of entropy is zero, or the entropy of the system is constant. Such a process is therefore called *isentropic*, meaning constant entropy. Observe that an isentropic process is not necessarily either adiabatic or reversible; isentropic means nothing but constant entropy. If, however, a process is *both* isentropic and reversible it must be adiabatic; if it is *both* adiabatic and reversible it must be isentropic; and if it is *both* adiabatic and isentropic it must be reversible.

In applications of thermodynamics to heat power engineering, the most common use for the entropy is to locate the path of a reversible adiabatic process. The reversible adiabatic is a convenient ideal or hypothetical process to use as a standard of comparison for many actual processes such as the expansion of a working fluid in an engine or turbine.

8-4 The temperature-entropy plot. Since entropy is a property, it may be used as a coordinate in a graphical representation of the property changes in a process or cycle. The temperature-entropy

* It can be shown that in an irreversible process the integral of dQ/T is less than the entropy change. See Keenan, J. H., *Thermodynamics*. New York: Wiley, 1941; chap. 8.

plot is particularly useful because, for reversible processes, areas upon this plot are directly proportional to heat quantities. Thus the temperature-entropy plot permits visualizing heat quantities, just as the pressure-volume plot permits visualizing work quantities.

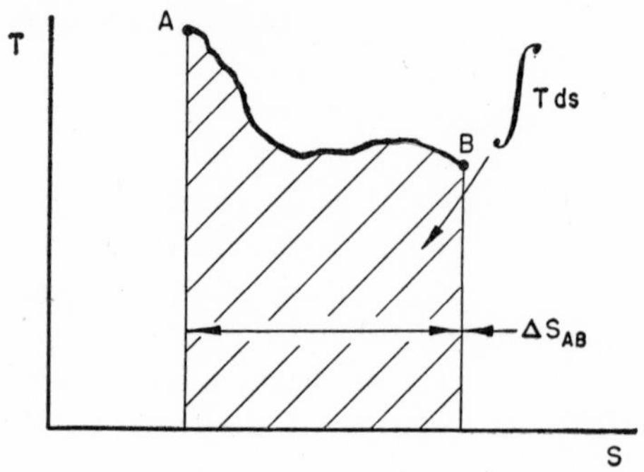

Fig. 8-4. Temperature-entropy plot of a process.

Consider a process A–B for which the path is plotted in Fig. 8-4 upon the temperature-entropy plane. The area under the curve A–B is proportional to $\int T\,ds$. By the definition of entropy, Eq. (7),

$$T\,ds = dQ_{\text{rev}} \tag{8}$$

or

$$\int T\,ds = Q_{\text{rev}} \tag{9}$$

Hence if A–B is a reversible process, the area under the curve is proportional to the heat transferred in the process. If A–B is an

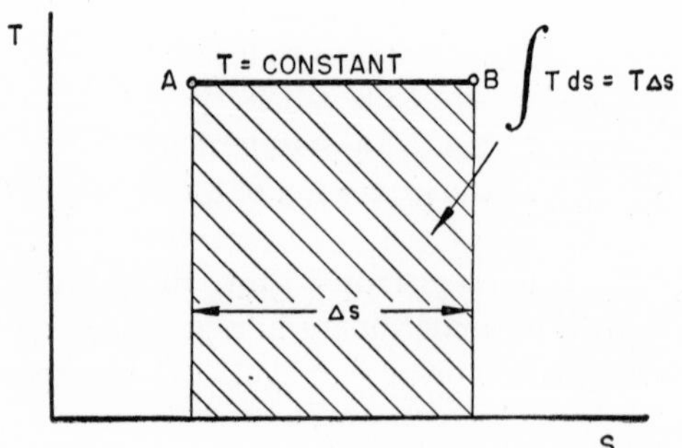

Fig. 8-5. Constant-temperature process.

irreversible process, the area under the curve does not represent the heat transferred.

Some examples of representation upon the T–s plane follow.

Reversible Constant-Temperature Process: In this case, Fig. 8-5, $Q = \int T\,ds = T\,\Delta s$

Reversible Adiabatic Process: In this case, Fig. 8-6,

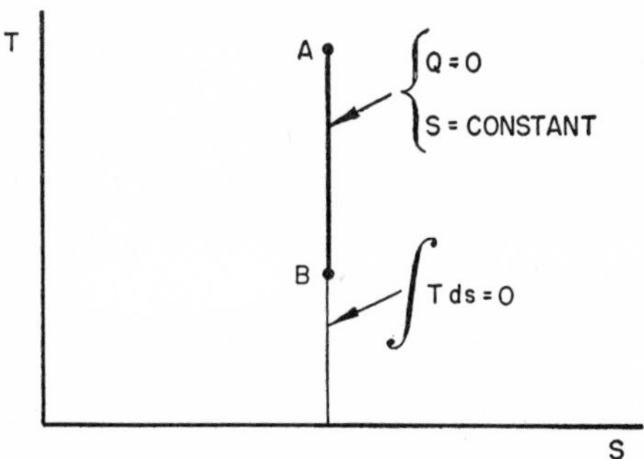

Fig. 8-6. Reversible adiabatic process.

$$Q = 0 = \int T\,ds \qquad \text{or} \qquad ds = 0$$

Carnot Cycle. The Carnot cycle is conveniently shown in the T–s plane because it always forms a rectangle in that plane. In Fig. 8-7 a–b–c–d is a Carnot cycle. The area a–b–m–n, under the reversible constant-temperature path a–b, represents the heat supplied. The area c–m–n–d represents the heat rejected. Hence the area a–b–c–d enclosed by the paths of the cycle represents the *net* heat transferred, which is equal to the net work of the cycle. The T–s plot of any reversible cycle will enclose an area proportional to the net work of the cycle. Such plots are useful in comparing the work outputs and efficiencies of various ideal reversible cycles.

For the Carnot cycle, Fig. 8-7,

$$Q_1 = T_1\,\Delta s, \qquad Q_2 = T_2\,\Delta s$$

$$W_{\text{net}} = Q_1 - Q_2 = (T_1 - T_2)\,\Delta s \tag{10}$$

8-5 The availability of energy. One of the basic general problems of heat engineering is to obtain the maximum work output from a cycle with the minimum heat input. The theoretical maximum work output of a heat engine supplied with a given quantity of heat

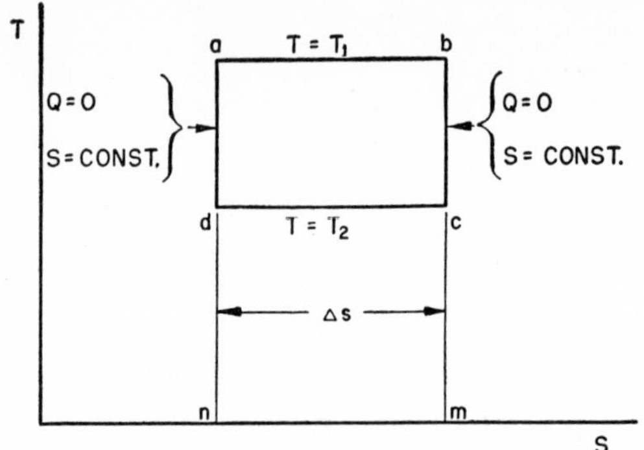

Fig. 8-7. Carnot cycle.

under given conditions is called the *available energy* supplied or the available part of the energy supplied. The remainder of the heat supplied is called *unavailable energy*. These terms refer solely to the possibility of work production by a cyclic heat engine.

The maximum efficiency of any heat engine working between the two fixed temperatures T_1 and T_2 is given by

$$\eta = \frac{T_1 - T_2}{T_1} \tag{6}$$

Therefore if T_1 is given, the maximum efficiency will be obtained by making T_2 as small as possible. But, for practical cases, T_2 must be the temperature of some naturally available body which can absorb a large amount of heat without appreciable temperature rise. Such bodies are the atmosphere, ocean, lakes, and rivers. The temperatures of these bodies vary from place to place and from time to time so it is customary to reason in terms of a "temperature of the surroundings" which is a hypothetical average temperature.

Consider a finite process *a–b*, Fig. 8-8, in which heat is supplied reversibly to a heat engine. Taking an elementary cycle, the engine receives heat dQ_1 at temperature T_1, rejects heat to the surroundings at temperature T_0, and goes between these temperatures by reversible adiabatic processes. For the elementary reversible cycle the work will be a maximum under the given conditions; it is given by

$$dW_{\text{max}} = \frac{T_1 - T_0}{T_1}\,dQ_1 = dQ_1 - \frac{T_0}{T_1}\,dQ_1 \tag{11}$$

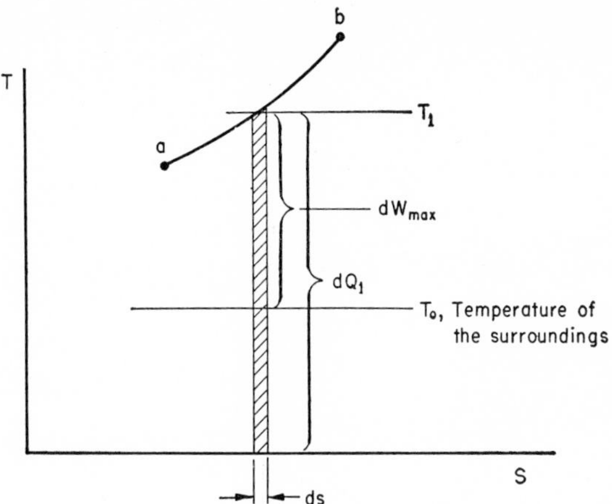

Fig. 8-8. Availability of energy.

The quantity of energy $(dQ_1 - dW_{\text{max}})$ which must necessarily be rejected as heat is said to be *unavailable* for work. The unavailable energy for the elementary cycle is then

$$dQ_1 - dW_{\text{max}} = \frac{T_0}{T_1} dQ_1 \tag{12}$$

If the heat engine receives heat through the whole process a–b, and rejects heat at T_0, the maximum work that can be obtained, or the *available energy*, for the finite cycle will be the sum of the elementary quantities of Eq. (11), or

$$W_{\text{max}} = \int_a^b \left(dQ_1 - \frac{T_0}{T_1} dQ_1 \right) = Q_{ab} - \int_a^b \frac{T_0}{T_1} dQ_1 \tag{13}$$

The unavailable part of the energy supplied is

$$Q_{ab} - W_{\text{max}} = \int_a^b \frac{T_0}{T_1} dQ_1 \tag{14}$$

Then, since T_0 is constant,

$$Q_{ab} - W_{\text{max}} = T_0 \int_a^b \frac{dQ_1}{T_1} = T_0(s_b - s_a) \tag{15}$$

The unavailable part of the heat supplied to a heat engine is equal to the product of the lowest temperature at which heat may be rejected and the entropy change of the system during the process of supplying heat. Figure 8-9 is a graphical presentation of this fact.

It is clear from the *t–s* plot that the available portion of the energy supplied to a heat engine will decrease as the temperature at which the heat is supplied decreases. It follows that when energy is transferred as heat to a region at a lower temperature than the source, some available energy is lost. Thus every heat transfer across a finite temperature difference involves an irrecoverable dissipation of available energy.

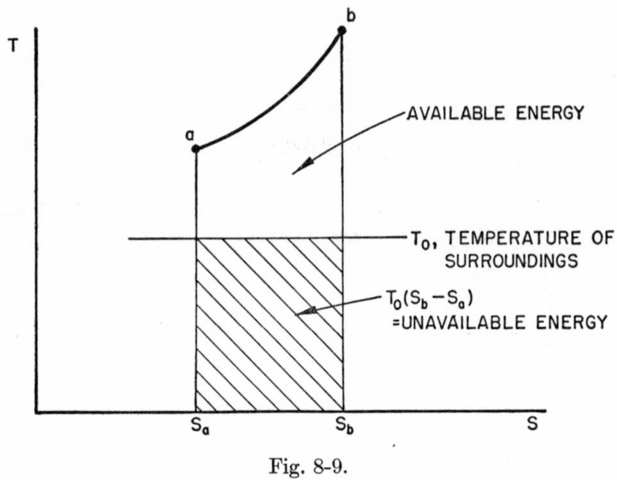

Fig. 8-9.

8-6 Entropy and reversibility. There is a direct relation between entropy change in an isolated system and the reversibility of processes in that system. An isolated system is a system which cannot exchange energy in any way with the surroundings. Obviously any process may be made to occur in an isolated system by including in the system everything affected by the process.

By definition every process in an isolated system must be adiabatic; hence for such processes

$$\int \frac{dQ}{T} = 0 \tag{16}$$

Also, it can be shown that for any process in any system

$$\frac{dQ}{T} \lesseqgtr ds \tag{17}$$

then for an isolated system

$$\Delta s \gtreqless 0 \tag{18}$$

Equation (18) is a statement of the *Principle of Increase of Entropy.* By this principle an isolated system in a given initial state can change only to states of the same or greater entropy; but by the First Law an isolated system must have constant internal energy, therefore an isolated system may change only to states having the same internal energy and the same or greater entropy than the initial state. The internal energy here considered is the most general internal energy and includes effects of motion and gravity (and surface tension, electricity and magnetism if they were to be permitted in any particular case).

The principle of increase of entropy provides a criterion of reversibility for, in an isolated reversible system, the entropy must be constant, all processes being both adiabatic and reversible. Therefore if the entropy of an isolated system increases, the process taking place must be irreversible.

8-7 Examples of second law problems. This section contains examples of the use of the ideas and principles presented in this and the preceding chapter.

Example 1. It is proposed to obtain power from the hot surface water of tropical seas using the cold water from the depths as a sink for heat rejection. The surface water is at 85°F, the deep water at 50°F. In the light of the Second Law is such a scheme possible? If so, what is the maximum thermal efficiency possible under the Second Law?

Solution: The scheme is possible since the Second Law permits net work to be produced by a heat engine if a temperature difference is available. The maximum possible efficiency is given by

$$\eta_{\text{max}} = \frac{T_1 - T_2}{T_1}$$

$$T_1 = 85 + 460 = 545^{\text{R}}$$

$$T_2 = 50 + 460 = 510^{\text{R}}$$

$$\eta_{\text{max}} = \frac{35}{545} = 0.064 \text{ or } 6.4 \text{ percent}$$

An actual plant, being subject to losses and irreversibility, would have a much lower efficiency.

Example 2. A proposed steam power plant will provide for supplying heat to steam at temperatures up to 1050°F. The temperature of heat rejection is about 90°F. It is stated that the thermal efficiency of the plant will approach 34 percent. Does this seem reasonable in the light of the Second Law?

Solution: The maximum possible efficiency for the given temperature range is

$$\eta_{\text{max}} = \frac{(T_1 - T_2)}{T_1} = \frac{960}{1510} = 0.63 = 63 \text{ percent}$$

The predicted efficiency does not seem unreasonably high since it is only a little more than half the maximum possible efficiency. Actual plants seldom approach the maximum efficiency, partly because most of the heat is actually supplied at temperatures well below the maximum temperature of the cycle.

Example 3. How much does the entropy of a pound of air change when the air is heated reversibly from 1 atm pressure, 50°F to 1 atm, 150°F, if the specific heat of air at constant pressure is 0.240 Btu/lb F?

Solution: Since the given states are at the same pressure, assume a reversible constant-pressure process between them and compute the entropy change for the assumed process. (Note that a constant-pressure process was not given in the example.)

$$\Delta s = \int_{\text{rev}} \frac{dQ}{T}$$

For a reversible constant-pressure process, from Eq. 3-2,

$$dQ = mc_p dt$$

then

$$\Delta s = \int \frac{mc_p\, dt}{T} = mc_p \int \frac{dT}{T} = mc_p(\ln T_2 - \ln T_1)$$

$$\Delta s = 1(0.240)(\ln 610 - \ln 510) = 0.0431 \text{ Btu/F}_{\text{abs}}$$

Example 4. Suppose the air in Example 3 is changed from the initial state to the final state of that problem by an adiabatic process involving work against friction. How much does the entropy of the air change in this irreversible process?

Solution: The same as in Example 3. If this is not clear, review the whole discussion of entropy.

Example 5. In a steam boiler, hot gases from a fire transfer heat to water which vaporizes at constant temperature. In a certain case the gases are cooled from 2000°F to 1000°F while the water evaporates at 400°F. The specific heat of the gases is 0.24 Btu/lb F, and the latent heat of the water is 826.0 Btu/lb. All the heat transferred from the gases goes to the water. How much does the total entropy of the combined system of gas and water increase as a result of the irreversible heat transfer? Obtain the result on the basis of one pound of water evaporated.

Solution: The entropy change of the combined system is the sum of the entropy changes of its parts. The individual entropy changes of the gas and water systems can be found by assuming in each case a reversible process between the actual end states of the individual system. For the water, since temperature is constant,

$$\Delta s_{H_2O} = \frac{Q}{T} = \frac{826}{860} = 0.965 \text{ Btu/R}$$

For the gas, per pound of water evaporated, $Q = -826$ Btu. Also

$$Q = m\int c_p\, dT = mc_p\, \Delta T_{\text{gas}}$$

since c_p is constant. Then, for the gas,

$$\Delta s_{\text{gas}} = \int_{\text{rev}} \frac{dQ}{T} = mc_p \int \frac{dT}{T} = mc_p(\ln T_2 - \ln T_1)$$

$$= \frac{Q}{\Delta T_{\text{gas}}}(\ln T_2 - \ln T_1)$$

$$= 0.826(\ln 1460 - \ln 2460) = -0.431 \text{ Btu/R}$$

The total change is

$$\Delta s = \Delta s_{H_2O} + \Delta s_{\text{gas}} = 0.534 \text{ Btu/R}$$

Observe that the total change of entropy is positive, as it must be by the principle of increase of entropy.

Example 6. For the process of Example 5 find the increase in unavailable energy due to the irreversible heat transfer. Assume the temperature of the surroundings is 80°F.

Solution: Sketch on the T–s plane the two assumed reversible processes, heat transfer from the gas and heat transfer to the water. Use a basis of one pound of water evaporated. The gas is cooled from 1 to 2 by transferring heat to the evaporating water which changes from a to b. If the heat from the gas had been transferred reversibly to a reversible heat engine the heat engine would have received the heat along the path 2–1 because reversible heat transfer requires zero temperature difference as each increment of heat is transferred; the heat engine would have rejected at T_0 a heat quantity $T_0\Delta s_{21}$ or $-T_0\Delta s_{\text{gas}}$. If a reversible heat engine received the heat reversibly at 400°F the heat engine would have rejected at T_0 a heat quantity $T_0\Delta s_{ab}$

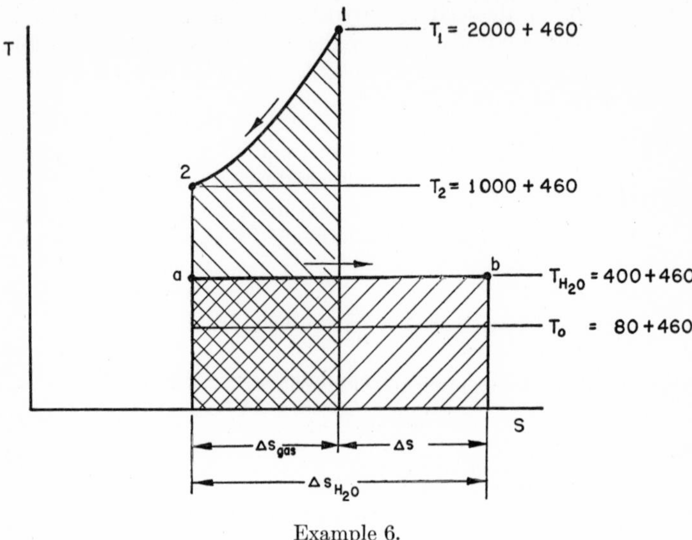

Example 6.

or $T_0\Delta s_{H_2O}$. The increase in unavailable energy due to the transfer of heat from the gas to the water is then

$$T_0\Delta s_{H_2O} - (-T_0\Delta s_{gas}) = T_0(\Delta s_{H_2O} + \Delta s_{gas}) = T_0\Delta s$$

where Δs applies to the combined system as in Example 5. The numerical result is then

$$T_0\Delta s = 540(0.534) = 289 \text{ Btu}$$

It is of interest to note that the available energy in the heat transferred from the gas was, at the temperature level of the gas,

$$Q - T_0\Delta s_{21} = 826 - 540(0.431) = 594 \text{ Btu}$$

Therefore the irreversible transfer of heat has resulted in a loss of 289/594, or almost 49 percent of the available energy. The situation here depicted is typical of what is happening constantly in even the best power plants. It shows that, despite the great advances so far made, there is still room for improvement in basic heat power operations.

8-8 Summary. No cyclic heat engine operating between fixed temperature levels can be more efficient than a reversible engine operating between the same temperatures.

The absolute thermodynamic temperature scale is defined by the relation

$$\frac{T_1}{T_2} = \frac{Q_1}{Q_2} \tag{5}$$

where Q_1 is the heat received from a source at T_1, and Q_2 is the heat rejected to a sink at T_2, by a reversible heat engine which exchanges heat only at T_1 and T_2. Equation (5) establishes the ratio of any two temperatures and the zero point of the scale, but does not establish the size of the degree of temperature, which is taken the same as the conventional Fahrenheit degree (or the Centigrade degree) for the Rankine scale (or the Kelvin scale). The absolute thermodynamic scale is identical with the ideal gas scale discussed in Chap. 3. In thermodynamic analysis temperature implies the absolute thermodynamic scale, although data are frequently presented with reference to the conventional temperature scales.

By the definition of the temperature scale, the efficiency of a reversible heat engine receiving heat solely at T_1 and rejecting heat solely at T_2 is given by

$$\eta = \frac{T_1 - T_2}{T_1} \tag{6}$$

This equation sets an upper limit on the possible efficiency of *any* cyclic heat engine working between T_1 and T_2.

The *entropy* s of a system is a quantity which satisfies the relation

$$ds = \frac{dQ_{\text{rev}}}{T} \tag{7}$$

The change in entropy between any two states of a system is measured by the integral of dQ/T for a *reversible process* between the two states. Entropy is a property of a system but has no definite value until a value is assigned for one state of the system; then the values at all other states become fixed. In an irreversible process the change of entropy is the same as for a reversible process between the same end states, but the integral of dQ/T is less for an irreversible process than for a reversible process between the same end states. Because entropy is a property, the integral of ds around a cycle must always be zero.

A process in which the entropy of a system remains constant is called *isentropic.* A process in which heat transfer is zero is called *adiabatic.* A *reversible* adiabatic process is isentropic but an adiabatic process is not necessarily isentropic, nor is an isentropic process necessarily reversible or adiabatic. Of the three characteristics, reversible, adiabatic, and isentropic, no one necessarily implies the others, but any two together necessarily require the third.

The plot of a reversible process upon the coordinates of temperature and entropy has the useful property that the area between the path of the process and the line of zero temperature is proportional to the heat transferred in the process. The area enclosed within the plot of a reversible cycle on the T–s plane is proportional to the work of the cycle.

Given a certain process in which heat is supplied to a cyclic engine and given that heat is to be rejected at the temperature of the surroundings, the Second Law determines the maximum possible work that can be obtained from the engine. This quantity of work is called the available part of the heat supplied, or the *available energy* supplied to the engine. The remainder of the heat supplied is called *unavailable.* If the lowest usable sink temperature (temperature of the surroundings) is T_0 the unavailable part of the heat supplied is given by $T_0 \, \Delta s$ where Δs is the increase in the entropy of the system during the process of *supplying* heat. The available portion of the heat supplied to a heat engine decreases as the temperature at which the heat is supplied decreases. Whenever heat flows through a finite temperature difference some energy becomes unavailable.

In any isolated system the entropy may increase or remain constant but it cannot decrease. This is the principle of increase of entropy. By this principle, if a system is so chosen as to include all bodies affected by a process, the entropy of the system will increase if the process is irreversible, and will remain constant if the process is reversible.

PROBLEMS

8-1. A cyclic heat engine receives heat solely from a source at 1000°F and rejects heat solely to cooling water at 100°F. What is the maximum possible efficiency of the heat engine?

8-2. On the basis of the First Law fill the blank spaces in the following table of hypothetical heat engine cycles; on the basis of the Second Law classify each cycle as either *irreversible, reversible,* or *impossible.*

Cycle	*Temperatures*		*Rates of Heat Flow*		*Rate of Work Output*	*Efficiency*
	Source	Sink	Supply	Rejection		
(a)	540°F	40°F	100 Btu/sec	55 Btu/sec	___ hp	___
(b)	1000°F	100°F	___ Btu/min	1000 Btu/min	___ kw	65%
(c)	1500°R	600°R	___ Btu/hr	___ Btu/hr	35 hp	60%
(d)	1200°R	530°R	10000 Btu/hr	___ Btu/hr	1 kw	___

8-3. The latent heat of fusion of water at 32°F is 143 Btu/lb. How much does the entropy of a pound of ice change as it melts to liquid in each of the following ways? (a) Heat is supplied reversibly to a mixture of ice and water at 32°F. (b) A mixture of ice and water at 32°F is stirred by a paddle wheel.

8-4. Liquid water flows through a hydraulic turbine in which friction causes the water temperature to rise from 65°F to 67°F. If there is no heat transfer in the turbine how much does the entropy of the water change in passing through the turbine? (Note that liquid water is essentially incompressible, then compute the entropy change for a reversible constant-volume process between 65°F and 67°F.)

8-5. Two pounds of water at 140°F are mixed with 3 pounds of water at 40°F in a constant-pressure process at 1 atmosphere. Find the increase of entropy of the total mass of water due to the mixing process.

8-6. In a Carnot cycle heat is supplied at 380°F and rejected at 80°F; the working fluid is water, which, while receiving heat, evaporates from liquid at 380°F to steam at 380°F. According to the steam tables the entropy change for this process is 1.0059 Btu/lb R. (a) If the cycle operates on a stationary mass of 1 lb of water, Fig. 7-4, how much work is done per cycle, and how much heat is supplied? (b) If the cycle operates in steady flow, Fig. 7-5, with a power output of 25 hp, what is the steam flow rate?

8-7. A heat engine receives reversibly 100 Btu of heat per cycle from a source at 540°F, and rejects heat reversibly to a sink at 40°F; there are no other heat transfers. For the three hypothetical amounts of heat rejected, in (a), (b), and (c), below, show which case is irreversible, which reversible, and which impossible. (a) 50 Btu/cycle rejected; (b) 75 Btu/cycle rejected; (c) 25 Btu/cycle rejected.

8-8. As shown in Sec. 7-8 a heat engine cycle may be operated in reverse, as a heat pump. Show, by the method used in Sec. 8-1, that no heat pump cycle working between two reservoirs at fixed temperatures can have a larger ratio of Q_1 to W_{net} than a reversible cycle.

8-9. Show that no heat pump cycle working between two reservoirs at fixed temperatures can require less net work per unit of heat received than a Carnot cycle.

8-10. In a certain process a vapor, while condensing at 800°F, transfers heat to water evaporating at 500°F. The resulting steam is used in a power cycle which rejects heat at 100°F. What fraction of the available energy in the heat transferred from the process vapor at 800°F is lost due to the irreversible heat transfer to the water at 500°F?

8-11. Exhaust gases leave an internal combustion engine at 1500°F and 1 atmosphere, after having done 400 Btu of work per pound of gas in the engine. The specific heat of the gases at 1 atmosphere is 0.26 Btu/lb F, and the temperature of the surroundings is 80°F. (a) How much available energy per pound of gas is lost by throwing away the exhaust gases? (The available energy is the work of a reversible heat engine which receives heat reversibly from the exhaust gases, while cooling them to 80°F, and rejects heat at 80°F). (b) What is the ratio of the lost available energy to the engine work?

Chapter 9

PROPERTIES OF PURE SUBSTANCES

In the foregoing chapters the basic laws of thermodynamics and some of their general applications have been presented. In subsequent chapters it is desired to apply this information to the analysis of specific engineering processes. However, to obtain quantitative results, knowledge of the properties of the substances involved is necessary. This chapter, therefore, contains a general description of the relations among properties, and the following chapters give the common methods of presenting and using data.

9-1 Methods of presenting property relations. The properties usually needed in applying thermodynamics to mechanical engineering problems are pressure, temperature, volume, internal energy, enthalpy, and entropy. If the portions of the internal energy due to gravity, observable motion, surface tension, electricity, or magnetism are separately accounted for, a pure substance has in general only two independent properties. If pressure and specific volume, for example, are fixed then all the other properties become fixed. This makes it possible to express any property of a substance as a function of two other properties or to plot lines of constant value for any one property on coordinates of two other properties.* Of the properties listed above, the most evident to the senses are pressure, volume, and temperature; therefore these properties are often used to describe the states that a particular substance may assume. The equation relating pressure, volume, and temperature is called an equation of state. Some other information in addition to the equation of state is necessary to determine the remaining properties.

* See the Appendix for tables of properties, and envelope charts at back of book for plots.

An example of a simple equation of state which is satisfactory for many gaseous substances is the *perfect gas equation*

$$pv = RT \tag{1}$$

in which p is the pressure, v the specific volume, T the temperature, and R a constant for the particular gas. The same information may be shown graphically, as in Fig. 9-1. In the general case of substances which may pass from solid to liquid and to gaseous states, no such simple equation is even approximately correct. The best way to describe the general nature of the property relations is by a graphical exposition. In order to make the discussion specific the important substance *water* will often be used as an example. The characteristics of other substances are similar to those of water but some peculiarities of water will be noted.

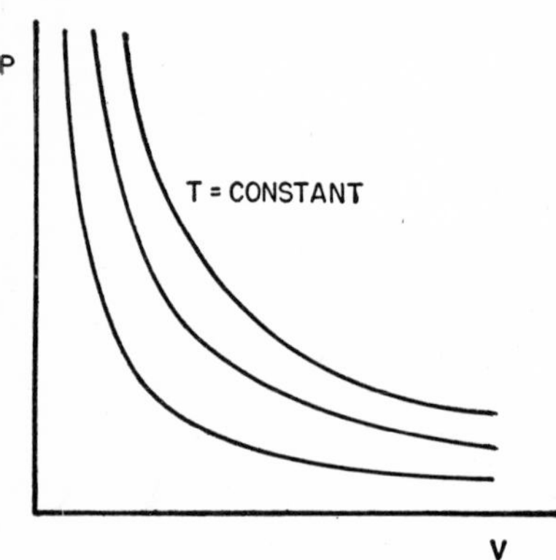

Fig. 9-1. Perfect gas equation of state.

9-2 Solid, liquid and gas phases. Assume that a unit mass of ice (solid water) at 0°F and 14.7 psia fills the clearance space of a cylinder and piston machine, Fig. 9-2. It is desired to trace the changes which occur in the mass of water as the temperature is increased while the pressure is held constant. Use will be made of two plots, Fig. 9-3, on coordinates of pressure vs. volume and temperature vs. volume respectively. Let heat flow slowly to the ice so that its temperature is always uniform; then changes in volume and temperature will be observed as plotted in Fig. 9-3 (data from Keenan and Keyes):

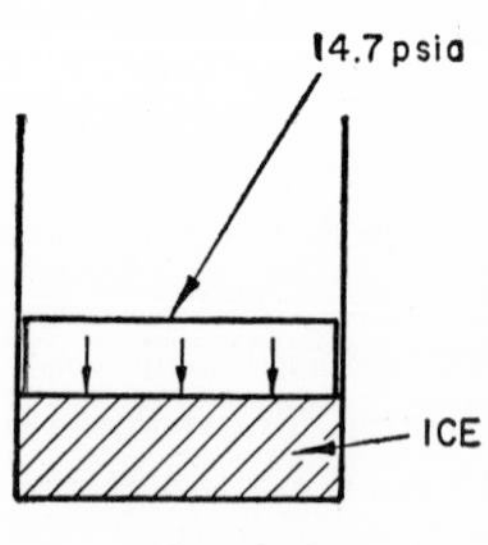

Fig. 9-2.

1–2 the temperature of the solid increases from 0°F to 32°F with a small increase in volume,

2–3 the solid melts to liquid at constant temperature with a small decrease in volume,

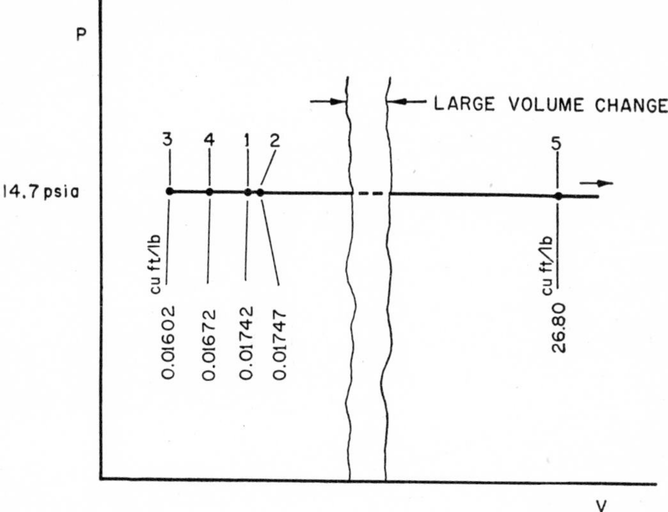

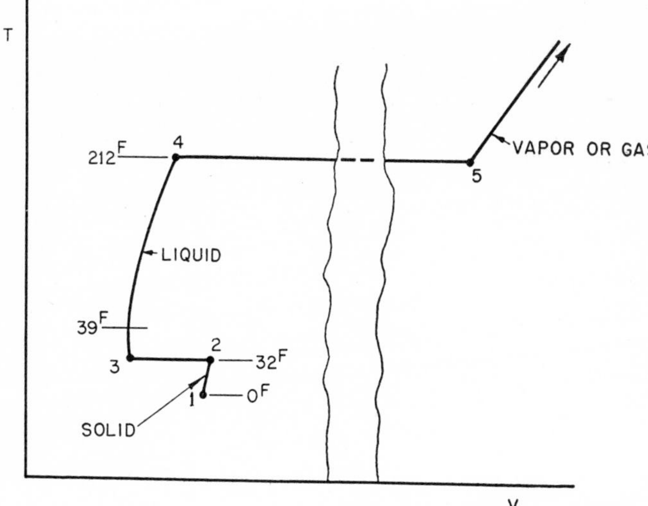

Fig. 9-3. Heating water at constant pressure.

3–4 the temperature of the liquid increases from 32°F to 212°F; the volume decreases to a minimum at 39°F and then increases,

4–5 the liquid boils to a vapor or gas at constant temperature with a large increase in volume.

Beyond 5 both the temperature and the volume of the vapor increase.* The decrease in volume during melting is a peculiarity of water. Most substances show a continuous increase in volume as indicated in Fig. 9-4. In a slow cooling process the paths of Figs. 9-3 and 9-4 would be followed in the reverse direction.

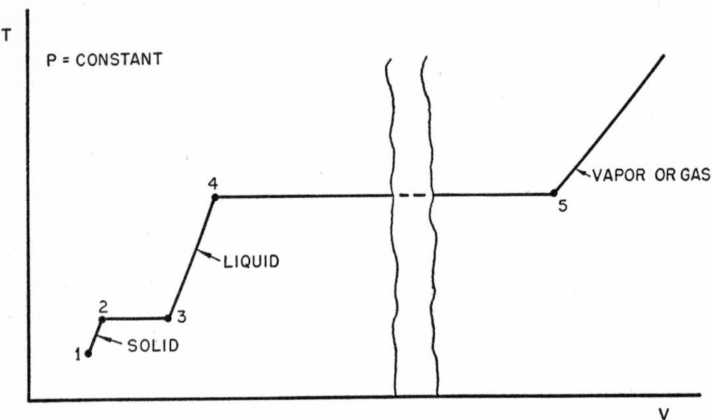

Fig. 9-4. Constant-pressure heating; general substance.

9-3 Saturation states. In the process described above the water existed in three different homogeneous states of aggregation, or phases; solid between 1 and 2; liquid between 3 and 4; and gas beyond 5. Between 2 and 3 the water was not homogeneous but consisted of a mixture of two phases, solid and liquid; between 4 and 5 it consisted of a mixture of liquid and gas. The states indicated by points 2, 3, 4, and 5 are known as *saturation* states. A saturation state is a state from which a change of phase may occur without change of pressure or temperature. Thus state 2 is a saturated solid state because the solid can change to liquid at constant pressure and temperature from state 2. State 1 is not a saturation state because in a constant pressure process from state 1 the temperature will change before melting occurs. States 3 and 4 are both saturated liquid states; in state 3 the liquid is saturated with respect to solidification, whereas

* The terms *vapor* and *gas* are often not rigorously distinguished. *Vapor* is generally used for states not far from saturation; *gas* is sometimes restricted to temperatures above the critical temperature. See Sec. 9-3 for the meaning of saturation and critical temperature.

in state 4 the liquid is saturated with respect to vaporization. State 5 is a saturated vapor state because, from state 5, the vapor can condense to liquid without change of pressure or temperature.

Experience shows that for any given substance, in saturation states at various pressures, there is a definite temperature corresponding to each pressure. Thus if the process 4–5 in Fig. 9-3, corresponding to vaporization, is plotted for a number of different pressures, a, b, c, etc., a series of saturation states 4 and 5 appears as in Fig. 9-5. A curve drawn through the points 4 is called the saturated liquid

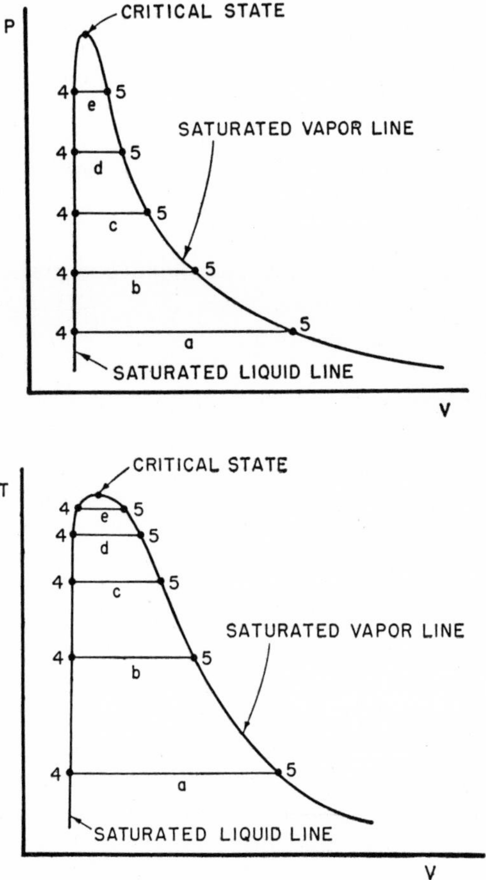

Fig. 9-5. Saturation curve on p–V and T–V coordinates.

line, and a curve drawn through the points 5 is called the saturated vapor line. It is found that when these two curves are extended to higher pressures and temperatures they do not continue indefinitely but turn toward each other and join in a continuous curve at a state called the critical state. Thus the two saturation lines are actually a single line forming what is often called the *saturation dome* or *vapor dome.*

In the plots of Fig. 9-5 each point in the plane represents a particular state of the substance considered. States to the right of the saturated vapor line are called *superheated vapor* states or gas states. States to the left of the saturated liquid line are called *compressed liquid* states. States between the saturation lines are called *mixture* states or *wet vapor* states. The pressure and temperature at the critical state are called the *critical pressure* and the *critical temperature* respectively.

It should be understood that at pressures higher than the critical pressure the phenomenon of boiling does not exist. At such pressures there is never any time in a constant-pressure heating process when the temperature remains constant, and liquid and gas become distinct phases in a mixture. Instead, the fluid, remaining always homogeneous, simply becomes gradually less dense, with no distinction between liquid and gas. Sometimes the arbitrary rule is followed that the liquid becomes a gas when it passes the critical temperature.

9-4 Equilibrium between phases. The saturation states are states in which two different phases of a substance can exist together in equilibrium. If liquid water at 1 atm and 212°F is brought into contact with water vapor at 1 atm and 212°F there is no disturbance of either phase by the other. Since pressures are equal no work is transferred; since temperatures are equal no heat is transferred. If liquid water at 1 atm and 200°F is brought into contact with water vapor at 1 atm and 212°F there is equilibrium of pressure but the difference of temperature will cause heat transfer until the two phases have equal temperatures. Assuming the pressure held constant by some external restraint, the only temperature at which the liquid and vapor can reach equilibrium is 212°F; therefore sufficient vapor must condense to raise the temperature of the liquid to 212°F. Similarly, if liquid and vapor at different pressures are placed in contact, they must assume the same pressure, exchanging work or heat as necessary to accomplish this.

It should be apparent that all this discussion of property relations involves an assumption that the substance is in equilibrium throughout its mass. A diagram or plot showing the conditions under which the various phases of water can be in equilibrium with each other is given in Fig. 9-6. This shows the relations between saturation temperatures and saturation pressures for water, plotted on coordinates of temperature vs. pressure. It is seen that the solid-liquid equi-

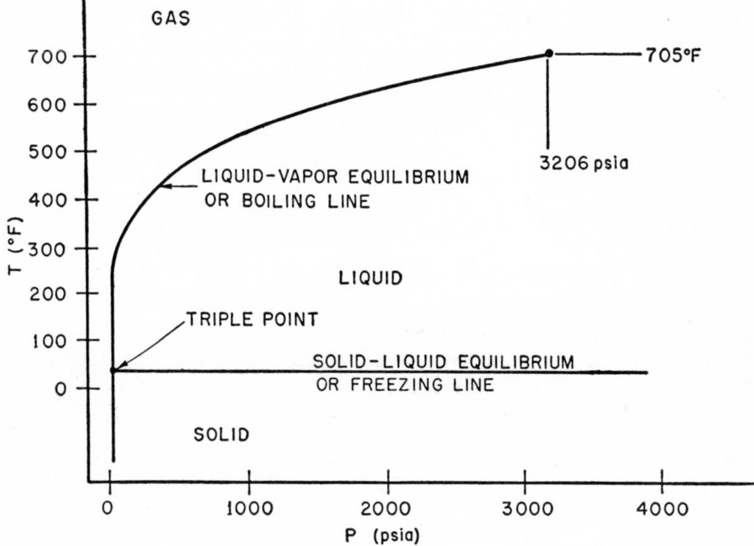

Fig. 9-6. Equilibrium among phases of water.

librium, or freezing, temperature is little affected by pressure; the variation is only about $-1°F$ per 1000 psi pressure increase up to pressures of several thousand psia.* Usually such a variation is negligible and the freezing temperature is considered independent of pressure. The liquid-vapor equilibrium, or boiling, temperature is seen to be very much a function of pressure; this is of great importance in the design of steam power apparatus.

The liquid-vapor saturation line does not extend indefinitely in either direction. It is limited at the upper end by the critical state,

* For most substances the variation is of opposite sign, but still of small magnitude. For water above 2000 atm the freezing temperature increases with pressure.

above which no boiling process occurs. At the lower end the circumstances are as shown in Fig. 9-7 which is an enlarged plot of part of Fig. 9-6. The enlarged plot shows that, whereas gaseous and liquid water can exist together in equilibrium above 32°F, gaseous and solid water can exist together in equilibrium below 32°F. The direct change from solid to gas is called *sublimation.* The point at the intersection of the boiling line and the freezing line is called the triple point because at that state the three phases of gas, liquid, and solid may exist together in equilibrium. Since the triple point for water is at a pres-

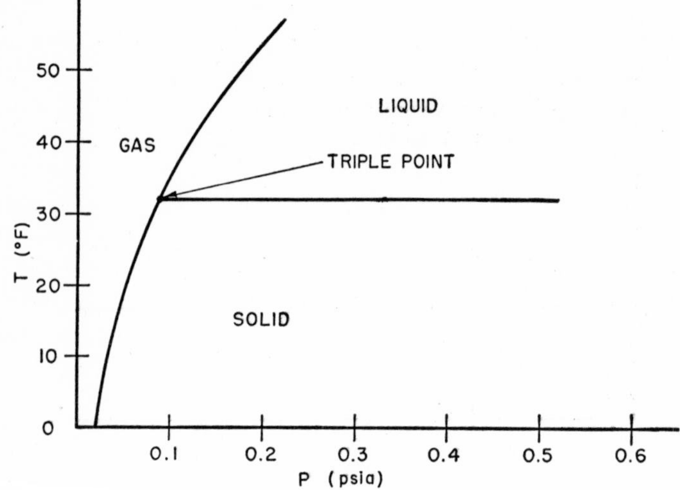

Fig. 9-7. Triple point for water.

sure of about 0.09 psia while the thermometric calibration point of 32°F is the melting point at 14.7 psia pressure, it follows that the triple point for water is not exactly at 32°F, but the difference is insignificant for engineering purposes.

The triple point for carbon dioxide is at about −70°F and about 5 atm pressure. Hence when solid CO_2 is exposed to 1 atm pressure it begins to change directly to gas, and cools toward an equilibrium temperature below −70°F. This phenomenon is applied commercially in the use of "dry ice" for refrigeration.

9-5 Phase diagram on the pressure-volume plane. A typical phase diagram on pressure-volume coordinates is shown in Fig. 9-8.

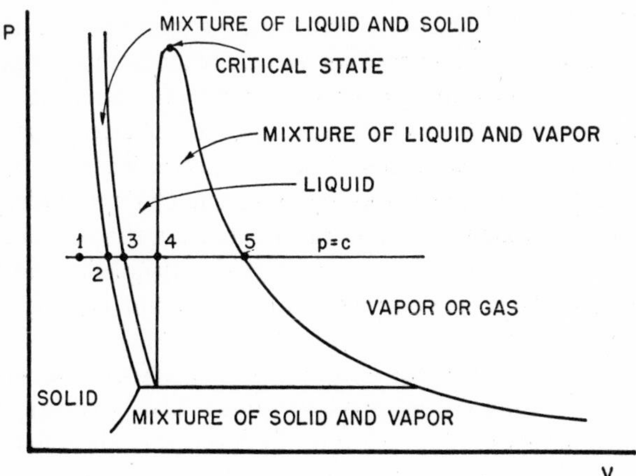

Fig. 9-8. Pressure-volume phase diagram.

The points 1, 2, 3, 4, and 5 correspond to like points in Fig. 9-4. The scales of the diagram have been distorted because in true scale the liquid and solid regions would be microscopic compared to the vapor regions.

In Fig. 9-9 is shown the liquid-vapor portion of Fig. 9-8 with

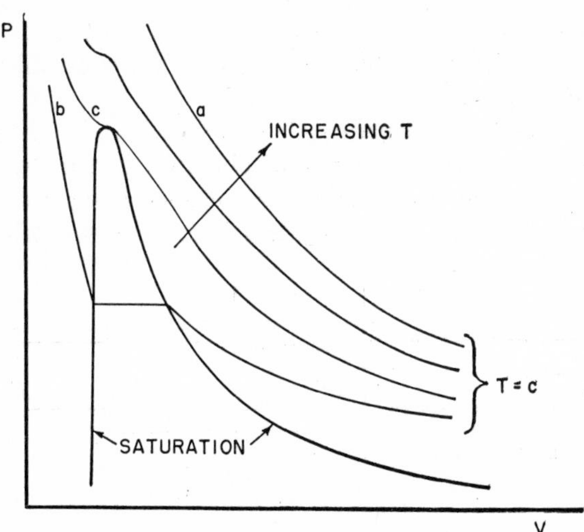

Fig. 9-9. Isothermal lines.

some typical lines of constant temperature superimposed. In the gas region, particularly with curves such as a, for temperatures well above the critical, the constant-temperature curve approximates a rectangular hyperbola according to Eq. (1). The curves like b, for temperatures below the critical, consist of two approximately hyperbolic segments in the liquid and vapor regions joined by a straight segment parallel to the v axis in the mixture region. As the temperature increases, the length of the straight segment decreases until at the critical temperature, curve c, there is only a point of tangency with the horizontal at the critical state.

9-6 Enthalpy, internal energy and entropy of the pure substance; temperature-entropy plot. The pressure, volume, temperature data discussed in the preceding sections must be supplemented by other data (from heat or work measurements, Joule-Thomson experiments) in order to compute enthalpy, internal energy, and entropy values corresponding to each point on the pressure-volume plot.

In the constant-pressure heating process described in Sec. 9-2 the heat supplied to the water is equal to the change of enthalpy of the water. Hence the variation of enthalpy at constant pressure can be obtained by heat measurements or from a knowledge of specific heats and latent heats; but to have a complete knowledge of the enthalpy it is necessary to know how it changes when the pressure changes. This information may be obtained in several ways, of which two will be described briefly.

The Joule-Thomson experiment, Fig. 9-10, consists in passing a fluid in steady flow through a porous plug in a duct under such conditions that heat transfer, kinetic energy change, and potential energy change are negligible. Such a process is called a *throttling* process.

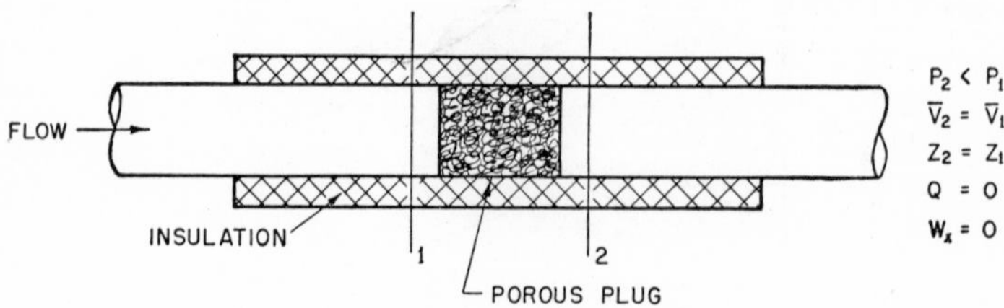

Fig. 9-10. Joule-Thomson experiment.

In the *steady flow energy equation* for this process all terms except the enthalpies will disappear leaving

$$h_1 = h_2$$

By the throttling experiment, states having the same enthalpy may be identified at different pressures.

Another method of obtaining the enthalpy difference between two states at different pressures is to carry out a frictionless constant-volume heating process in which the heat supplied and the temperature change are measured. Such an experiment will yield the change in internal energy at constant volume since, for such a process

$$(\Delta u)_v = Q_v$$

If the p, v, T relation is known, the pressure change can be found, and the enthalpy change follows from the definition $h = u + pv$. When the enthalpy is known as a function of p and t it is possible to compute specific heats, internal energy, and entropy from the known relations among these properties. The property diagram may then be replotted on coordinates of temperature vs. entropy, Fig. 9-11, or other coordinates as desired. In the back cover envelope are such plots for water, air, carbon dioxide, and Freon-12.

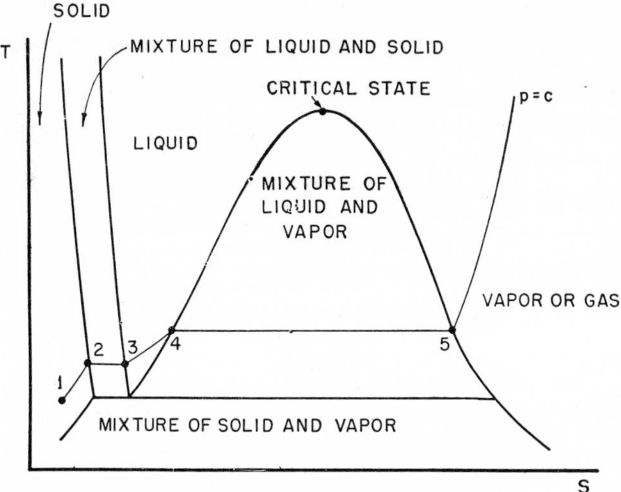

Fig. 9-11. Temperature-entropy phase diagram.

9-7 General property relations for pure substances. Since a pure substance has only two independent properties, it must be possible to express any property as a function of two others. But this general fact provides no clue as to the nature of the relations for any particular substance. Hence the p, v, T relation for a substance is determined by making measurements of pressure, volume, and temperature. Other properties such as specific heats may also be measured as functions of pressure and temperature. However, not all properties need be measured directly as functions of other properties because there exist certain general relations which apply to the properties of all pure substances. These general relations, when combined with empirical data, permit the computation of properties not directly measured.

Of the many relations which must hold among the properties of all pure substances some are simply definitions; for example

$$h = u + pv$$

and

$$c_v = \left(\frac{du}{dT}\right)_v$$

Other relations are derived from the application of the First and Second Laws to a generalized process in a pure substance. Consider any reversible process in a system of unit mass of a pure substance in the absence of effects of motion and gravity. From the First Law

$$Q = \Delta u + W$$

or for a small change

$$dQ = du + dW \tag{2}$$

Now from the Second Law development, Chap. 8, since the process is reversible

$$dQ = T\,ds \tag{3}$$

and, since motion and gravity effects are absent,

$$dW = p\,dv \tag{4}$$

Then for a reversible process in the given system

$$T\,ds = du + p\,dv \tag{5}$$

Equation (5) is a relation solely among the properties of a system. If the system passes from one given state to another given state, the property changes are the same for all processes by which the system could pass from the one state to the other. Hence Eq. (5) is true for *all processes* between states which could hypothetically be connected by a reversible process. A pure substance can pass by a hypothetical reversible process from any equilibrium state to any other equilibrium state. Therefore Eq. (5) is a true relation among the properties of a pure substance for all processes during which the system is in equilibrium throughout its mass. This restriction is obvious from the viewpoint of physical measurement of the properties for, if equilibrium did not exist, there would not exist any particular properties which could be called *the* pressure and *the* temperature of the system. The equation would then be a relation among non-existent entities.

Note that Eqs. (2) and (5) are applicable to both reversible and irreversible processes; Eq. (2) because it is the First Law, and Eq. (5) because it is a relation solely among properties, and is therefore independent of the type of process. Equation (5) is *not* the First Law; it is a property relation that is dependent upon *both* the First and Second Laws.

From Eq. (5) and the definitions of properties, numerous useful property relations can be derived. As an example, a simple derivation of a very useful relation follows:

By definition $$h = u + pv$$

or $$dh = du + p\,dv + v\,dp$$

Substituting in (5) $$T\,ds = dh - v\,dp \tag{6}$$

Other relations will be derived in subsequent chapters. For general discussion of such relations and examples of their use see the references below.

PROBLEMS

9-1. From the temperature-entropy diagram for carbon dioxide (in the back cover envelope) find the phase (liquid, solid, vapor, or mixture) in which carbon dioxide exists in equilibrium at each of the following states: (a) $t = 180°\text{F}$, $p = 1200$ psia; (b) $t = -100°\text{F}$, $v = 10$ cu ft/lb; (c) $p = 100$ psia, $h = 180$ Btu/lb; (d) $p = 1200$ psia, $t = 0°\text{F}$; (e) $t = -80°\text{F}$, $p = 75$ psia; (f) $v = 0.2$ cu ft/lb, $p = 400$ psia.

9-2. In the temperature-entropy diagram for carbon dioxide find the data needed to answer the following questions: (a) What are the critical pressure and temperature of carbon dioxide? (b) How much does the volume of carbon dioxide change in going from saturated liquid at 0°F to saturated vapor at 0°F? (c) Find the saturation temperature of carbon dioxide at 100 psia and at 1000 psia. (d) At what temperature will solid carbon dioxide be in equilibrium with its vapor under a pressure of 1 atm? (This is the temperature of "dry ice.") (e) At the triple-point temperature, how much does the enthalpy of carbon dioxide change in going from saturated solid to saturated liquid; from saturated liquid to saturated vapor; from saturated solid to saturated vapor?

9-3. In the temperature-entropy diagram for air (in the back cover) find the data to answer the following: (a) What are the critical pressure and temperature of air? (b) Find the enthalpy of superheated air at −100°F and 100 psia. (c) Find the constant-pressure specific heat of air at −200°F, 14.7 psia, and at −200°F, 1000 psia. (Find how much h changes for a small temperature change along the constant-pressure line, and compute $\Delta h/\Delta t$.) (d) Find c_p for air at 240°F, 14.7 psia, and at 240°F, 1000 psia. (e) At 100 psia find the saturation temperatures for vapor and for liquid. (The saturation temperature is not constant at constant pressure because air is a mixture of oxygen and nitrogen; oxygen-rich liquid condenses first, and a progressively lower temperature is required to condense the vapor as it becomes richer in nitrogen.)*

9-4. From the pressure-enthalpy diagram for Freon-12 (in the back cover envelope) find the data to answer the following: (a) What are the critical pressure and temperature of Freon-12? (b) In what phase is Freon-12 at 200 psia, 400°F? (c) At $h = 55$ Btu/lb, $t = 40$°F, what is the state of aggregation (liquid, vapor, mixture)? (d) What is the normal boiling point (at 1 atm pressure) for Freon-12? (e) What is the density of Freon-12 saturated vapor at 0°F?

9-5. A Joule-Thomson experiment (Fig. 9-10) is carried out with Freon-12; the initial state is at 100 psia, 100°F, and the final pressure is 20 psia. Using the chart find (a) the final temperature of the fluid; (b) the change of internal energy during the process.

9-6. Using either the enthalpy-entropy diagram, or the temperature-entropy diagram in Keenan and Keyes' steam tables, find: (a) the saturation pressure for water at 400°F; (b) the enthalpy of saturated water vapor at 100 psia; (c) the approximate specific heat at constant pressure for water vapor at 1 atm, 300°F. (Find the ratio of Δh to Δt for a small temperature change at constant pressure.)

9-7. From the temperature-entropy diagram for carbon dioxide find the change in the entropy and the temperature of carbon dioxide when it goes through a throttling process beginning at 800 psia, 100°F, and ending at

* See Keenan, *Thermodynamics*, chap. 22, for a discussion of binary mixtures.

100 psia. Sketch the area under the path of the process in the T–s plane, and explain why the area is not zero even though the process is adiabatic.

REFERENCES

Keenan, J. H., *Thermodynamics.* New York: Wiley, 1941, chap. 19.

Zemansky, M. W., *Heat and Thermodynamics.* New York: McGraw-Hill, 1943, chap. 13.

Keenan, J. H. and F. G. Keyes, *Thermodynamic Properties of Steam.* New York: Wiley, 1936, Introduction, pp. 11–24.

Keenan, J. H. and J. Kaye, *Gas Tables; Thermodynamic Properties of Air.* New York: Wiley, 1948, Sources and Methods, pp. 199–203.

Sears, F. W., *Introduction to Thermodynamics, Kinetic Theory of Gases, and Statistical Mechanics.* Cambridge: Addison-Wesley, 1950, chap. 9.

See also the list of references at the end of Chap. 10, p. 139.

Chapter 10

TABULATED PROPERTIES: STEAM TABLES

In this chapter the use of tabulated property data will be explained with particular reference to the steam tables.* The properties of water are arranged in the steam tables as functions of pressure and temperature. Separate tables are provided to give the properties of the substance in the liquid and vapor phases and in the saturation states. The several tables will be considered separately. In all the tables the zero points for both enthalpy and entropy are taken at 32°F saturated liquid.

10-1 Saturation states. Tables 1 and 2 give the properties of saturated liquid and saturated vapor. Either the pressure or the temperature is sufficient to identify a saturation state so Tables 1 and 2 need only one independent variable. In Table 1 the independent variable is temperature; on each line the following data are tabulated:

t, °F, temperature
p, psia, saturation pressure corresponding to t
v_f, cu ft/lb, specific volume of saturated liquid
v_{fg}, cu ft/lb, change in v during evaporation
v_g, cu ft/lb, specific volume of saturated vapor
h_f, Btu/lb, specific enthalpy of saturated liquid
h_{fg}, Btu/lb, change in h during evaporation (latent heat)
h_g, Btu/lb, specific enthalpy of saturated vapor
s_f, Btu/lb°R, specific entropy of saturated liquid
s_{fg}, Btu/lb°R, change in s during evaporation
s_g, Btu/lb°R, specific entropy of saturated vapor

* Keenan, J. H. and F. G. Keyes, *Thermodynamic Properties of Steam.* New York: Wiley, 1936. Extracts from these tables appear in the Appendix.

The following relations will obviously hold:

$$v_{fg} = v_g - v_f, \qquad h_{fg} = h_g - h_f, \qquad s_{fg} = s_g - s_f$$

Table 2 is similar to Table 1 but has pressure as the independent variable. Table 2 also contains columns for internal energy:

u_f, Btu/lb, specific internal energy of saturated liquid
u_{fg}, Btu/lb, change in u during evaporation
u_g, Btu/lb, specific internal energy of saturated vapor

$$u_{fg} = u_g - u_f$$

Tables 1 and 2 may be used interchangeably. The reason for two tables is to reduce the amount of interpolation required. The use of the tables is best illustrated by examples.

Example 1. What is the heat of vaporization of water at 100 psia?

The heat of vaporization is the change of enthalpy from saturated liquid to saturated vapor,

$$\Delta h = h_g - h_f = h_{fg}$$

At 100 psia, from Table 2, $h_{fg} = 888.8$ Btu/lb.

How much work is done by 1 lb of water as it evaporates at 100 psia?

For a stationary system with no friction

$$W = \int p\, dv$$

At constant pressure $\qquad W = p\,\Delta v = p(v_g - v_f)$

From Table 2,

$$v_g = 4.432 \text{ cu ft/lb} \qquad \text{and} \qquad v_f = 0.01774 \text{ cu ft/lb}$$

Then

$$\begin{aligned} W &= (100)(144)(4.432 - 0.01774) \\ &= 63{,}700 \text{ ft lb/lb} = 81.8 \text{ Btu/lb} \end{aligned}$$

The work may also be obtained by use of the First Law:

$$\begin{aligned} W &= Q - \Delta u \\ \Delta u &= u_{fg} = 807.1 \text{ Btu/lb} \qquad \text{(Table 2)} \\ W &= 888.8 - 807.1 = 81.7 \text{ Btu/lb} \end{aligned}$$

Since the process of evaporation at constant pressure also takes place at constant temperature, the heat of vaporization could have been obtained by assuming a reversible process for which

$$Q = \int T\, ds = T\,\Delta s = Ts_{fg}$$

From Table 2, at 100 psia, $t = 327.8°F$ ($T = 788°R$), and $s_{fg} = 1.1286$ Btu/lb°R. Then

$$Q = (788)(1.1286) = 889 \text{ Btu/lb}$$

Example 2. Saturated steam has entropy of 1.6315 Btu/lb°R; what are its pressure, temperature, and enthalpy?

In Table 1 are found entropy values of 1.6326 at 302°F and 1.6302 at 304°F. By linear interpolation t is found to be very close to 303°F, $h = 1180.6$ Btu/lb, and p is close to 70 psia. Checking in Table 2 the exact entropy value of 1.6315 Btu/lb°R happens to appear in the table at 70 psia. Then from Table 2,

$$p = 70 \text{ psia} \qquad t = 302.92°\text{F} \qquad \text{and} \qquad h = 1180.6 \text{ Btu/lb}$$

Example 3. What is the internal energy of saturated water vapor at 250°F?

Table 1 gives no values of u; however, $u = h - pv$. At 250°F $h_g = 1164.0$ Btu/lb, $p = 29.825$ psia, and $v_g = 13.821$ cu ft/lb. Then

$$u_g = 1164.0 - \frac{(29.825)(144)(13.821)}{778} = 1087.6 \text{ Btu/lb}$$

Another method is to interpolate in Table 2. At 29 psia $t = 248.40$ and $u_g = 1087.3$. At 30 psia $t = 250.33$ and $u_g = 1087.8$. Using linear interpolation, when t is 250.00°F u_g will be 1087.7 Btu/lb

10-2 Liquid-vapor mixtures. When two identical masses of substance are placed together to form a single system there are some properties (called *intensive* properties) which have the same value for the combined system as for the individual masses. For example, the pressure and temperature are of this class. Other properties (called *extensive* properties) are twice as great for the single system of two identical masses as for either mass alone; for example, the volume, internal energy, enthalpy and entropy are of this class. If any two masses of a given substance having equal pressures and temperatures are placed in contact to form a single system, the pressure and the temperature of the combined system will be the same respectively as the pressure and temperature of the separate systems. The volume of the combined system will however be equal to the sum of the volumes of the separate systems and a similar summation will apply to the internal energy, enthalpy and entropy.

Consider a mixture of saturated liquid water and water vapor in equilibrium at pressure p and temperature t. The pressure and temperature of both components of the mixture must also be p and t.

Let the composition of the mixture be given as the fraction, by mass, of vapor in the mixture; the symbol x and the name *quality* are used for this fraction. Then if the mixture is 10 percent liquid by mass the quality, x, is 0.90; if the mixture is equal parts by mass of liquid and vapor the quality is 0.50. In a mixture of total mass m and quality x the mass of liquid, m_f, is $(1 - x)m$ and the mass of vapor m_g is xm. Now as stated above, each extensive property of the mixture will be equal to the sum of the values of the same property for the two components. Thus, letting subscript x denote the property of the mixture at quality x, for total volumes

$$V_x = V_f + V_g$$

or, since total volume is mass times specific volume,

$$mv_x = m_f v_f + m_g v_g$$

Then

$$mv_x = (1 - x)mv_f + xmv_g$$

or

$$v_x = (1 - x)v_f + xv_g$$

similarly

$$\begin{aligned} u_x &= (1 - x)u_f + xu_g \\ h_x &= (1 - x)h_f + xh_g \qquad \text{(a)} \\ s_x &= (1 - x)s_f + xs_g \end{aligned}$$

The relations (a) can be changed in form as follows:

$$\begin{aligned} v_x &= (1 - x)v_f + xv_g \\ &= v_f + x(v_g - v_f) \\ &= v_f + xv_{fg} \qquad \text{(b)} \end{aligned}$$

or alternatively

$$\begin{aligned} v_x &= (1 - x)v_f + xv_g \\ &= (1 - x)v_f - (1 - x)v_g + v_g \\ &= v_g - (1 - x)v_{fg} \qquad \text{(c)} \end{aligned}$$

The physical interpretation of (b) is that the volume of one pound of mixture at quality x is equal to the volume of one pound of liquid plus the increase of volume due to the evaporation of x pounds. The physical interpretation of (c) is that the volume of one pound of mixture of quality x is equal to the volume of one pound of vapor minus the decrease of volume due to condensation of $(1 - x)$ pounds. Precisely analogous equations and interpretations can be given for internal energy, enthalpy and entropy. The quantity $(1 - x)$ is called the *moisture* fraction.

Example 4. Find the properties of a mixture of steam and liquid water at 100 psia, containing 40 percent liquid.

The quality, x, is 0.60.

Using (a) $v_x = 0.40v_f + 0.60v_g$

From Table 2, $v_f = 0.01774, \qquad v_g = 4.432$

$$v_x = (0.4)(0.01774) + (0.6)(4.432)$$
$$= 2.666 \text{ cu ft/lb}$$

Using (b) $h_x = h_f + (0.6)h_{fg}$

$h_f = 298.4, h_{fg} = 888.8 \qquad$ (Table 2)

$$h_x = 298.4 + 533.3 = 831.7 \text{ Btu/lb}$$

Using (c) $s_x = s_g - (0.4)s_{fg}$

$s_g = 1.6026, s_{fg} = 1.1286 \qquad$ (Table 2)

$$s_x = 1.6026 - 0.4514 = 1.1512 \text{ Btu/lb°R}$$

In this example it makes little difference whether the equation form (a), (b) or (c) is used; however, when the quality is close to unity the precision of computation is improved if the form (c) is used, because the larger part of the result then comes directly from the tables and only a small subtractive quantity is computed.

Example 5. Find the enthalpy of wet steam, 0.97 quality, at 100 psia.

Using (b) $h_x = h_f + xh_{fg}$

$$h_x = 298.4 + (.97)(888.8)$$

By slide rule $h_x = 298.4 + 862 = 1160$ Btu/lb

Precision achieved is not better than to the nearest whole Btu/lb.

Using (c) $h_x = h_g - (1 - x)h_{fg}$

$$h_x = 1187.2 - (0.03)(888.8)$$

By slide rule $h_x = 1187.2 - 26.6 = 1160.6$ Btu/lb

Precision here is good to the first decimal place, assuming that the tables are precise.

It may seem unnecessary to obtain such precision as $\frac{1}{10}$ in 1100 but it must be remembered that most calculations involve differences between tabulated values. If a difference of 10 Btu/lb were involved, precision to the first decimal place would be highly desirable, even though the actual tabulated values are of the order of 1000.

In many problems a mixture is described not by its quality but by a pair of properties such as entropy and pressure. The procedure in obtaining the other properties is to find the quality from the given properties, and then find the other properties from the quality.

Example 6. Water at 15 psia has entropy of 1.7050 Btu/lb°R. Find its enthalpy and volume. To find the composition of the mixture,

$$s_x = s_g - (1 - x)s_{fg}$$

From Table 2 $1.7050 = 1.7549 - (1 - x)(1.4415)$

$$1 - x = 0.0346 \text{ moisture fraction}$$

Then $h_x = h_g - (1 - x)h_{fg}$

From Table 2 $h_x = 1150.8 - (0.0346)(969.7) = 1117.2$ Btu/lb

Similarly $v_x = v_g - (1 - x)v_{fg}$

$$= 26.29 - (0.0346)(26.27) = 25.38 \text{ cu ft/lb}$$

10-3 Superheated vapor. When the temperature of a vapor is greater than the saturation temperature corresponding to the existing pressure, the vapor is said to be *superheated.* The difference between the existing temperature and the saturation temperature corresponding to the existing pressure is called the superheat; thus steam at 14.7 psia and 220°F has 8 degrees superheat. In a superheated vapor at a given pressure the temperature may have any value greater than the saturation temperature; therefore the superheated vapor table (Table 3 in Keenan and Keyes) is arranged with two independent variables, pressure and temperature. Three properties, volume, enthalpy and entropy are given for each tabulated pair of values of pressure and temperature.

Example 7. Find the enthalpy of steam at 150 psia, 400°F. Inspection of the saturation data shows that the given state is superheated. Then in Table 3 at the intersection of the 150 psia line with the 400°F column find $h = 1219.4$ Btu/lb.

Example 8. Find the temperature of steam having enthalpy 1303.7 Btu/lb and pressure 300 psia. First determine whether the steam is a mixture or is superheated. Since h_g at 300 psia is found to be less than the given h, the steam is superheated. In Table 3 at 300 psia the given value of h appears in the column for 580°F. If the exact value of h had not appeared in the table it would have been necessary to interpolate to obtain the temperature.

In some cases it may be necessary to interpolate in two directions in the tables. For example, the properties at 296 psia and 565°F can be found only by interpolating between 295 and 300 psia and between 560 and 580°F. It is desirable to construct an auxiliary table such as follows to facilitate the work. The values tabulated in Table 3 are filled in first,

p	560°F	565°F	580°F
295	from Table 3	______	from Table 3
296	______	______	______
300	from Table 3	______	from Table 3

and the 565° column (or the 296 row) may be filled in two end spaces by interpolation. A second interpolation will then fill the center space. As an example, the entropy for the state 296 psia and 565°F may be found; the value is 1.6099 Btu/lb°R.

10-4 Compressed liquid. When the temperature of a liquid is less than saturation temperature for the existing pressure, the liquid is called *compressed liquid.* The pressure and temperature of compressed liquid may be varied independently; consequently a table of properties could be arranged like the superheated vapor table to give the properties at any pressure and temperature. However, the properties of liquids change little with pressure; hence for pressures of a few hundred psi, little error is made if the properties are assumed independent of pressure. In this case properties are taken from the saturation tables at the *temperature* of the compressed liquid.

Example 9. Find the enthalpy of liquid water at 200 psia and 70°F. From Table 1 at 70°F, $h_f = 38.04$ Btu/lb; therefore h at 200 psia, 70°F is approximately 38.0 Btu/lb.

When the pressures are high, or great precision is desired, the properties of compressed liquid water may be obtained from Table 4 of Keenan and Keyes. Table 4 does not give property values directly; instead it gives the properties in terms of differences from the properties of saturated liquid at the same temperature. Such a table of correction factors is less convenient but much more compact than a complete table.

Example 10. Find the volume and enthalpy of liquid water at 100°F and 1000 psia.

From Table 4 at 100°F and 1000 psia

$$v - v_f = -5.1 \times 10^{-5}$$
$$h - h_f = +2.70$$

From Table 1 at 100°F $v_f = 0.01613$

$h_f = 67.97$

Then at 100°F and 1000 psia

$$v = 0.01608 \text{ cu ft/lb}$$
$$h = 70.67 \text{ Btu/lb}$$

In using Table 4, care must be taken with signs and decimal points. Another method of obtaining the enthalpy of compressed liquid water will be found in the discussion of the Rankine cycle, Sec. 13-2, Example 1.

10-5 Charts of thermodynamic properties. There are certain advantages in the presentation of the properties of substances in graphical form. The manner of variation of the properties is made more evident and the problems of interpolation are simplified. However, it is not usually convenient to use a chart large enough to give the same precision as tables. Charts are commonly used to determine the general nature of a problem before proceeding with a detailed computation and to provide data for computation when the required precision can be obtained with the charts.

Keenan and Keyes present two charts for steam, a temperature-entropy plot and an enthalpy-entropy plot. The temperature-entropy plot shows the vapor dome and lines of constant pressure, constant volume, constant enthalpy, constant quality and constant superheat. The scale of this plot is so small that its use is limited to the general orientation of problems. For example, given the volume and enthalpy of a mass of steam, to find the exact state in the tables is a tedious task since neither property is an independent variable in the tables. With the chart, however, the desired state can easily be located approximately, thus reducing the search in the tables to a narrow range. Processes and cycles may be plotted on the chart for convenient visualization of property relations.

The enthalpy-entropy chart, commonly called a *Mollier* chart, contains the same data as the T–s chart with the exception of the constant-volume lines. The scale of the h–s chart is large enough to provide data suitable for many computations.

Some authors use coordinates other than enthalpy-entropy for charts to be used in computation. Enthalpy and volume have been used for a steam chart and the most common coordinates for charts relating to refrigerants are logarithm of pressure vs. enthalpy.*

* Ellenwood, F. O. and C. O. Mackey, *Thermodynamic Charts*. New York: Wiley, 1944. US Bureau of Standards, *Thermodynamic Properties of Ammonia*,

10-6 Measurement of steam quality. In general all the properties of a pure substance are fixed if two independent properties are given. The properties which are most convenient to measure are pressure and temperature; therefore whenever the pressure and temperature are independent properties it is customary to measure them to determine the state of the substance. As shown in Fig. 10-1, measured values of pressure and temperature which fall in the superheated vapor region or in the compressed liquid region will fix definite points on the plot of properties. In these regions pressure and temperature are independent. Measured values of pressure and temperature which correspond to saturation conditions, however, could apply equally well to saturated liquid point f, saturated vapor point g, or to mixtures of any quality, points x_1, x_2 or x_3. In order to fix the state, some other property such as specific volume, enthalpy or composition of the mixture (quality) must be measured. Specific volume measurements are usually impractical in engineering work; devices known as calorimeters are used for determining the enthalpy or composition of mixtures.

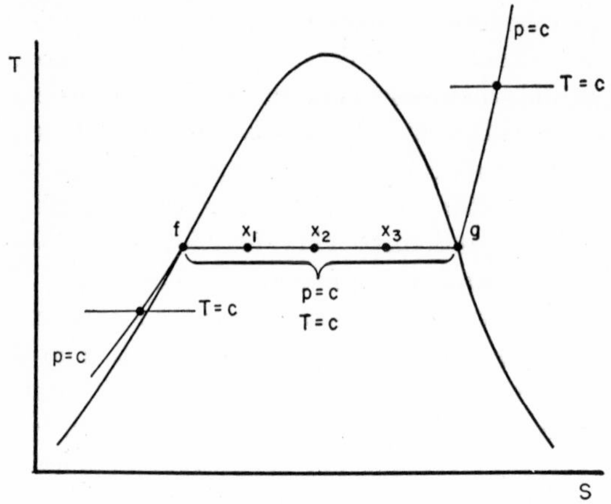

Fig. 10-1.

In the separating calorimeter a stream of fluid is separated mechanically into liquid and vapor by a sudden change in direction of

---

Circular 142. Amer. Soc. Refrigerating Engineers, *Refrigerating Data Book*, published periodically.

flow. The liquid and vapor are measured separately by a calibrated receiver and a calibrated orifice. This device is used when the moisture fraction is more than a few percent.

The throttling calorimeter, Fig. 10-2, is a steady flow device in which a sample of steam is put through a throttling process.

In this process the fluid flows in a well-insulated duct through a localized restriction (such as a partially closed valve or a small orifice) and then resumes its original velocity downstream from the restriction. For this process the steady flow energy equation reduces to

$$h_1 = h_2$$

In the use of the throttling calorimeter the sample of wet steam entering the device becomes superheated in the throttling process so that the final state can be determined by measurements of pressure and temperature. Since the enthalpy of the initial state is the same as the enthalpy of the final state the initial state can then be determined. For example, if a sample of steam at 100 psia containing 3 percent moisture is expanded to 15 psia in a throttling calorimeter, the location of the initial and final states, 1 and 2, would appear on the h–s and t–s diagrams as shown in Fig. 10-3. This may be checked on the diagrams in the steam tables. The process in the calorimeter is not a constant-enthalpy process; nevertheless $h_2 = h_1$ and the determination of state 2 fixes state 1.

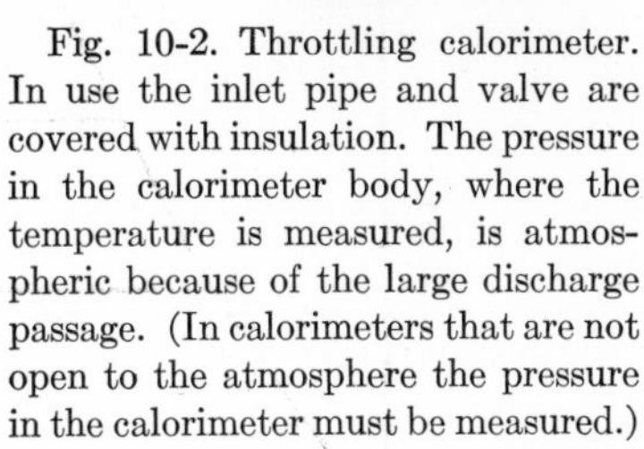

Fig. 10-2. Throttling calorimeter. In use the inlet pipe and valve are covered with insulation. The pressure in the calorimeter body, where the temperature is measured, is atmospheric because of the large discharge passage. (In calorimeters that are not open to the atmosphere the pressure in the calorimeter must be measured.) The steam leaving through the annular jacket space reduces heat loss from the inner chamber. (Courtesy Ellison Draft Gage Company.)

Example 11. A sample of steam at 200 psia flows to a throttling calorimeter in which the pressure is 15 psia and the temperature 280°F. Find the quality of the sample. At 15 psia and 280°F the enthalpy is found in Table 3

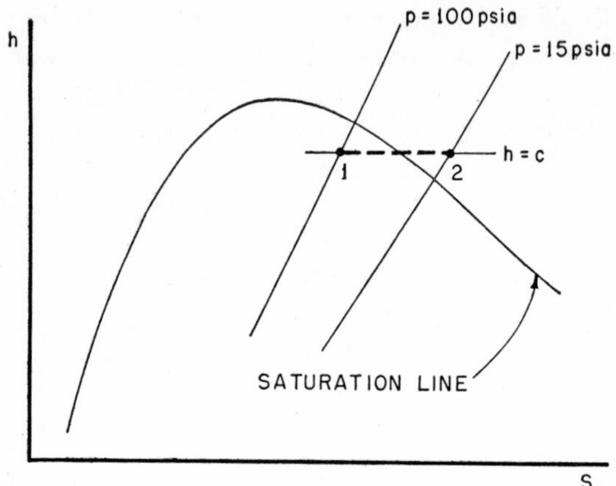

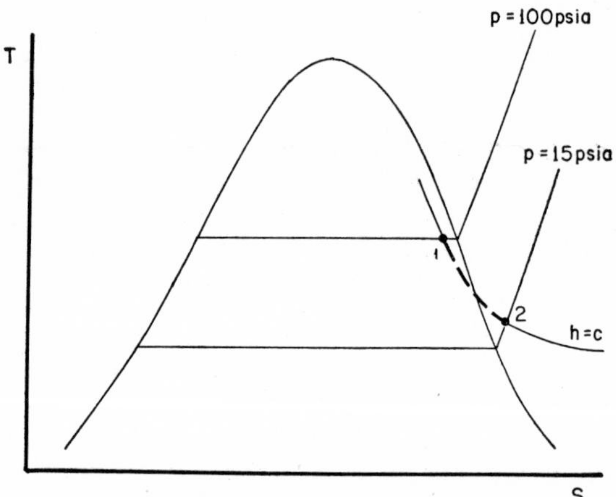

Fig. 10-3. Throttling calorimeter process.

to be 1183.2 Btu/lb. Therefore $h_1 = 1183.2$ Btu/lb. To determine the quality, x,

$$h_1 = h_g - (1 - x)h_{fg}$$

$$1183.2 = 1198.4 - (1 - x)843.0 \qquad \text{(Table 2)}$$

$$1 - x = 0.018$$

$$x = 0.982$$

A solution may be made on the h–s chart: find point 2 at 15 psia, 280°F, and proceed on a constant h line to 200 psia; there read moisture 1.7 percent.

The range of usefulness of the throttling calorimeter is limited by the fact that for given values of p_1 and p_2 there is a maximum moisture content at 1 for which the steam at 2 will be superheated; with greater moisture the method fails. This situation is illustrated in Fig. 10-4. For the pressures p_1 and p_2 the method will work with a sample at state 1, but will fail with a sample at state 1′ or any state of greater moisture. If the pressure in the calorimeter can be reduced to p_3, the moisture range is increased as is evident from the diagram. The throttling calorimeter also has a limitation on initial pressure; if p_2 is fixed at 1 atm, the moisture range increases for increased p_1

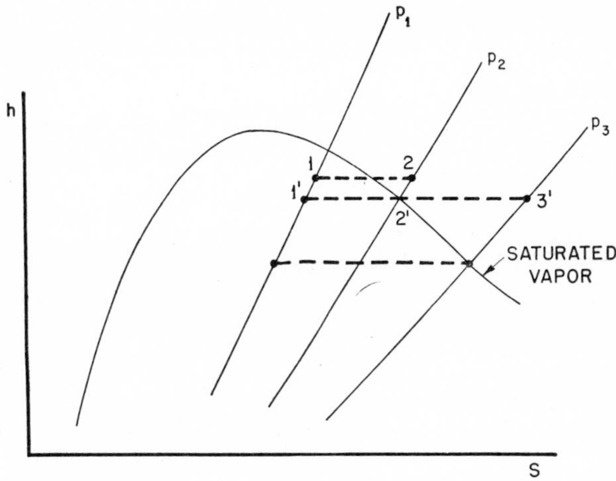

Fig. 10-4.

up to about 700 psia and then decreases to zero at about 1800 psia.

In the use of a throttling calorimeter the measured superheat in the calorimeter should be sufficient to give reasonable assurance that the apparent superheat is not the result of errors in measurement. If the steam were not really superheated in the calorimeter the results would be worthless. An apparent superheat of 10°F is a desirable minimum.

Calorimeters and their connections should be well insulated and have sufficient steam flow through so that external heat transfer will be negligible. They must also be run long enough to warm up thoroughly before use. In any case a calorimeter can give only the quality of the sample it receives; the most difficult part of the problem of determining steam quality is to get a sample which truly represents the fluid whose quality is desired.

Every problem in which tables are used is to some degree a cut-and-try problem. The search for data, at first, appears laborious but with tables, as with any other tool, experience leads to greater facility.

In the use of tables the units should be checked, as no standards are generally followed.

In many cases for which complete tabulations of data are not available, it is possible to find specific heats, coefficients of expansion, and the other data from which the needed properties can be computed.

PROBLEMS*

10-1. Complete the following table for water (liquid, vapor, or mixtures). (Insert a dash for irrelevant items.)

	p (psia)	t (°F)	v (cu ft/lb)	x %	*Superheat* (F)	h (Btu/lb)	s (Btu/lb R)
(a)	_____	100	350.4	_____	_____	_____	_____
(b)	_____	_____	0.01700	_____	_____	218.48	_____
(c)	_____	400	_____	0.9	_____	_____	_____
(d)	14.696	_____	_____	_____	_____	_____	1.4500
(e)	140	600	_____	_____	_____	_____	_____
(f)	60	_____	9.403	_____	_____	_____	_____
(g)	60	_____	10.000	_____	_____	_____	_____
(h)	_____	800	_____	_____	_____	1416.4	_____
(i)	_____	_____	4.002	_____	_____	1476.2	_____

* In solving the problems of this chapter use the abridged tables in the Appendix or the complete tables if available; use charts for preliminary orientation and for check solutions.

(j)	280	_____	_____	_____	100	_____	_____
(k)	225	_____	_____	_____	_____	1300.0	_____

10-2. A pound of water in a closed container is one-third liquid and two-thirds vapor, by volume; the temperature is 300°F. Find the pressure, the quality, the volume, and the enthalpy of the mixture.

10-3. Three pounds of liquid water at 1 atm is heated at constant pressure until it becomes saturated vapor. Find the changes in volume, temperature, and internal energy during the process.

10-4. Steam initially at 400°F, 40 psia, is cooled at constant volume. (a) At what temperature does the steam become saturated vapor? (b) What is the quality when the temperature reaches 200°F? (c) How much heat is transferred, per pound of steam, in the process between 400°F and 200°F?

10-5. Steam from a pipe where the pressure is 110 psia flows steadily through an electric calorimeter and comes out at 100 psia, 430°F; the electric power input to the calorimeter is 1 kw, and the amount of steam that flows through the calorimeter in 5 min is 4.1 lb. Find the moisture fraction or the superheat of the steam taken from the pipe.

10-6. A sample of steam from a boiler drum at 390 psia is put through a throttling calorimeter in which the pressure and temperature are found to be 14.7 psia and 318°F. Find the quality of the sample taken from the boiler.

10-7. A sample of steam taken from the supply pipe for a turbine, where the pressure is 235 psia, goes to a throttling calorimeter where the pressure is 20 psia. (a) If the calorimeter temperature is 325°F what is the state of the sample taken from the pipe? (b) If the calorimeter must show at least 10°F superheat for acceptable results, what is the maximum moisture content that could be measured under the given pressures?

REFERENCES

International Critical Tables. New York: McGraw-Hill. (For general use.)

Smithsonian Physical Tables. (General use.)

Keenan, J. H. and F. G. Keyes, *Thermodynamic Properties of Steam.* New York: Wiley, 1936. (For data on water.)

Dorsey, Noah E., *Properties of Ordinary Water Substance.* New York: Reinhold, 1940. (Data on water.)

Keenan, J. H. and J. Kaye, *Gas Tables; Thermodynamic Properties of Air.* New York: Wiley, 1948. (For data on gases.)

Ellenwood, F. O. and others, *Specific Heats of Certain Gases.* Cornell University Engineering Experiment Station Bulletin 30, 1942. (Data on gases.)

Hottel, H. C. and others, *Thermodynamic Charts for Combustion Processes.* New York: Wiley, 1949. (Data on gases.)

Amer. Soc. Heating and Ventilating Engineers, *Heating and Ventilating Guide.* Published annually. (Data on gases, especially on moist air.)

Amer. Soc. Refrigerating Engineers, *Refrigerating Data Book.* Published periodically. (For data on refrigerants.)

Spiers, H. M. (Ed.), *Technical Data on Fuel.* London: Brit. Nat. Comm. World Power Conf., 1950. Available through Amer. Soc. Mech. Engrs., New York. (Data on fuels and many other substances.)

US Bureau of Standards, *Tables of Selected Values of Properties of Hydrocarbons Circular C461.* Washington: Govt. Printing Office, Nov. 1947. (Data on hydrocarbons.)

Chapter 11

PROPERTIES OF GASES

The properties of gases are frequently correlated by means of certain simple rules called the *Perfect Gas Laws.* These rules were originally formulations of the best experimental data, but more accurate experiments showed them to be only approximations. It is often of great convenience, however, to assume that a gas is a so-called "perfect gas," a hypothetical substance which obeys the perfect gas laws. The two characteristics that define a perfect gas are the form of the equation of state and the fact that the specific heats are constant.

11-1 p, v, T relations for the perfect gas. The equation of state of a perfect gas is

$$pv = RT \tag{1}$$

where p is absolute pressure, v is specific volume, T is absolute temperature, and R is an experimental constant called the *gas constant.* In the usual case, R is in units of ft lbf/lbm °R. The gas constant has a different value for each gas (see Table A-1 on p. 386).

Equation (1) is satisfactory for real gases at high temperatures (more than twice the critical temperature) or at low pressures (of the order of 1 atmosphere for temperatures below the critical). For example, the critical temperature of nitrogen is 227°R; at room temperature (530°R), nitrogen agrees with Eq. (1) to within 1 percent at pressures as high as 10 atmospheres. Water (critical temperature 1165°R) deviates from Eq. (1) by about 3 percent at 1 atm, 800°R, and by about 1 percent at 1 psia, 800°R.

Equation (1) may be used with gas mixtures such as air so long as the composition of the mixture is not changed by condensation, evaporation, or chemical reactions (see Chap. 12).

If both sides of (1) are multiplied by the mass m of a gaseous system, the result is

$$pV = mRT \tag{2}$$

where V, which equals mv, is the volume of the mass m.

Another form of the equation of state, much used by physicists and chemists, is based on the *mol* as a unit of quantity of gas. A mol of substance has a mass numerically equal to the molecular weight of the substance. Thus a pound mol of oxygen is 32 lb, and a gram mol of oxygen is 32 gm. Since the molecular weight is a number proportional to the mass of a molecule it follows that 1 mol contains the same number of molecules for any substance. Avogadro's law states that a given volume will contain the same number of molecules of any gas at given pressure and temperature. (This law is based on the volumes entering into chemical reactions.) Then for all gases, given the same p, V and T, the number of mols is the same. This is valid for real gases under the same conditions as is the equation of state, (1). Now for n mols of gas the mass is

$$m = nM$$

where M is the molecular weight. Substituting in (2)

$$pV = nMRT$$

If p, V and T are fixed at p_0, V_0, and T_0, n will have a fixed value n_0 the same for all gases. Then

$$MR = \frac{p_0 V_0}{n_0 T_0}$$

or MR is the same for all gases at p_0, V_0 and T_0. But for any one gas both M and R are constant for all states, therefore MR is a constant for all gases in all states. The name *universal gas constant* and the symbol $\bar{R}$ are used for the product MR. Then

$$pV = n\bar{R}T \tag{3}$$

The value of the universal gas constant is 1545 ft lb/lb mol °R; in heat units this is 1545/778 = 1.986 Btu/lb mol °R.

The gas constant for any perfect gas may then be obtained from

$$R = \frac{\bar{R}}{M} = \frac{1545}{M} \text{ ft lbf/lbm °R} \tag{4}$$

or

$$R = \frac{1.986}{M} \text{ Btu/lb °R}$$

Another characteristic number for all gases is the volume of a mol of gas at standard pressure and temperature, 1 atm and 32°F. This is called the molecular volume. The pound molecular volume is

$$V_M = \frac{V_0}{n} = \frac{\overline{R}T_0}{p_0} = \frac{(1545)(492)}{(14.7)(144)} = 359 \text{ cu ft}$$

The gram molecular volume, used by scientists, is 22.4 liters.

The volume of n mols of perfect gas at any temperature and pressure, T and p, is given by

$$V = V_M n \frac{T}{T_0} \frac{p_0}{p} = 359 \frac{m}{M} \frac{T}{492} \frac{14.7}{p} \tag{5}$$

where p is in psia and T is in °R.

The actual values of R, $\overline{R}/M$, and V_M for certain gases are listed in Table A-1, p. 386. It will be observed that gases of simpler molecular structure show better agreement with Eq. (4) at 1 atm, 32°F, because of their low critical temperatures.

The graphical representation of Eqs. (1), (2) and (3) upon the pressure-volume plane is a series of rectangular hyperbolas for lines of constant temperature, Fig. 11-1.

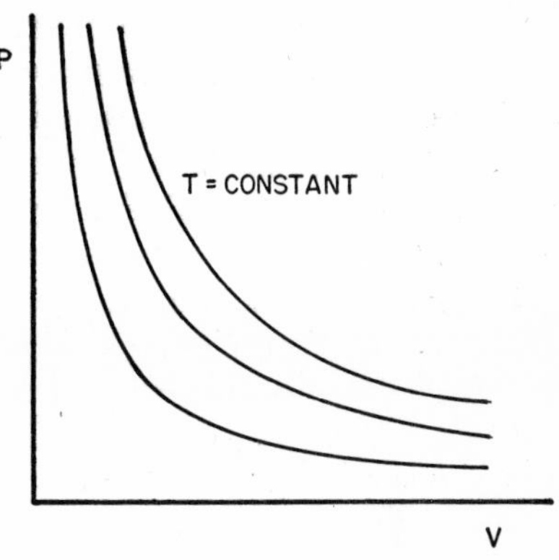

Fig. 11-1. P–v–T relation for a perfect gas.

Example 1. A certain piece of apparatus of constant volume is filled with nitrogen at 15 psia, 80°F. From a nitrogen bottle on a weighing scale exactly 3 lb of nitrogen is added to the apparatus. The final pressure and temperature are 25 psia, 75°F. Find the volume of the apparatus.

In the initial state $p_1V_1 = m_1RT_1$

In the final state $p_2V_2 = m_2RT_2$

Since volume is constant $V_1 = V_2$

It is given that $m_2 = m_1 + 3$

Then $\dfrac{p_1V_1}{p_2V_2} = \dfrac{m_1RT_1}{m_2RT_2}$

$$\frac{m_1}{m_2} = \frac{p_1T_2}{p_2T_1} = \frac{15}{25}\frac{535}{540} = 0.594$$

$$m_2 = 0.594m_2 + 3 = 7.39 \text{ lb}$$

$$V_2 = \frac{m_2RT_2}{p_2} = \frac{(7.39)(55.16)(535)}{(25)(144)} = 60.9 \text{ cu ft}$$

Note how the problem of units is simplified by dealing as far as possible with ratios in which units cancel.

11-2 Specific heats, internal energy and enthalpy. A perfect gas not only satisfies the equation of state (1), but its specific heats are constant. Real gases deviate appreciably from this rule, as shown in Tables A-4–A-6. The variation of specific heats is primarily with temperature change, but there is a small variation with pressure. For many purposes the use of constant specific heats is satisfactory, particularly if suitable average values are chosen. When this is not satisfactory, the Gas Tables may be used (see Sec. 11-6).

A corollary of the perfect gas equation of state is that the internal energy of a gas is solely a function of temperature. This is approximately true for real gases (Joule's law, based on experiment) and it can be proved analytically for any pure substance having the equation of state (1).

In general, if u is a function of T and v,

$$du = \left(\frac{\partial u}{\partial T}\right)_v dT + \left(\frac{\partial u}{\partial v}\right)_T dv$$

But if u is a function of T only,

$$\left(\frac{\partial u}{\partial v}\right)_T = 0 \tag{6}$$

and by definition

$$c_v = \left(\frac{\partial u}{\partial T}\right)_v \tag{5-6}$$

Therefore for the perfect gas

$$du = c_v\, dT \tag{7}$$

not only for a constant-volume process but for any process. Since c_v is constant for the perfect gas, it follows that

$$\Delta u = c_v\, \Delta T \tag{8}$$

The enthalpy of any substance is given by $h = u + pv$. Then for the perfect gas

$$h = u + RT$$

or h is a function of temperature only. By definition, for any substance,

$$c_p = \left(\frac{\partial h}{\partial T}\right)_p \tag{5-15}$$

Then for the perfect gas, since h is a function of T only,

$$dh = c_p\,dT \tag{9}$$

for any process. Since c_p is constant for the perfect gas

$$\Delta h = c_p\,\Delta T \tag{10}$$

If lines of constant internal energy or of constant enthalpy are plotted on the p–v plane, they will be identical with the lines of constant temperature in Fig. 11-1.

Observe that Eqs. (7) and (9) depend solely upon the assumption that $pv = RT$; (8) and (10) depend upon the further assumption that the specific heats are constant. Starting with (7) and (9) (i.e. assuming only that $pv = RT$) an important relation between the specific heats can be derived. From the definition of enthalpy it follows that

$$dh = du + d(pv)$$

Then for a gas

$$dh = du + R\,dT$$

or

$$c_p\,dT = c_v\,dT + R\,dT$$

Then

$$c_p - c_v = R \tag{11}$$

This relation will hold even if specific heats are permitted to vary with temperature. Hence for a substance satisfying the perfect gas equation of state a value of one of the specific heats leads immediately to the value of the other at the same temperature. It is well to recall here that the usual units for specific heats are Btu/lb degree whereas the usual units for R are ft lb/lb degree. It should be obvious that consistent units are necessary in Eq. (11); it is customary to use the heat units.

The ratio of specific heats c_p/c_v is of importance in perfect gas computations; it will be designated by the symbol k. According to the classical kinetic theory of gases the ratio k should have the values 5/3, 7/5, and 4/3 respectively for monatomic, diatomic and polyatomic gases. This theory is not accurate, as is indicated by the variation in measured values of k shown in Tables A-4–A-6. Nevertheless, at ordinary temperatures, for monatomic and diatomic gases the actual values of k are not far from the values 1.67 and 1.40 respectively. For polyatomic gases there is considerable deviation from the value 1.33. It may be observed that, if k can be predicted in any way or measured experimentally (several common phenomena are functions of k), the specific heats of a perfect gas can be determined without making heat measurements. From (11)

$$\frac{R}{c_v} = \frac{c_p}{c_v} - 1$$

so
$$c_v = \frac{R}{k - 1} \tag{12}$$

Also from (4)
$$R = \frac{1.986}{M}$$

so
$$c_v = \frac{1.986}{M(k - 1)}$$

Only the molecular weight and k are needed to determine the specific heats.

Modern theories of the structure of gases can predict specific heats accurately, but such computations are outside the realm of thermodynamics. Moreover the results are far too complicated for use in gas law computations; they are used in the preparation of tables or charts of properties.

Example 2. Given that oxygen has a gas constant of 48.3 ft lb/lb °R, and is a diatomic gas, compare its actual specific heats at 100°F, 500°F, and 1500°F with the values computed from the simple kinetic theory value of k.

$$c_p - c_v = 48.3/778 = 0.0622 \text{ Btu/lb °R}$$

$$c_p = \frac{k}{k-1}(c_p - c_v) = \frac{1.4}{0.4}(0.0622)$$

$$= 0.218 \text{ Btu/lb °R}$$

$$c_v = 0.218/1.4 = 0.156 \text{ Btu/lb °R}$$

From the plot, Table A-4,

t°F =	100	500	1500
c_p =	0.219	0.234	0.263
c_v =	0.157	0.172	0.201
k =	1.40	1.36	1.31

Example 3. Show that on a molal basis the specific heats of all gases having the same value of k are identical.

From (12),
$$c_p = \frac{k(c_p - c_v)}{k - 1} = \frac{kR}{k - 1}$$

On a molal basis c_p is M times as large as on a pound basis because each mol of gas contains M pounds.

$$\text{molal } c_p = \frac{kR}{k-1} M$$

But $MR = \overline{R}$ which is the same for all gases; therefore the molal c_p is a function of k only.

11-3 Entropy of the perfect gas. The entropy change between any two states of a substance may be determined by assuming a convenient reversible process connecting the states, and computing $\int dQ/T$ for the reversible process. For a perfect gas a convenient

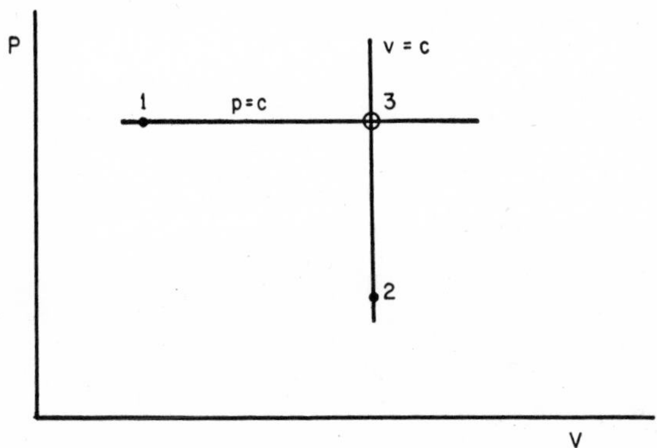

Fig. 11-2.

process is a combination of a constant-pressure process and a constant-volume process. In Fig. 11-2 any two states 1 and 2 are connected by the constant-pressure and constant-volume paths intersecting at 3. Assuming the processes 1–3 and 3–2 reversible, for a unit mass of gas

$$\Delta s_{12} = \int_1^3 \frac{dQ}{T} + \int_3^2 \frac{dQ}{T} = \int_1^3 c_p \frac{dT}{T} + \int_3^2 c_v \frac{dT}{T} \qquad \text{(a)}$$

Since c_p and c_v are constant

$$\Delta s_{12} = c_p \ln \frac{T_3}{T_1} + c_v \ln \frac{T_2}{T_3} \qquad \text{(b)}$$

It is desirable to eliminate the propertles at state 3; since the same gas is dealt with at each state,

$$\frac{p_1v_1}{T_1} = R = \frac{p_2v_2}{T_2} = \frac{p_3v_3}{T_3} \tag{c}$$

Then since $p_3 = p_1$ and $v_3 = v_2$,

$$\frac{T_3}{T_1} = \frac{v_2}{v_1} \quad \text{and} \quad \frac{T_2}{T_3} = \frac{p_2}{p_1}$$

Finally

$$\Delta s_{12} = c_p \ln \frac{v_2}{v_1} + c_v \ln \frac{p_2}{p_1} \tag{13}$$

By a simple exercise in algebra, using the relations (11) and (c) above, this can be converted to the forms

$$\Delta s_{12} = c_p \ln \frac{T_2}{T_1} - R \ln \frac{p_2}{p_1} \tag{14}$$

$$\Delta s_{12} = c_v \ln \frac{T_2}{T_1} + R \ln \frac{v_2}{v_1} \tag{15}$$

The same results may be arrived at in a more elegant way starting with the general property relation

$$T\,ds = du + p\,dv \tag{9-5}$$

or

$$ds = \frac{du}{T} + \frac{p}{T}\,dv$$

Then for the perfect gas

$$ds = c_v \frac{dT}{T} + R \frac{dv}{v}$$

or

$$\Delta s_{12} = c_v \ln \frac{T_2}{T_1} + R \ln \frac{v_2}{v_1}$$

11-4 The reversible adiabatic process. The reversible adiabatic process may be described in terms of properties by $\Delta s = 0$. From the general property relation (9-5) for a pure substance it follows that in a reversible adiabatic process

$$T\,ds = 0 = du + p\,dv$$

or

$$du = -p\,dv \tag{a}$$

for any pure substance. For the perfect gas, then,

$$c_v\,dT = -RT\frac{dv}{v}, \qquad c_v \frac{dT}{T} = (c_v - c_p)\frac{dv}{v}, \qquad \frac{dT}{T} = (1 - k)\frac{dv}{v}$$

Integrating, $\ln T = (1 - k) \ln v + \text{constant}$

$$\ln Tv^{k-1} = \text{constant}$$

$$Tv^{k-1} = \text{constant} \tag{16}$$

Substituting from $pv = RT$,

$$pv^k = \text{constant} \tag{17}$$

and

$$Tp^{(1-k)/k} = \text{constant} \tag{18}$$

The internal energy change and enthalpy change for a reversible adiabatic process in a perfect gas can be derived from (a) above.

$$\Delta u_{12} = -\int_1^2 p\, dv \tag{b}$$

Substituting from (17)

$$\Delta u_{12} = -\int_1^2 p_1 v_1^k \frac{dv}{v^k}$$

$$= \frac{p_2 v_2 - p_1 v_1}{k - 1}$$

$$= \frac{R}{k - 1} (T_2 - T_1) \tag{c}$$

From Eq. (12) it is seen that this result agrees with the general rule for the perfect gas, $\Delta u = c_v\, \Delta T$. Rearranging (c)

$$\Delta u_{12} = \frac{RT_1}{k - 1} \left(\frac{T_2}{T_1} - 1 \right) \tag{d}$$

and substituting from (18)

$$\Delta u_{12} = \frac{RT_1}{k - 1} \left[\left(\frac{p_2}{p_1} \right)^{(k-1)/k} - 1 \right] \tag{19}$$

From the definition of h,

$$dh = du + d(pv) = du + p\, dv + v\, dp$$

Substituting from (a)

$$dh = v\, dp \tag{e}$$

for a reversible adiabatic process in any pure substance. Then

$$\Delta h_{12} = \int_1^2 v\, dp \tag{f}$$

From (f), by a development similar to that for the internal energy, the result is obtained

$$\Delta h_{12} = \frac{kRT_1}{k-1}\left[\left(\frac{p_2}{p_1}\right)^{(k-1)/k} - 1\right] \tag{20}$$

Equations (19) and (20) are consistent with (8) and (10) from which, for any perfect gas process,

$$\frac{\Delta h}{\Delta u} = \frac{c_p}{c_v} = k \tag{21}$$

Example 4. Nitrogen, in a frictionless adiabatic process, expands from an initial state of 100 psia, 140°F to a final pressure of 10 psia. How much does the enthalpy change?

One method is to use Eq. (20).

$$\begin{aligned}\Delta h &= \frac{(1.4)(55.16)(600)}{1.4-1}\left[\left(\frac{10}{100}\right)^{.4/1.4} - 1\right]\\ &= 116{,}000(0.518 - 1)\\ &= -56{,}000 \text{ ft lb/lb} = -72 \text{ Btu/lb}\end{aligned}$$

Another method is to use the relation

$$\Delta h = c_p\,\Delta T$$

which applies to any perfect gas process. From (18), for a reversible adiabatic process,

$$T_2 = T_1\left(\frac{p_1}{p_2}\right)^{(1-k)/k} = T_1\left(\frac{p_2}{p_1}\right)^{(k-1)/k}$$

$$T_2 = 600(0.518) = 311°\text{R}$$

$$\Delta h = 0.248(311 - 600) = -71.8 \text{ Btu/lb}$$

11-5 The polytropic process. Many processes which occur in practice can be described approximately by an equation of the form pv^n = constant, where n is a constant. Such a process is called a *polytropic* process. The constant-temperature process and the reversible adiabatic process in a perfect gas are obviously special cases of the polytropic. By substitution from $pv = RT$ the following relations will hold for the *perfect gas* polytropic process:

$$Tv^{n-1} = \text{constant}$$

$$Tp^{(1-n)/n} = \text{constant}$$

The paths of various perfect gas polytropic processes on the pressure-volume plane and on the temperature-entropy plane are shown in Fig. 11-3.

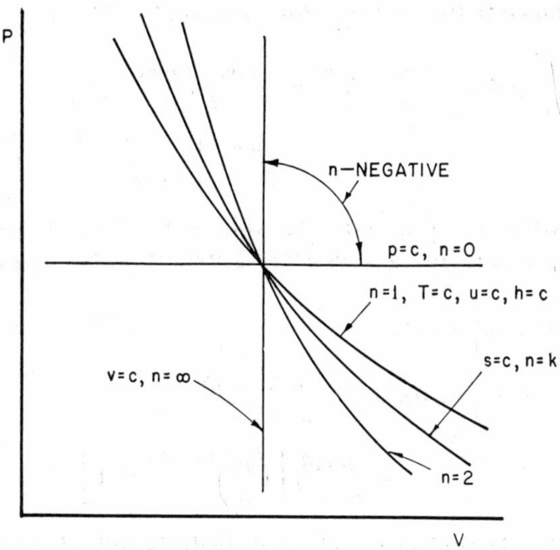

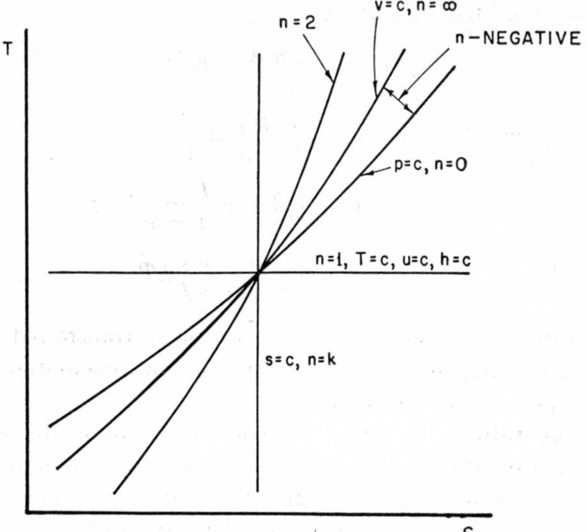

Fig. 11-3. Polytropic paths on p–V and T–s coordinates.

It was shown in Sec. 2-4 that for a process in which pv^n = constant

$$\int_1^2 p\,dv = \frac{p_2v_2 - p_1v_1}{1-n} = \frac{p_1v_1}{1-n}\left(\frac{p_2v_2}{p_1v_1} - 1\right)$$

$$= \frac{p_1v_1}{1-n}\left[\left(\frac{p_2}{p_1}\right)^{(n-1)/n} - 1\right] \qquad (22)$$

The similarity to Eq. (19) is apparent. Note, however, that $\Delta u = -\int p\,dv$ *only* for a reversible adiabatic; (22) does not give $-\Delta u$ unless $n = k$.

Also for the polytropic

$$\int_1^2 v\,dp = \frac{n(p_2v_2 - p_1v_1)}{n-1}$$

$$= \frac{p_1v_1n}{n-1}\left[\left(\frac{p_2}{p_1}\right)^{(n-1)/n} - 1\right] \qquad (23)$$

This equation is similar to (20) but $\int v\,dp$ is not Δh unless $n = k$.

It is possible to derive what is called a "polytropic specific heat" as follows: for a reversible polytropic process with a unit mass of perfect gas the First Law gives

$$Q = \Delta u + \int p\,dv$$

Now

$$\Delta u = c_v\,\Delta T \quad \text{and} \quad \int p\,dv = \frac{\Delta(pv)}{1-n} = \frac{R\Delta T}{1-n} = \frac{c_v(k-1)\,\Delta T}{1-n}$$

Then

$$Q = c_v\left(1 + \frac{k-1}{1-n}\right)\Delta T$$

$$= \left(c_v\frac{k-n}{1-n}\right)\Delta T \qquad (24)$$

The quantity in brackets in (24) is the heat transferred per unit temperature change in a *reversible* polytropic process and is therefore called the polytropic specific heat.

11-6 Gas tables. In many problems encountered by engineers the equation of state $pv = RT$ is satisfactory but the assumption of constant specific heats does not give the desired accuracy. The Gas Tables of Keenan and Kaye* provide data for variable specific heat computations. These tables will be discussed briefly. Since the per-

* Keenan, J. H. and J. Kaye, *Gas Tables; Thermodynamic Properties of Air.* New York: Wiley, 1948. Also see Appendix Table A-7, p. 394.

fect gas equation of state is assumed valid, the enthalpy and internal energy are functions of temperature only and a table for these data need have only one independent variable, temperature. Entropy change, however, is a function of two independent properties, for example pressure and temperature; allowing for variable specific heats, Eq. (14) becomes:

$$\Delta s_{12} = s_2 - s_1 = \int_0^{T_2} c_p \frac{dT}{T} - \int_0^{T_1} c_p \frac{dT}{T} - R \ln \frac{p_2}{p_1}$$

In the tables the quantity

$$\phi = \int_0^T c_p \frac{dT}{T}$$

is tabulated as a function of temperature.

The Gas Tables give two other functions of temperature, the *relative pressure* p_r, and the *relative volume* v_r, which are useful in finding the end state of an isentropic process. For an isentropic process between states 1 and 2 the following will hold:

$$p_2/p_1 = p_{r2}/p_{r1}$$
$$v_2/v_1 = v_{r2}/v_{r1}$$

Therefore if the initial state and the final pressure (or volume) are given, the final temperature may be found, as illustrated in the following example:

Example 5. Air initially at 100 psia, 2000°F, expands reversibly and adiabatically to 15 psia. Find the change of enthalpy and of specific volume by perfect gas laws, and by variable specific heats using the gas tables.

(1) Specific heats constant at room temperature values.

$$p_1 v_1^k = p_2 v_2^k$$

$$v_2 = v_1 \left(\frac{p_1}{p_2}\right)^{1/k} = \frac{RT_1}{p_1}\left(\frac{p_1}{p_2}\right)^{1/k} = \frac{(53.34)(2460)}{(100)(144)}\left(\frac{100}{15}\right)^{1/1.4}$$

$$= 9.12 \frac{1}{0.258} = 35.4 \text{ cu ft/lb}$$

$$\frac{p_1 v_1}{T_1} = \frac{p_2 v_2}{T_2}$$

$$T_2 = T_1 \frac{p_2 v_2}{p_1 v_1} = 2460 \frac{15}{100} \frac{35.4}{9.12} = 1425°\text{R} = 965°\text{F}$$

$$\Delta h = c_p \, \Delta T = (0.240)(965 - 2000) = -249 \text{ Btu/lb}$$

$$\Delta v = 35.4 - 9.12 = 26.3 \text{ cu ft/lb}$$

(2) Specific heats constant at average values suitable to the process. From the plot, Table A-4, take values at 1500°F as reasonable average values. Then $c_p = 0.276$, $c_v = 0.207$ Btu/lb °R, $k = 1.33$. Solving as before,

$$v_2 = (9.12)\frac{1}{0.241} = 37.8 \text{ cu ft/lb}$$

$$T_2 = 1520°\text{R} = 1060°\text{F}$$

$$\Delta h = 0.276(1060 - 2000) = -259 \text{ Btu/lb}$$

$$\Delta v = 37.8 - 9.12 = 28.7 \text{ cu ft/lb}$$

(3) Variable specific heats (gas tables). At T_1, 2460°R, from Table 1,

$$h_1 = 634.34$$

$$p_{r1} = 407.3$$

For a *reversible adiabatic* process

$$p_{r2} = p_{r1}\frac{p_2}{p_1} = 407.3\frac{15}{100} = 61.1$$

In Table 1, $p_r = 61.1$ at 1535°R. Then

$$T_2 = 1535°\text{R} = 1075°\text{F}$$

$$h_2 = 378.44$$

$$\Delta h = h_2 - h_1 = 378.44 - 634.34 = -255.90 \text{ Btu/lb}$$

$$v_2 = \frac{RT_2}{p_2} = \frac{(53.34)(1535)}{(15)(144)} = 37.9 \text{ cu ft/lb}$$

$$\Delta v = 37.9 - 9.12 = 28.8 \text{ cu ft/lb}$$

For situations in which the gas laws are not applicable but there are no detailed tables of thermodynamic data available, chemists and chemical engineers have developed methods of obtaining approximate properties from a minimum of experimental data. Such methods are described in texts on chemical thermodynamics and chemical engineering.

PROBLEMS

11-1. Nitrogen is to be stored at 2000 psia, 80°F, in a steel flask of 1.5 cu ft volume. The flask is to be protected against excessive pressures by a fusible plug which will melt and allow the gas to escape if the temperature rises too high. (a) How many pounds of nitrogen will the flask hold at the designated conditions? (b) At what temperature must the fusible plug melt in order to limit the pressure of a full flask to a maximum of 2400 psia?

11-2. Tanks of 200 cu ft volume are to be used for storing gases at 250 psia, 100°F. The gases are oxygen, nitrogen, helium, hydrogen, and carbon dioxide. Assuming these are all perfect gases, find: (a) the number of pounds of each gas that can be stored in a tank; (b) the number of pound-mols of each gas that can be stored in a tank.

11-3. For each of the following cases find the volume by the perfect gas rules, and compare with the true volume shown in the tables or charts for the substance: (a) water vapor at 1 psia, 300°F; (b) water vapor at 60 psia, 300°F; (c) water vapor at 60 psia, 1000°F; (d) carbon dioxide at 10 psia, 0°F; (e) carbon dioxide at 75 psia, 160°F; (f) carbon dioxide at 1100 psia, 160°F.

11-4. A tank contains dry air at 15 psia, 70°F. A flask of dry air containing 2 lb of air at 150 psia and 70°F is connected to the tank by a tube of negligible volume, and the pressures are allowed to equalize. Eventually the pressure is 30 psia and the temperature is 70°F in both containers. What is the volume of the tank?

11-5. A certain gas has specific heats as follows: $c_p = 0.47$ Btu/lb F, $c_v = 0.36$ Btu/lb F. Find the molecular weight and the gas constant for the gas.

11-6. The molecular weight of acetylene, C_2H_2, is 26; an experimental determination of the specific heat ratio gives $k = 1.26$. Find the two specific heats.

11-7. Nitrous oxide, N_2O, is a triatomic gas of molecular weight 44; estimate its specific heat at constant pressure from this information. The actual value at ordinary temperatures is about 0.21 Btu/lb F.

11-8. It was shown in Example 3, Sec. 11-2, that the molal specific heats of a perfect gas depend only upon the specific heat ratio. Find, to two significant figures, the molal specific heat at constant pressure and at constant volume for perfect gases as follows: (a) monatomic, $k = 5/3$; (b) diatomic, $k = 7/5$; (c) polyatomic, $k = 4/3$.

11-9. A perfect gas has a molecular weight of 20 and $c_v = 0.30$ Btu/lb F. The gas enters a pipe line at 100 psia and $1000°F_{abs}$, and flows steadily to the end of the pipe where the pressure is 75 psia and the temperature is $500°F_{abs}$. No shaft work is done in the pipe, and velocities are small. How much heat is transferred per pound of gas, and in which direction?

11-10. A constant-volume chamber of 10 cu ft capacity contains 2 lb of air at 40°F. Heat is transferred to the air until its temperature is 300°F. Find the work done by the air, the heat transferred, the change in internal energy, the change in enthalpy, and the change in entropy.

11-11. A constant-pressure process begins with 2 lb of air at 40°F, occupying a volume of 10 cu ft. Heat is transferred to the air until its temperature is 300°F; there is no work other than $p\,dv$ work. Find the work, the heat transferred, the change in internal energy, the change in enthalpy, and the change in entropy.

11-12. A quantity of nitrogen is cooled in a reversible constant-pressure process in which 300 Btu of heat are transferred. Find: (a) the work; (b) the change of internal energy.

11-13. A mass of hydrogen originally having a volume of 1 cu ft and a pressure of 180 psia expands in a reversible process at constant temperature of 400°F; the final pressure is 15 psia. Find: (a) the work of the process; (b) the heat transferred; (c) the entropy change of the gas.

11-14. Solve Problem 11-13 if the gas is nitrogen having: (a) the same original gas volume as in Problem 11-13; (b) the same gas mass as in Problem 11-13.

11-15. Air flows steadily through a porous plug in a Joule-Thomson experiment, Fig. 9-10. The initial pressure and temperature are 30 psia and 80°F, and the final pressure is 15 psia. Find the change of temperature, of internal energy, and of entropy, per pound of air.

11-16. Oxygen expands in a reversible adiabatic process from 45 psia, 60°F, to 30 psia. If the process occurs in a stationary system find the final temperature, the change of specific enthalpy, the heat transferred per pound of gas, and the work done per pound of gas.

11-17. If the process of Problem 11-16 occurs in steady flow find the final temperature, the change of specific internal energy, the heat transferred per pound of gas, and the shaft work done per pound of gas; changes in elevation and in kinetic energy are negligible.

11-18. Helium is compressed reversibly in a non-conducting cylinder from 70°F, 10 cu ft, 15 psia, to 1 cu ft. Find the change of enthalpy, the work done, and the final temperature.

11-19. In the compression of air in a certain cylinder cooled by a water jacket the indicator diagram shows that pressure and volume are related by $pv^{1.30}$ = constant. The process begins at 80°F, 14 psia, and the pressure rises to 80 psia. If the process is reversible how much heat is transferred per pound of air?

11-20. A perfect gas cycle of three processes uses argon (c_p = 0.125 Btu/lb F, k = 1.67) as a working substance. Process 1–2 is a reversible adiabatic expansion from 0.5 cu ft, 100 psia, 540°F, to 2 cu ft. Process 2–3 is a reversible isothermal process. Process 3–1 is a constant-pressure process in which the heat transfer is zero. Sketch the cycle in the p–v and T–s planes, and find: (a) the work for process 1–2; (b) the work for process 2–3; (c) the net work of the cycle.

11-21. Is the cycle of Problem 11-20 reversible? Give proof.

11-22. A Carnot cycle uses 1 lb of nitrogen as a working fluid. The maximum and minimum temperatures of the cycle are 1000°R and 500°R, the maximum pressure of the cycle is 100 psia, and the volume of the gas doubles during the isothermal expansion. Taking nitrogen as a perfect gas (specific heats constant) show by detailed computation of the heat supplied and the work done that the efficiency of the Carnot cycle agrees with that obtained in terms of temperature in Chapter 8.

11-23. A mass of air is initially at 500°F and 100 psia, and occupies 1 cu ft.

The air is expanded at constant pressure to 3 cu ft; a polytropic process with $n = 1.50$ is then carried out, followed by a constant-temperature process which completes a cycle. All of the processes are reversible. (a) Sketch the cycle in the p–v and T–s planes. (b) Find the heat received and the heat rejected in the cycle. (c) Find the efficiency of the cycle. (d) Describe qualitatively any differences that would have been found in (a), (b), and (c), if the gas had been helium instead of air.

11-24. Air is to be heated in steady flow from 60°F to 1300°F. Find the heat required per pound of air by each of three methods: (a) use room-temperature specific heats; (b) use a constant specific heat chosen from Table A-4, p. 391, as a good average value for the given temperature range; (c) use enthalpy values from the air table.

11-25. Solve Problem 11-10 by air tables.

11-26. Solve Problem 11-11 by air tables.

11-27. For nitrogen, oxygen, carbon dioxide, water vapor, and hydrogen find the enthalpy change between 500°F_{abs} and 2500°F_{abs} by: (a) the relation $\Delta h = c_p \, \Delta T$, using room-temperature c_p; (b) the same relation, using c_p as an average value estimated from the plots of c_p vs. T in the Appendix, Tables A-4 and A-6; (c) the plots of h/T vs. T in the Appendix, Tables A-5 and A-6.

11-28. One-tenth pound of air is compressed reversibly and adiabatically from 12 psia, 140°F, to 60 psia, and is then expanded at constant pressure to the original volume. (a) Sketch these processes on the p–v and T–s planes. (b) Compute the heat transfer and the work for the total path by simple gas-law methods. (c) Compute the heat transfer and the work for the total path by the air tables.

11-29. Solve Problem 11-19 by air tables.

11-30. One pound of air, initially at 80°F, is heated reversibly at constant pressure until the volume is doubled, and is then heated reversibly at constant volume until the pressure is doubled. (a) For the total path find the work, the heat transfer, and the change of entropy by simple gas-law methods. (b) For the total path find the work, the heat transfer, and the change of entropy by the air tables.

11-31. A Carnot cycle with 0.2 lb of air as working fluid operates between 2000°R and 1000°R. The volume of the air at the beginning of isothermal expansion is 1 cu ft, and at the end of isothermal expansion is 2 cu ft. (a) Find the work per cycle and the maximum volume of the gas by simple gas-law methods. (b) Find the work per cycle and the maximum volume of the gas by the air tables.

11-32. Solve Problem 11-23(a), (b), and (c) by air tables.

REFERENCES

For properties of gases, see the list of references at the end of Chap. 10, p. 139.

Chapter 12

PROPERTIES OF GASEOUS MIXTURES

Since a pure substance is a substance of constant and uniform chemical composition, a homogeneous mixture of gases would appear from the thermodynamic or macroscopic viewpoint to be a pure substance, and its properties could be formulated in the same way as for a single gas.* One important example of this type of pure substance is dry air. Mixtures cease to be pure substances when, by reason of phase changes or chemical reactions, the chemical composition of the mixture ceases to be constant and uniform. For example, moist air is not a pure substance in any process in which condensation or vaporization of moisture occurs.

For certain mixtures of great importance (dry air, moist air, certain combustion products) thermodynamic data are available, but for many mixtures with which the engineer has to deal it is necessary to compute the properties of the mixture from the composition of the mixture and the properties of the components. In this chapter relations will be given for mixtures of components which can be treated as perfect gases. Gas mixture properties are used in combustion calculations (Chap. 14) and in humidity calculations.

12-1 Basic rules for mixture properties. From experiments on the mixing of gases, Dalton's Law of Partial Pressures was developed. *Dalton's Law: The pressure of a mixture of gases is equal to the sum of the pressures of the individual components taken each at the temperature and volume of the mixture.*

When gases at equal pressures and temperatures are mixed adiabatically without work, as by inter-diffusion in a constant-volume container, the First Law requires that the internal energy of the gaseous system remain constant, and experiments show that the temperature remains constant. From this fact this rule may be deduced

* Even chemically pure gases are often mixtures of isotopes.

for internal energy: *The internal energy of a mixture of gases is equal to the sum of the internal energies of the individual components taken each at the temperature and volume of the mixture.*

It is possible to devise hypothetical reversible mixing processes for which the entropy change during mixing may be computed.* *Gibbs's Theorem* may thereby be deduced: *The entropy of a mixture of gases is equal to the sum of the entropies of the individual components taken each at the temperature and volume of the mixture.*

The three rules above are sometimes presented as deductions from a more general theorem known as the Gibbs-Dalton Law; hence properties obtained by these rules are said to be obtained by the Gibbs-Dalton Law.

12-2 Perfect gas mixtures: p, V, T relationships. The pressure of an individual component of a mixture, taken at the temperature and volume of the mixture, is called the *partial pressure* of the component. For a mixture of three perfect gas components A, B, and C let p_A, p_B and p_C be the respective partial pressures; then

$$\left.\begin{aligned} p_A V &= m_A R_A T = n_A \bar{R} T \\ p_B V &= m_B R_B T = n_B \bar{R} T \\ p_C V &= m_C R_C T = n_C \bar{R} T \end{aligned}\right\} \tag{1}$$

where subscripts indicate properties of the individual components, and the absence of a subscript indicates a property of the mixture. By Dalton's Law

$$p = p_A + p_B + p_C \tag{2}$$

In the absence of chemical reactions the mols of mixture will be the sum of the mols of components;

$$n = n_A + n_B + n_C \tag{3}$$

From (1), (2) and (3)

$$pV = n\bar{R}T$$

Thus if Dalton's Law is satisfied, a mixture of perfect gases follows the perfect gas equation of state. Writing this equation in terms of the mass and gas constant of the mixture, $pV = mRT$. Also, from (1) and (2),

$$pV = m_A R_A T + m_B R_B T + m_C R_C T$$

* See Zemansky, M. W., *Heat and Thermodynamics.* New York: McGraw-Hill, 1943, chap. 16.

Then the gas constant for the mixture is given by

$$mR = m_A R_A + m_B R_B + m_C R_C \tag{4}$$

The gas constant of the mixture is thus the weighted mean, on a mass basis, of the gas constants of the components.

Strictly a mixture has no molecular weight but an equivalent molecular weight is defined by

$$M = \frac{m}{n} = \frac{\overline{R}}{R} \tag{5}$$

Then from (3) and (5)

$$\frac{M}{m} = \frac{1}{\dfrac{m_A}{M_A} + \dfrac{m_B}{M_B} + \dfrac{m_C}{M_C}} \tag{6}$$

A quantity called the *partial volume* of a component of a mixture is the volume that the component alone would occupy at the pressure and temperature of the mixture. Designating the partial volumes by V_A, V_B and V_C,

$$pV_A = m_A R_A T, \quad pV_B = m_B R_B T, \quad pV_C = m_C R_C T \tag{7}$$

From (1), (2) and (7)

$$V = V_A + V_B + V_C \tag{8}$$

Observe that *in the mixture* each component occupies the total volume;* therefore the specific volume of each component is the *volume of the mixture* divided by the mass of the component. *In the mixture*

$$v_A = \frac{V}{m_A}, \qquad v_B = \frac{V}{m_B}, \qquad v_C = \frac{V}{m_C} \tag{9}$$

therefore

$$v = \frac{V}{m} = \frac{V}{m_A + m_B + m_C}$$

and

$$\frac{1}{v} = \frac{1}{v_A} + \frac{1}{v_B} + \frac{1}{v_C} \tag{10}$$

Mixture relations may be shown in diagrams; Fig. 12-1 shows the mixture and the components each at the temperature and volume of the mixture.

* In a gas which behaves like a perfect gas the molecules are so thinly distributed through the volume occupied that no component of a mixture interferes appreciably with the freedom of the molecules of another component to roam throughout the volume.

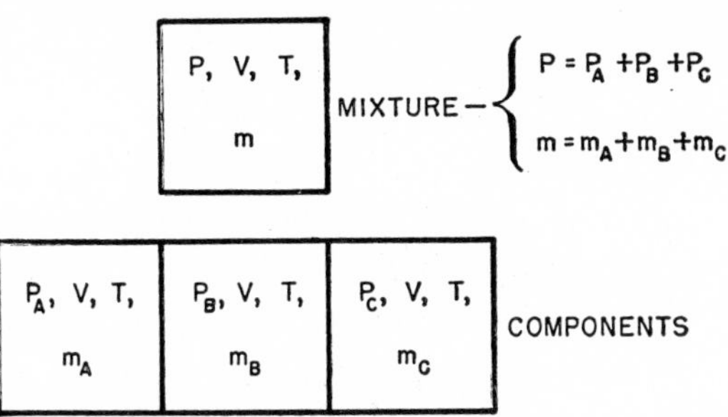

Fig. 12-1. Partial pressures.

If each of the components is compressed at constant temperature until its pressure becomes p the diagram will be as in Fig. 12-2.

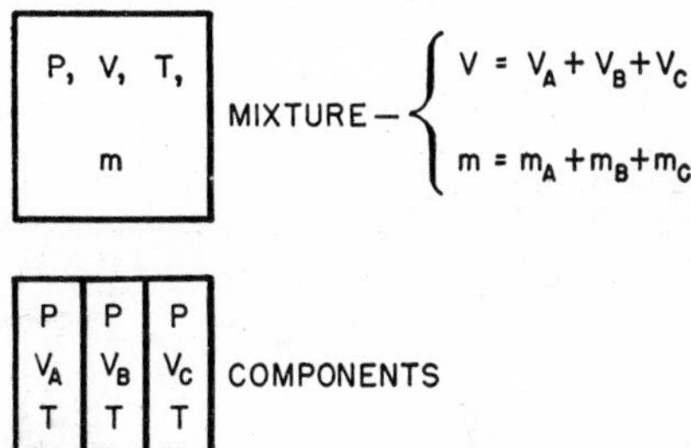

Fig. 12-2. Partial volumes.

12-3 Parts by mass; parts by volume; mol fractions. Three common ways of specifying the composition of a mixture are: (1) in parts by mass; (2) in parts by volume; (3) in mol fractions. The part by mass of a component is the ratio of the mass of the component to the mass of the mixture. The sum of all the parts by mass is unity.

$$\frac{m_A}{m} + \frac{m_B}{m} + \frac{m_C}{m} = \frac{m}{m} = 1 \tag{11}$$

The part by volume of a component of a mixture is the ratio of the partial volume of the component to the volume of the mixture. The sum of the parts by volume is unity.

$$\frac{V_A}{V} + \frac{V_B}{V} + \frac{V_C}{V} = \frac{V}{V} = 1 \tag{12}$$

The mol fraction of a component is the ratio of the number of mols of the component to the number of mols of mixture. The sum of the mol fractions is unity.

$$\frac{n_A}{n} + \frac{n_B}{n} + \frac{n_C}{n} = \frac{n}{n} = 1 \tag{13}$$

Parts by volume are related to parts by mass as follows:

$$pV_A = m_A R_A T = m_A \frac{\overline{R}}{M_A} T$$

and

$$pV = mRT = m \frac{\overline{R}}{M} T$$

then

$$\frac{V_A}{V} = \frac{m_A}{m} \frac{M}{M_A} \tag{14}$$

Both the part by volume and the mol fraction of a component are equal to the ratio of its partial pressure to the pressure of the mixture.

$$pV_A = p_A V = n_A \overline{R} T$$

and

$$pV = n\overline{R}T$$

then

$$\frac{n_A}{n} = \frac{p_A}{p} = \frac{V_A}{V} \tag{15}$$

Parts by volume are encountered in analyses of flue gases or exhaust gas. Mol fractions are frequently of convenience in dealing with mixtures because the use of mols in gas computations reduces the number of different constants and properties that have to be dealt with.

The above results depend upon Dalton's Law which is not exact. However, in cases for which the perfect gas law is a satisfactory representation of the properties of the components, Dalton's Law will be satisfactory for mixtures of uniform constant composition.

12-4 Internal energy, enthalpy, and specific heats of gas mixtures. The rule for the internal energy of a gas mixture, for a mixture of three components, leads to the equation

$$mu = m_A u_A + m_B u_B + m_C u_C \tag{16}$$

From (2), (9), (16), and the definition of enthalpy

$$mh = m_A h_A + m_B h_B + m_C h_C \tag{17}$$

From the definitions of the specific heats it follows that

$$mc_v = m_A c_{vA} + m_B c_{vB} + m_C c_{vC} \tag{18}$$

and

$$mc_p = m_A c_{pA} + m_B c_{pB} + m_C c_{pC} \tag{19}$$

All of the equations above apply to components at the temperature and volume of the mixture.

12-5 Entropy of gas mixtures. From the Gibbs Theorem,

$$ms = m_A s_A + m_B s_B + m_C s_C \tag{20}$$

where the components are at the temperature and volume of the mixture. It is instructive to consider the entropy change of a system during adiabatic mixing at constant volume. Let two gases A and B be contained in the two halves of an insulated rigid chamber divided by a thin wall, Fig. 12-3. With the pressures and temperatures respectively equal in A and B the wall is broken and the gases are allowed to mix. Experience shows that with perfect gases there will be no temperature or pressure change. Taking both gases as the system, the entropy before mixing is

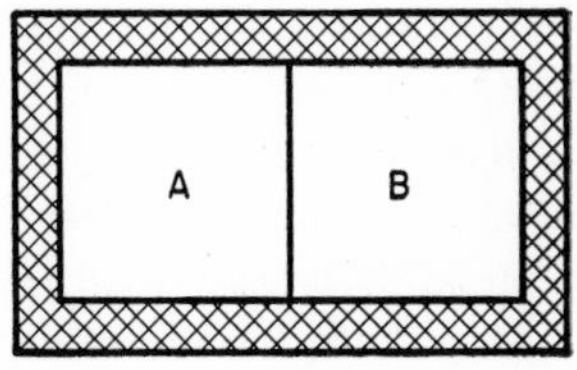

Fig. 12-3. Adiabatic mixing at constant total volume.

$$ms_1 = m_A s_{A1} + m_B s_{B1}$$

and the entropy after mixing is

$$ms_2 = m_A s_{A2} + m_B s_{B2}$$

At state 1 each component has the pressure and temperature of the mixture; at state 2 each component has the volume and temperature of the mixture. For each component

$$m\,\Delta s_{12} = m\left(c_p \ln \frac{T_2}{T_1} - R \ln \frac{p_2}{p_1}\right)$$

Then the change of entropy for the system comprising both components, in the constant-temperature process, is

$$m\,\Delta s = -m_A R_A \ln \frac{p_A}{p} - m_B R_B \ln \frac{p_B}{p}$$

Since the partial pressures are less than the total pressure, the change of entropy is positive. Recall from Sec. 8-6 that an increase of entropy in an isolated system is a criterion of an irreversible process. It is well known from experience that the mixing of gases by diffusion is irreversible. Hence the result conforms to the Second Law.

12-6 Mixtures of gases and water vapor. Superheated water vapor at pressures less than one atmosphere is approximately a perfect gas. Therefore the properties of gas mixtures (for example, atmospheric air) containing superheated water vapor at low pressure may be obtained by the Gibbs-Dalton Law. However it is also necessary to consider mixtures of water and gas in which the water vapor is not superheated. If a mixture of superheated water vapor and air comes into contact with a body of liquid water at the same temperature, the partial pressure of the superheated vapor will be less than the saturation pressure of the liquid, and liquid will evaporate until the vapor pressure (partial pressure in this case) equals the saturation pressure for the existing temperature. The mixture is then said to be a *saturated* mixture.* So long as a gas mixture is in contact with a body of liquid water, the partial pressure of the water vapor will be determined by the temperature of the system and the pressure-temperature relation for saturation as given in the steam tables. It is assumed here that time is permitted for equilibrium to be established; it may take much longer for water vapor to distribute itself uniformly throughout a space full of air than to fill to saturation pressure the same space empty of air.

In a saturated mixture in contact with liquid water the Gibbs-Dalton relations will hold for any given state, but the composition of the mixture will be a function of the temperature and pressure, and must be determined at each state by reference to the properties of saturated vapor.

If a gas mixture containing superheated water vapor is cooled at constant pressure, the mixture will eventually reach the saturation temperature corresponding to the partial pressure of the water vapor. This is called the *dew-point* temperature because it is often associated with the condensation of liquid drops or dew from a gas mixture. A common example of this type of process is the clouding of a window on a cold day, when the temperature of the surface in contact with the air becomes less than the dew-point temperature. Such a process is indicated in Fig. 12-4 which is a temperature-entropy plot for the water vapor alone. If the initial state of the vapor is at 1 (partial pressure p_w, temperature T_1), constant-pressure cooling will change the state eventually to d. The partial pressure of the vapor at d is still p_w because the mixture remains at constant composition during

* There is a small difference between the saturation pressure of the pure substance and the partial pressure of the vapor in the saturated mixture, but this will be neglected here.

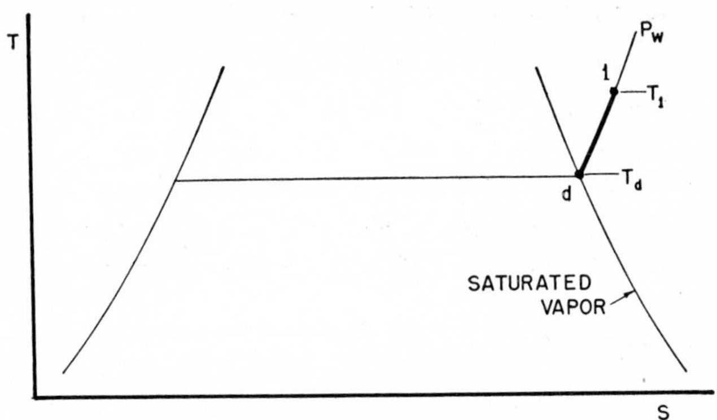

Fig. 12-4. Dew point.

cooling from 1 to d. But the partial pressure at d is also the saturation pressure corresponding to the dew-point temperature T_d; therefore T_d is the saturation temperature corresponding to p_w.

An apparatus in which a solid surface is cooled until condensation appears may be used to determine the water-vapor content of moist air as shown in the following example.

Example 1. In a dew-point apparatus a metal beaker is cooled by gradually adding ice water to the water initially in the beaker at room temperature. The moisture from the room air circulating around the beaker begins to condense on the beaker when its temperature is 55°F. If the room temperature is 70°F and the barometric pressure is 15.0 psia, find: (1) the partial pressure of the water vapor in the room air and (2) the parts by mass of water vapor in the room air.

Solution: The partial pressure of the water vapor in the room air is equal to the saturation pressure corresponding to the dew-point temperature 55°F; from the steam tables this is 0.2141 psia. By Dalton's Law

$$p = p_A + p_w$$

where p_A is the partial pressure of the air and p_w that of the water vapor. Then

$$p_A = 15.00 - 0.2141 = 14.79 \text{ psia}$$

By (15) the mol fractions are

$$\frac{n_A}{n} = \frac{p_A}{p}, \qquad \frac{n_w}{n} = \frac{p_w}{p}$$

But the mass of each component is given by

$$m = nM$$

therefore the parts by mass are

$$\frac{m_A}{m} = \frac{n_A M_A}{nM} = \frac{p_A M_A}{pM}, \qquad \frac{m_w}{m} = \frac{n_w M_w}{nM} = \frac{p_w M_w}{pM}$$

Since the molecular weight of the mixture is unknown, eliminate the mixture properties by taking the mass ratio of the two components; combining the two equations,

$$\frac{m_A}{m_w} = \frac{p_A M_A}{p_w M_w} = \frac{(14.79)(29)}{(0.2141)(18)} = 111$$

Then
$$\frac{m_A}{m} = \frac{111}{111 + 1} = 0.991, \qquad \frac{m_w}{m} = \frac{1}{112} = 0.00893$$

The composition of a mixture of air and water vapor is often given in terms of *specific humidity*, which is the ratio of mass of water vapor to mass of air in the mixture. The composition is also often stated in terms of *relative humidity*, which is the ratio of the partial pressure of the water vapor in the mixture to the saturation pressure for water vapor at the temperature of the mixture. In Example 1 the partial pressure of the water vapor was 0.2141 psia; at 70°F the saturation pressure p_g is 0.3631 psia. Therefore in Example 1 the relative humidity ϕ was

$$\phi = \frac{p_w}{p_g} = \frac{0.2141}{0.3631} = 0.59$$

the specific humidity γ was

$$\gamma = \frac{m_w}{m_A} = \frac{p_w M_w}{p_A M_A} = 0.0090 \text{ lb vapor/lb air}$$

Example 2. Air supplied to a furnace has relative humidity 75 percent, temperature 80°F, and gage pressure 10 inches of water. The barometer reads 29.50 inches of mercury. How many pounds of water vapor enter the furnace per pound of dry air?

Solution: As in Example 1,

$$\frac{m_w}{m_A} = \gamma = \frac{p_w M_w}{p_A M_A}$$

$$p_w = \phi p_g = (0.75)(0.5069) = 0.380 \text{ psi}$$

$$p_A = p - p_w$$

$$p = \left(29.50 + \frac{10}{13.6}\right)(0.491) = 14.85 \text{ psia}$$

$$p_A = 14.47 \text{ psi}$$

$$\frac{m_w}{m_A} = \frac{0.380}{14.47}\frac{18}{29} = 0.0163 \text{ lb vapor/lb air}$$

12-7 Humidity relations. Assuming that the mixtures follow the perfect gas rules and Dalton's Law, the relations between specific humidity and relative humidity may be developed. Specific humidity γ is given by

$$\gamma = \frac{m_w}{m_a} = \frac{v_a}{v_w} \tag{21}$$

where m_w is the mass of water vapor in a given space, m_a is the mass of dry air in the same space, v_a is the specific volume of the dry air *in the mixture*, v_w is the specific volume of the water vapor *in the mixture.*

Relative humidity is given by

$$\phi = \frac{p_w}{p_g} = \frac{v_g}{v_w} \tag{22}$$

where the subscript w refers to the water vapor in the actual mixture and the subscript g refers to the properties of saturated water vapor at the same temperature. Figure 12-5 is a temperature-entropy diagram for the water vapor alone. Combining Eqs. (21) and (22),

$$\gamma = \phi \frac{v_a}{v_g} \tag{23}$$

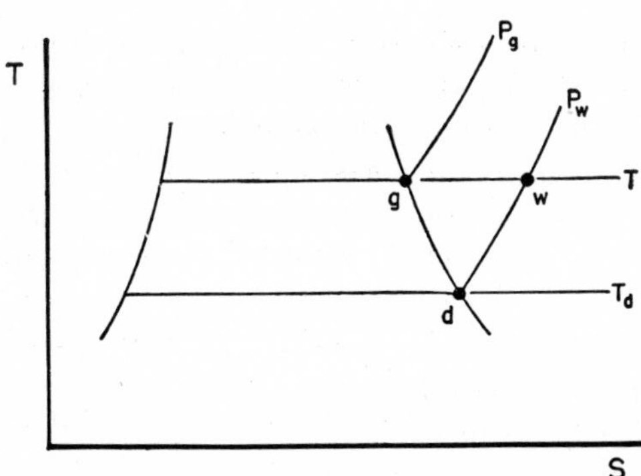

Fig. 12-5. States of water vapor in mixture.

By the perfect gas rules and Eq. (21),

$$\gamma = \frac{R_a T/p_a}{R_w T/p_w} = \frac{R_a p_w}{R_w p_a} = 0.622 \frac{p_w}{p_a} \tag{24}$$

Combining (22) and (24),

$$\phi = \frac{\gamma p_a}{0.622 p_g} \tag{25}$$

By Dalton's Law,

$$p = p_a + p_w \tag{26}$$

where p is the total pressure of the mixture.

In humidity computations the saturation pressure p_g is taken from the steam tables; the saturation volume may be taken from the tables or computed by the gas laws.

Example 3. A mixture of air and water vapor at 75°F and 14.7 psia has relative humidity 0.50; find its specific humidity and its dew-point temperature.

Solution: From (24),

$$\gamma = 0.622 \frac{p_w}{p_a}$$

From (22) $\quad p_w = \phi p_g = 0.5(0.4298) = 0.2149$ psi

From (26) $\quad p_a = p - p_w = 14.49$ psi

Then $\quad \gamma = 0.622 \dfrac{0.2149}{14.49} = 0.00925$ lb water/lb dry air

The dew-point temperature is the saturation temperature corresponding to p_w; this is found in the tables to be 55°F. The dew point is indicated as point d in Fig. 12-5.

12-8 Wet-bulb temperature. The usual method of determining experimentally the humidity of a mixture of air and water vapor is to obtain dry-bulb and wet-bulb temperatures. *Dry-bulb temperature* is the actual temperature of a gas; the designation "dry bulb" is used merely to avoid ambiguity. *Wet-bulb temperature* is the temperature indicated by a thermometer having its bulb covered by a film of water, when the thermometer is exposed to a stream of air in turbulent flow. Experimental correlations of these temperatures with humidity are available as formulas, tables, and charts. A typical psychrometric chart is folded in the back cover envelope of this book.

In the usual wet-bulb thermometer a water film is maintained around the bulb by a cotton wick saturated with water; for intermittent use the wick may be dipped in water before taking the reading, while for continuous use the wick may extend to a reservoir of water. The *sling psychrometer*, a common instrument for intermittent readings, consists of a pair of thermometers mounted on a holder which can be whirled through the air manually. One of the thermometers has a wet bulb, the other a dry bulb. A continuous psychrometer with a fan for drawing air over the thermometer bulbs is shown in Fig. 12-6; such devices are used in permanent installations.

In a psychrometer the dry-bulb thermometer gives the actual air temperature. At the wet bulb, if the air is not saturated, evaporation will occur from the liquid film on the bulb. The immediate effect is to lower the liquid temperature, but as soon as the liquid temperature

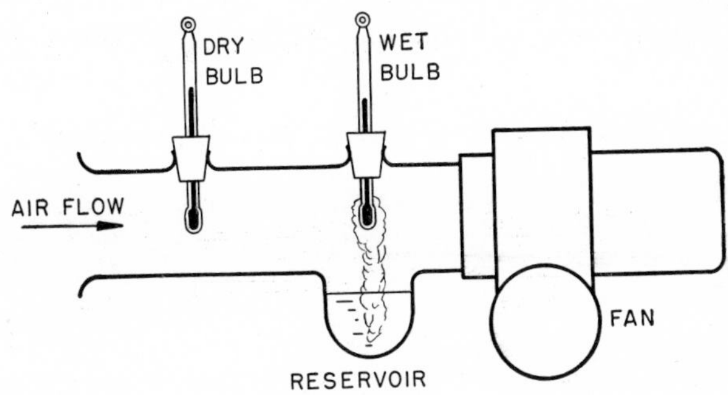

Fig. 12-6. Continuous wet- and dry-bulb psychrometer.

falls below the air temperature there will be heat transfer from the air to the liquid film. When steady state conditions are reached there will be a balance between energy removed from the liquid film by vaporization, and energy supplied to the liquid film by heat transfer. If the conditions are such that the heat transfer to the film is solely by convection from air in turbulent flow the temperature of the liquid film is the wet-bulb temperature.

It is clear that the wet-bulb temperature is not a property of the gas mixture since it depends upon heat transfer rates and mass transfer rates between the liquid film and the air. These rates depend upon the geometry of the bulb and the air velocity, neither of which is a thermodynamic property of the mixture.

12-9 Adiabatic saturation process. A process which is similar in some respects to the process at the wet bulb, but in which the air and the liquid water come to a true equilibrium state, may be accomplished as follows. Assume that a mixture of air and water vapor passes in steady flow through an adiabatic saturating chamber, Fig. 12-7. This is a long duct in which the air is in contact with a body of water, the whole apparatus being well insulated. The mixture entering is not saturated, but if the duct is long enough the mixture leaving is saturated. It is assumed that the water level is kept constant by make-up water supplied at the temperature of the saturated air leaving the chamber. Such a process is called an *adiabatic saturation process* and the temperature of the air leaving is the *adiabatic saturation temperature.*

The adiabatic saturation process differs from the wet-bulb process in that the liquid and the mixture of air and vapor are kept in contact until they come to equilibrium and the final state is not dependent

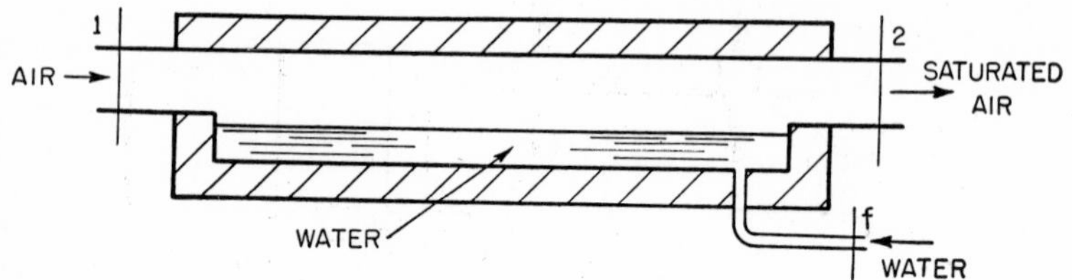

Fig. 12-7. Adiabatic saturation process.

in any way upon the *rates* of heat transfer and evaporation. It happens, however, that for water-air mixtures in the range of ordinary atmospheric pressures and temperatures the wet-bulb temperature and the adiabatic saturation temperature are practically identical.* This does not hold true at other states, or for substances other than air and water.

12-10 The psychrometric chart. Chart E-5 (see envelope inside back cover) is an example of a psychrometric chart suitable for air conditioning computations. This chart is based on the energy equation for the adiabatic saturation process at a total pressure of 1 atmosphere (29.92 in Hg); corrections for other barometric pressures are tabulated on the chart.† The principal coordinates of the chart are

* Carrier and Lindsay, "The Temperature of Evaporation of Water into Air," *Trans. ASME*, Vol. 44, p. 325, 1922.

† References to charts for other pressures are given at the end of the chapter.

specific humidity and dry-bulb temperature. Specific humidity is plotted in grains per pound of dry air (1.0 grain = 1/7000 lb), but a scale of pounds per pound of dry air is also given.

12-11 Humidifying processes. A simple humidifying process is often used with a heating process in comfort air conditioning; Fig. 12-8 is a diagram of such a process. For this process the steady flow equations may be written as follows:

$$w_{a1} = w_{a2} = w_a, \qquad w_{w1} + w_l = w_{w2}$$
$$w_a h_1 + w_l h_l + Q = w_a h_2$$

where h_1 and h_2 are in Btu/lb of dry air.

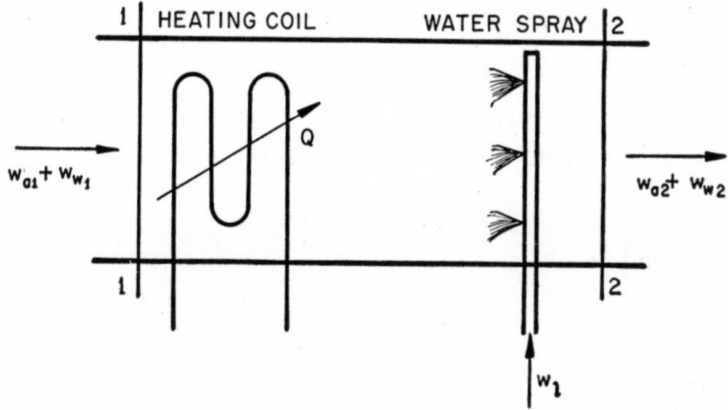

Fig. 12-8. Humidifying process; spray cooler with heating coil.

Example 4. In the process shown in Fig. 12-8 the air is received at 1 atm, 40°F, relative humidity 60 percent, and it is desired to discharge it at 70°F, relative humidity 50 percent. How much heat and how much water at 45°F must be supplied per pound of dry air passing through the apparatus?

Solution: From the chart, at state 1 the enthalpy is 12.9 Btu/lb of dry air and specific humidity is 0.0032 lb water/lb dry air. At state 2 the enthalpy is 25.33 Btu/lb of dry air and the specific humidity is 0.0078 lb water/lb dry air. The enthalpy of the liquid water h_l is h_f at 45°F or 13 Btu/lb from the steam tables. The water to be supplied is the difference between the specific humidities of the air:

$$w_l = \gamma_2 - \gamma_1 = 0.0046 \text{ lb/lb of dry air}$$

Then the energy equation on the basis of one pound of air is

$$12.9 + 0.0046(13) + Q = 25.33$$
$$Q = 12.4 \text{ Btu/lb of dry air}$$

The problem may also be solved by the mixture rules as follows,

$$\gamma_1 = 0.622 \frac{p_g}{p_a} \phi$$

From the steam tables at 40°F, $p_g = 0.1217$ psia. Since $p_a = p - p_g$

$$\gamma_1 = 0.622 \frac{0.1217}{14.7 - 0.12} (0.60) = 0.00311 \text{ lb/lb of dry air.}$$

By the same method $\gamma_2 = 0.00788$ lb/lb of dry air

Then $w_l = 0.00477$ lb/lb of dry air

The energy equation is

$$c_{pa}t_1 + \gamma_1 h_{w1} + w_l h_l + Q = c_{pa}t_2 + \gamma_2 h_{w2}$$

$$Q = 0.240(70 - 40) + 0.00788(1092.6) - 0.00311(1079.6) - 0.00477(13)$$

$$Q = 12.39 \text{ Btu/lb of dry air}$$

12-12 Dehumidifying processes. Dehumidifying, or the removal of moisture from air, is often required in comfort air condition-

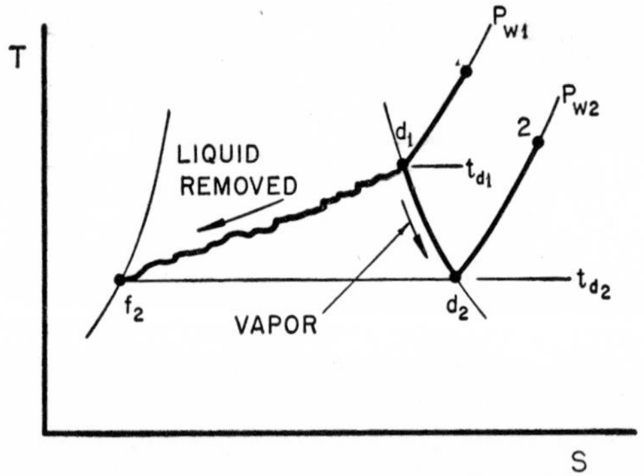

Fig. 12-9. State paths of water in a dehumidifying process.

ing and in industrial processes. Moisture may be removed by absorption in liquids or solids (so-called chemical dehumidifying), or by cooling below the dew point. Only the latter method will be considered here. The removal of moisture by cooling is shown on the temperature-entropy diagram, Fig. 12-9, which shows the states of the vapor only. Figure 12-10 is a flow diagram for the process.

If a mixture originally containing water vapor at state 1 is to have its specific humidity reduced to that of state 2 by cooling, the mixture must first be cooled to the dew-point temperature t_{d2} corresponding to state 2. The cooling will take place in two steps, first, cooling of the original mixture to its dew point t_{d1} and then cooling of the saturated mixture, with condensation of water as the temperature

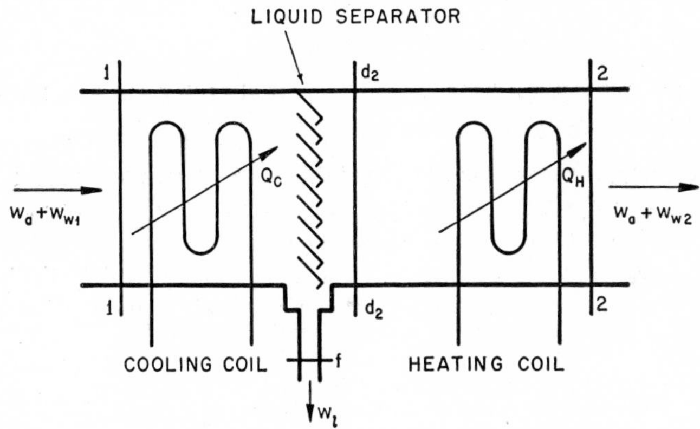

Fig. 12-10. Dehumidifying process; flow diagram.

falls. If the condensed liquid is separated from the mixture, the saturated air at state d_2 may then be heated to the desired final state, 2. In Fig. 12-11 the path of the process is shown on the psychrometric chart.

It will be observed that the *net* heat transfer for process 1–2 is of little significance, since to accomplish the desired effect the two

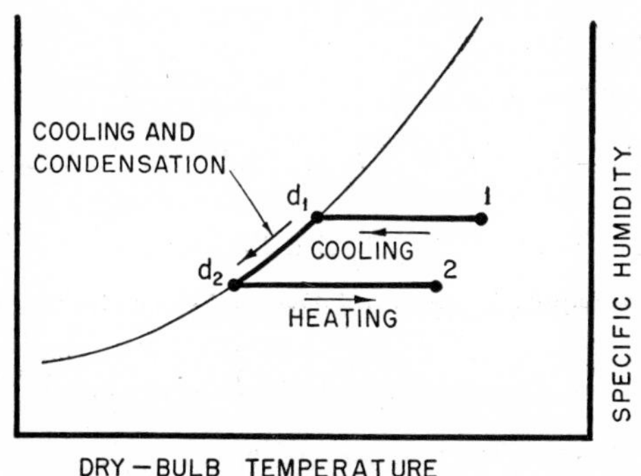

Fig. 12-11. Dehumidifying process on the psychrometric chart.

individual transfers Q_C and Q_H are both essential. Therefore the energy equation, to be useful, must be written for the two separate processes of cooling and heating. The mass and energy equations are

$$w_a + w_{w1} = w_a + w_{w2} + w_l$$
$$w_a h_1 + Q_C = w_a h_{d2} + w_l h_f$$
$$w_a h_{d2} + Q_H = w_a h_2$$

where the mixture enthalpies are per pound of dry air, and both heat quantities are taken positive for heat flow to the mixture.

Example 5. Air at 1 atm, 75°F, 70 percent relative humidity is to be brought to 70°F, 60 percent relative humidity by the process of Fig. 12-10. To what temperature must the mixture be cooled? How much heat must be removed by the cooling coil and how much must be supplied by the heating coil per pound of dry air? What fraction of the heat removed in the cooling coil is required to cool and condense the water removed?

Solution: Using the psychrometric chart, and referring to Fig. 12-10 and 11,

$$\gamma_1 = 0.0131 \text{ lb water/lb dry air}, \qquad \gamma_2 = 0.0093$$
$$h_1 = 32.36 \text{ Btu/lb dry air}, \qquad h_2 = 27.03$$

The mixture must be cooled to the dew-point temperature corresponding to state 2; this is found on the chart to be 55.3°F. Then from the chart and the steam tables

$$h_{d2} = 23.40 \text{ Btu/lb of dry air}, \qquad h_f = 23.4 \text{ Btu/lb of liquid}$$

From the mass equation,

$$\frac{w_l}{w_a} = \gamma_1 - \gamma_2 = 0.0038 \text{ lb water/lb dry air}$$

By the energy equations,

$$Q_C = h_{d2} - h_1 + \frac{w_l}{w_a} h_f = -8.87 \text{ Btu/lb dry air}$$

$$Q_H = h_2 - h_{d2} = 3.63 \text{ Btu/lb dry air}$$

The heat transfer required to cool and condense the water removed is given by

$$\frac{w_l}{w_a}(h_{w1} - h_f)$$

taking h_{w1} as h_g at 75°F, the heat transfer is

$$0.0038(1094.5 - 23.4) = 4.07 \text{ Btu/lb dry air}$$

Then the fraction of the heat removed is

$$\frac{4.07}{8.87} = 0.46$$

PROBLEMS

12-1. One cu ft of helium at 1 atm, 70°F, is mixed with one cu ft of nitrogen at 1 atm, 70°F. For the mixture at 1 atm, 70°F, find: (a) the volume of the mixture; (b) the partial volumes of the components; (c) the partial pressures of the components; (d) the parts by mass of the components; (e) the mol fractions of the components; (f) the specific heats of the mixture; (g) the gas constant of the mixture; (h) the adiabatic exponent, k, of the mixture.

12-2. Show that the molal specific heat (Btu/lb mol F) of a gas mixture is the weighted mean, on a mol basis, of the molal specific heats of the components. Check the numerical results of Problem 12-1 against this rule.

12-3. One pound of helium and one pound of nitrogen are mixed at 70°F and at a total pressure of 1 atm. Find all the quantities called for in Problem 12-1.

12-4. The mixture of Problem 12-1 is compressed reversibly and adiabatically from 20 psia, 140°F, to 40 psia. Find: (a) the final specific volume of the mixture; (b) the entropy change of the mixture; (c) the entropy change of the helium; (d) the entropy change of the nitrogen. What should be the relation among the answers to (b), (c), and (d)?

12-5. A stream of oxygen is mixed adiabatically in steady flow with a stream of hydrogen, the mass ratio of oxygen to hydrogen being 8 to 1. The velocities of the mixed stream and the two component streams are equal. For one pound of the total stream of oxygen and hydrogen find the change, during the mixing process, in: (a) enthalpy; (b) internal energy; (c) entropy.

12-6. The products of combustion of a hydrocarbon fuel have the following composition by mass: N_2, 0.72; CO_2, 0.18; H_2O, 0.07; O_2, 0.03. Using the charts in the Appendix for properties of the components, find c_p and k for the mixture at 1500°F.

12-7. The products of combustion from a furnace having the following composition in fractions by volume: N_2, 0.77; CO_2, 0.13; O_2, 0.05; H_2O, 0.05. How much heat must be transferred, per pound of mixture, to cool the gas from 2000°F to 500°F in steady flow at constant velocity? Take the properties of the components from the chart of h/T in the Appendix.

12-8. (a) Find the composition by volume of the gas in Problem 12-6. (b) Find the dew point of the gas at a total pressure of 14.5 psia.

12-9. A mixture of air and water vapor containing 0.015 lb of water vapor and 0.985 lb of dry air occupies a constant-volume tank at 14.7 psia and 80°F. (a) What is the dew point of the mixture? (b) What is the relative humidity?

(c) To what temperature must the mixture be cooled at *constant volume* to reach saturation?

12-10. An air compressor takes in air from the atmosphere at 75°F, 14.0 psia, and 75 percent relative humidity. The air is discharged from the compressor at 70 psia, 280°F. At the compressor discharge find: (a) the specific humidity; (b) the relative humidity.

12-11. The air discharged from the compressor in Problem 12-10 is passed through an aftercooler in which heat is transferred to metal tubes cooled by circulating water. Some of the moisture in the air condenses on the tubes, and the liquid is drained off separately from the air. The air then leaves the cooler as saturated air at 68 psia, 100°F. How much water is removed from each pound of mixture flowing into the cooler?

12-12. Atmospheric air is at a pressure of 14.7 psia and temperature 75°F. Plot the mixture density vs. relative humidity as the latter varies from 0 to 1.00. What percentage error would be made in using dry-air specific volume for a calculation involving air of 80 percent relative humidity at 75°F?

12-13. Condensation on cold water pipes often occurs in warm humid rooms. If the water temperature may reach a minimum of 45°F and the room temperature is kept at 75°F, what is the limit of relative humidity in the room to avoid condensation at normal atmospheric pressure? What is the limit at a barometer of 26 in. Hg?

12-14. Air at 1 atm has a dry-bulb temperature of 80°F and a wet-bulb temperature of 70°F. Find: (a) the specific humidity; (b) the relative humidity; (c) the enthalpy, Btu/lb of dry air.

12-15. The air of Problem 12-14 is heated in steady flow to 135°F, 1 atm. Find: (a) the final specific humidity; (b) the final relative humidity; (c) the heat transferred, Btu/lb of dry air; (d) the heat transferred, Btu/lb of mixture; (e) the heat transferred, Btu/1000 cu ft of mixture.

12-16. Three thousand cfm of air at 1 atm, 69°F, 60 percent relative humidity, are to be humidified by passing through a water spray chamber. The water in the spray chamber is continuously recirculated through the sprays and back to a reservoir; heat is supplied to keep the reservoir temperature constant, and make-up water is supplied at 60°F to compensate for the evaporation. Assuming that external heat losses and the work of the spray pump are negligible, if the air comes out at 69°F, 80 percent relative humidity; find: (a) the rate of make-up water supply; (b) the rate of heat supply.

12-17. Solve Problem 12-16 if the air comes out at 60°F, saturated.

12-18. Solve Problem 12-16 if the air comes out at 55°F, saturated.

12-19. Four thousand cfm of air at 75°F, 70 percent relative humidity, are to be cooled and dehumidified to 68°F, 60 percent relative humidity by cooling and reheating the entire flow at 1 atm pressure. Find: (a) the temperature to which the air must be cooled; (b) the tons of refrigeration required; (c) the Btu/hr of heat required. One ton of refrigeration is 200 Btu/min.

12-20. The requirements of Problem 12-19 can also be satisfied by a system in which only a portion of the total air flow is cooled and dehumidified, and is then mixed with the remainder of the air to obtain the final state desired. In such a process, if the cooling is carried to 45°F find: (a) the fraction of the entire flow which passes through the cooler; (b) the tons of refrigeration required; (c) the Btu/hr of heat required. Compare with results of Problem 12-19.

REFERENCES

American Society of Heating and Ventilating Engineers, *Heating and Ventilating Guide*. Published annually.

American Society of Refrigerating Engineers, *Refrigerating Data Book*. Published periodically.

Palmatier and Wile, "A New Psychrometric Chart." *Refrigerating Engineering*, v. 52, no. 1, 1946, p. 31.

Karig, H. E., "Psychrometric Charts for High Altitude Calculations." *Refrigerating Engineering*, v. 52, no. 5, 1946, p. 434.

Rohsenow, W. M., "Psychrometric Determination of Absolute Humidity at Elevated Pressures," *Refrigerating Engineering*, v. 51, no. 5, 1946, p. 423.

Carrier, W. H. and others, *Modern Air Conditioning, Heating, and Ventilating*. New York: Pitman, 1950.

Jordan, R. C. and G. B. Priester, *Refrigeration and Air Conditioning*. New York: Prentice-Hall, 1948.

See also the references at the end of Chap. 10, p. 139.

Chapter 13

VAPOR CYCLES

Beginning with this chapter the principles and methods developed in the preceding chapters will be applied to engineering processes. One of the first and most important applications of thermodynamics is the study of the production of power from the combustion of fuel. Thermal power plants may utilize the energy released by combustion in either of two ways: (1) by transfer of heat from the combustion products to a cyclic heat engine which produces work, as in steam power plants, or (2) by the extraction of work directly from the combustion products, as in the internal combustion engine. Heat power cycles will be discussed first.

Up to the present time the only important cyclic power plants have been those using condensable vapor, almost invariably steam, as the working fluid. Internal combustion engines are not cyclic heat engines; hot-air engines or closed-cycle gas turbine plants have yet to achieve an important role in practice. First consideration will therefore be given to vapor cycles.

13-1 Thermal power plants and ideal cycles. A cyclic thermal power plant utilizes a series of processes each of which involves some unnecessary or undesirable effects in addition to the intended effect of the process. For example, in a steam power plant the purpose of the boiler or steam generator is to convert liquid water to superheated vapor at constant pressure. Actually, because of friction, the pressure of the superheated vapor may be appreciably less than the pressure of the liquid. Similarly the purpose of the turbine is to obtain shaft work from the expansion of steam to a lower pressure; frictional effects and heat transfer are not desired but they always occur. For each process in the power plant it is possible to specify a hypothetical or ideal process which represents the basic intended operation and involves no extraneous effects. For the steam boiler this would be a reversible constant-pressure heating process; for the turbine the ideal process would be a reversible adiabatic expansion. A cycle composed of such processes is called an ideal cycle.

The efficiency and capacity (work output) of an ideal cycle, when compared with the efficiency and capacity of the actual plant, give a measure of the perfection of the plant apparatus. On the other hand, comparison of different ideal cycles gives a measure of the perfection of the thermodynamic design. Thus the ideal cycle is an invaluable aid in understanding existing plants and in developing improved plants.

13-2 The Rankine cycle. The basic operations involved in the simplest type of steam power plant are (1) pumping liquid water

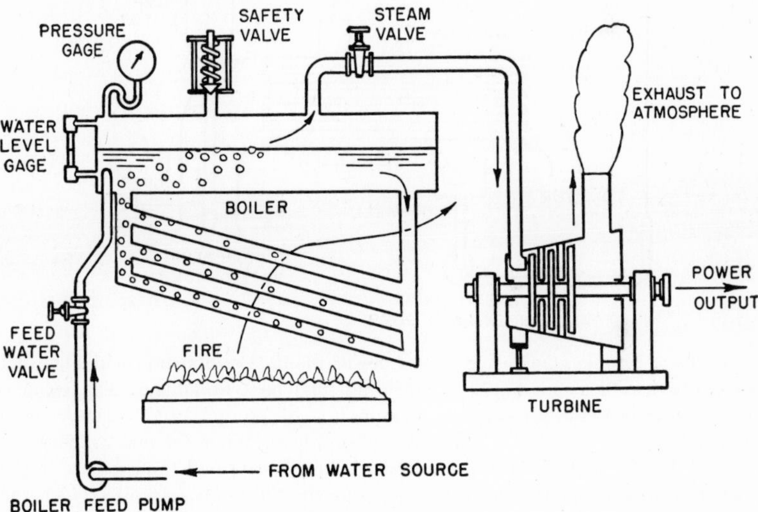

Fig. 13-1. Simple non-condensing steam power plant. The valves and gages shown are those which are essential for safe operation and control.

into a steam boiler or steam generator, (2) evaporation of the water at a high pressure by heat transfer in the boiler, (3) expansion of the vapor through an engine or turbine to a low pressure, producing shaft work. In the simplest case the vapor leaving the engine at low pressure (*exhaust steam*) may be discharged into the atmosphere and a fresh supply, or *makeup*, of *boiler feed water* must be used constantly to keep the plant in operation (see Fig. 13-1). It is possible, however, to condense the exhaust steam by heat transfer, and to return the resultant liquid (*condensate*) to the boiler feed pump for return to the boiler (see Fig. 13-2). When this is done the plant operates in a cycle according to the flow diagram, Fig. 13-3. The two types of plant are called respectively *non-condensing* and *condensing* plants. Although the non-condensing plant does not operate

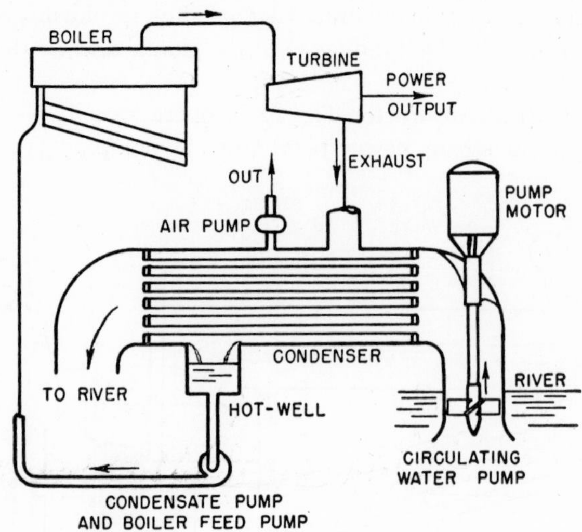

Fig. 13-2. Simple condensing steam power plant. Details shown in Fig. 13-1 are omitted. The condenser is a shell-and-tube heat exchanger. The exhaust pressure may be below atmospheric pressure if an air pump is used to remove non-condensable gas which leaks in from the atmosphere. When the non-condensable gas content of the steam in the condenser is kept very small the pressure in the condenser approaches the saturation pressure for the temperature to which the steam is cooled. This may be 20 or 30° above the cooling water temperature. When operating with exhaust pressure below atmosphere, a condensate pump is used to extract the condensate from the hot well and supply it to the suction of the boiler feed pump.

in a complete cycle it may be analyzed as a cyclic plant because the make-up feed water is the same, thermodynamically, as the condensate that might have been obtained by completing the cycle. The natural surroundings may be considered as the condenser, receiving the vapor and returning an equal amount of condensate.

The simple steam plant is analyzed by the aid of the following ideal cycle: (a) reversible adiabatic expansion, starting at the state at which steam enters the turbine and ending at the pressure of the exhaust steam leaving the turbine; (b) reversible constant-pressure heat transfer from the steam, ending at the saturated liquid state; (c) reversible adiabatic compression of the liquid, ending at the initial pressure; (d) reversible constant-pressure heat transfer to the fluid, ending at the initial state. Such a cycle, called a Rankine cycle, is plotted on the p–v and T–s planes in Fig. 13-4; the numbers on the plots correspond to the numbers on the flow diagram, Fig. 13-3. For any given pressures the steam approaching the turbine may have

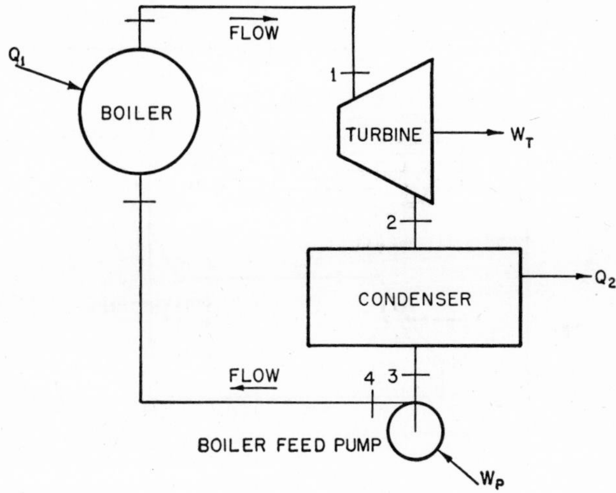

Fig. 13-3. Elementary steam plant flow diagram.

any of a variety of states as indicated by points 1, 1′, and 1″, but the fluid approaching the boiler feed pump may, in the ideal cycle, have only one state, that of saturated liquid.

For purposes of analysis the Rankine cycle is assumed to be carried out in a steady flow operation. Applying the steady flow energy equation to the processes of the cycle the equations for work and heat quantities may be set up on a basis of unit mass of fluid, assuming negligible changes of potential and kinetic energy.

Process 1–2, the turbine or engine process, is adiabatic; then

$$h_1 = h_2 + W_{x1-2}$$

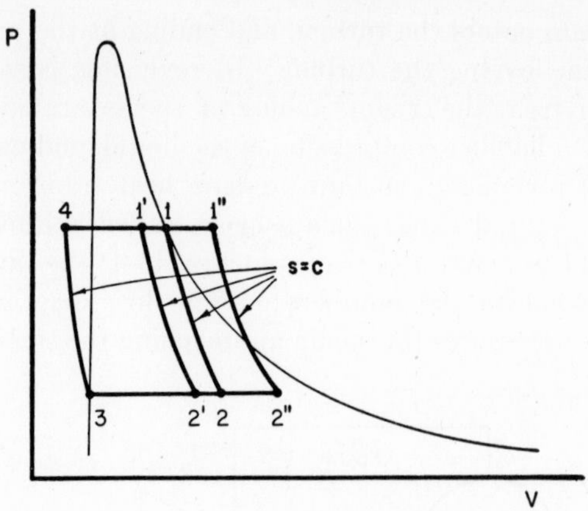

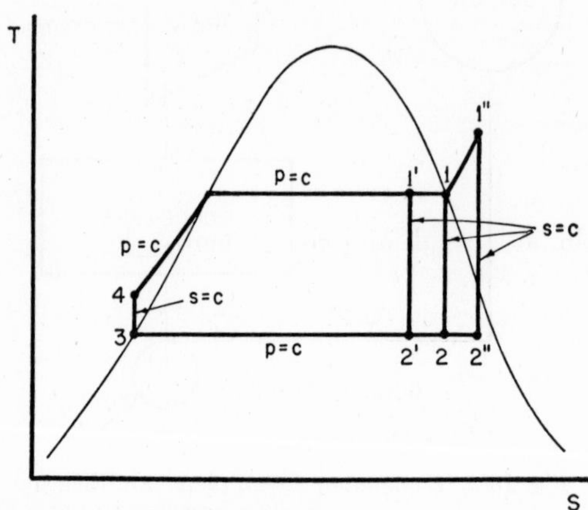

Fig. 13-4. Rankine cycle.

and the turbine work is given by

$$W_T = W_{x1-2} = h_1 - h_2 \tag{1}$$

Process 2–3, the condenser process, involves no shaft work; then

$$h_2 + Q_{2-3} = h_3$$

The heat rejected Q_2, is taken of opposite sign from Q_{2-3} (see Sec. 7-2); then

$$Q_2 = -Q_{2-3} = h_2 - h_3 \tag{2}$$

Process 3–4, the pump process, is adiabatic; then

$$h_3 = h_4 + W_{x3-4}$$

and the pump work is

$$W_P = -W_{x3-4} = h_4 - h_3 \tag{3}$$

Process 4–1, the boiler process, involves no shaft work; then

$$h_4 + Q_{4-1} = h_1$$

and the heat supplied is given by

$$Q_1 = Q_{4-1} = h_1 - h_4 \tag{4}$$

The efficiency of the Rankine cycle is given by

$$\eta = \frac{W_{\text{net}}}{Q_1} = \frac{W_T - W_P}{Q_1} = \frac{h_1 - h_2 - (h_4 - h_3)}{h_1 - h_4} \tag{5}$$

Cycle efficiency may be expressed alternatively as *heat rate;* this is the heat supplied per unit work output, Btu/hphr or Btu/kwhr.

Since the compression process is carried out with liquid water, the specific volume is small and consequently the work of compression is small and is sometimes neglected. Assuming the pump work negligible is equivalent to setting h_4 equal to h_3: If this is done Eq. (5) reduces to

$$\text{approximate } \eta = \frac{h_1 - h_2}{h_1 - h_3} \tag{5a}$$

In practice it is important to know not only the efficiency but also the capacity of a power plant. One measure of capacity is the *steam rate,* the pounds of fluid that flow to the turbine or engine per unit time per unit power output. Steam rate is usually given in pounds per horsepower-hour or pounds per kilowatt-hour. For the ideal Rankine cycle the steam rate will be

$$w_R = \frac{\text{Btu of work in one hphr}}{\text{Btu of work done per lb of steam}} = \frac{2545}{W_T - W_P} \frac{\text{lb}}{\text{hphr}} \tag{6}$$

or

$$w_R = \frac{\text{Btu of work in one kwhr}}{\text{Btu of work done per lb of steam}} = \frac{3413}{W_T - W_P} \frac{\text{lb}}{\text{kwhr}} \tag{6a}$$

Example 1. A Rankine cycle operates with steam conditions 200 psia, 750°F, and exhaust pressure 1 psia. Find the heat supplied, the turbine work, and the pump work per pound of steam. Find the cycle efficiency and steam rate.

Solution: Obtain the enthalpy at states 1, 2, 3, and 4, Fig. 13-4. From the tables at 200 psia, 750°F, $h_1 = 1399.2$ Btu/lb, $s_1 = 1.7448$ Btu/lb °R. By the definition of the cycle $s_2 = s_1$. At 1 psia, $s = 1.7448$, find from the tables $x_2 = 0.874$, $h_2 = 976$ Btu/lb. At 1 psia saturated liquid, $h_3 = 69.70$ Btu/lb. By the definition of the cycle $s_4 = s_3$; in the liquid region, however, it is inconvenient to use the tables to trace a constant-entropy path. A convenient method of obtaining the enthalpy change in a reversible adiabatic process in a liquid is by use of the general property relation from Chap. 9,

$$T\,ds = dh - v\,dp \tag{9-6}$$

since ds is zero,

$$dh = v\,dp$$

In the liquid the volume change due to compression to 200 psia is negligible, as can be verified in Table 4 of Keenan and Keyes. Assuming v is constant

$$\Delta h = v\,\Delta p$$

or $$h_4 - h_3 = v_3(p_4 - p_3) = 0.01614\,\frac{144}{778}\,(200 - 1) = 0.595 \text{ Btu/lb}$$

$$h_4 = 69.70 + 0.60 = 70.3 \text{ Btu/lb}$$

Then $$Q_1 = h_1 - h_4 = 1329 \text{ Btu/lb}$$

$$W_T = h_1 - h_2 = 423 \text{ Btu/lb,} \qquad W_P = h_4 - h_3 = 0.595 \text{ Btu/lb}$$

$$\eta = \frac{W_T - W_P}{Q_1} = 0.318$$

$$w = \frac{2545}{422} = 6.01 \text{ lb steam per hphr}$$

The efficiency of the Rankine cycle may be presented graphically by the use of the temperature-entropy plot, Fig. 13-5. Because the Rankine cycle is a reversible cycle, the areas representing the integral of $T\,ds$ for the various processes are proportional to the heat quantities transferred. Thus Q_1 is proportional to area 4156, Q_2 is proportional to area 3256, and the net work, being equal to the net heat, is proportional to the area 1234 enclosed by the cycle diagram. The efficiency can readily be estimated if the plot is made to scale. Such scale plots are useful in comparing the efficiencies of cycles working at different pressures and temperatures, and in comparing different types of cycles.

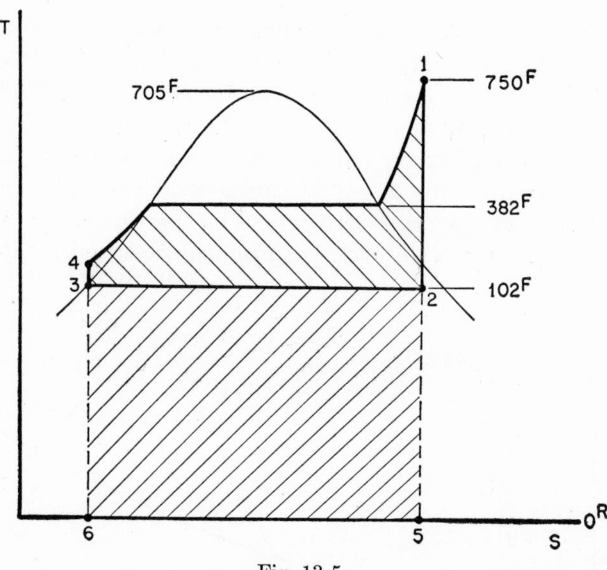

Fig. 13-5.

13-3 Rankine cycle—influence of pressure and temperature. The efficiency and steam rate of the Rankine cycle, as deduced above, are dependent upon the initial steam state and the exhaust pressure. It has been shown by the Second Law, however, that the maximum

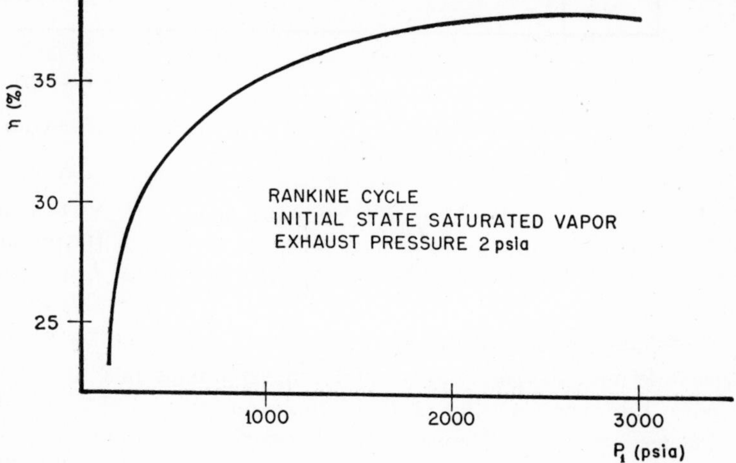

Fig. 13-6. Efficiency of Rankine cycle vs. initial pressure for cycles using saturated steam.

efficiency of a heat engine cycle is limited by the temperature range employed. Therefore it is of interest to investigate the effect of an increase in the initial temperature upon the efficiency of the Rankine cycle. Assuming a reference cycle using saturated steam, two possible ways of increasing the initial temperature would be (1) by going to higher pressure with saturated steam, or (2) by going to superheated steam at the same pressure. The corresponding efficiency characteristics for certain cycle conditions are plotted in Figs. 13-6 and 13-7.

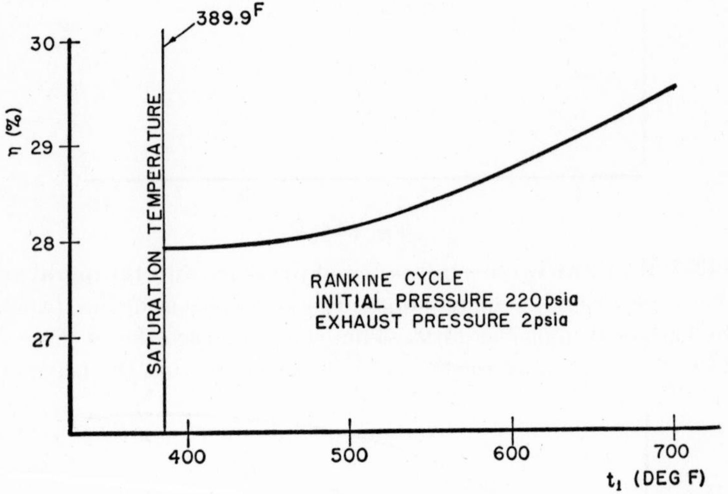

Fig. 13-7. Efficiency of Rankine cycle vs. initial temperature for cycles using a constant initial pressure.

The efficiency of a Rankine cycle may also be expected to increase with decreasing temperature of heat rejection, that is, with decreasing exhaust pressure. This effect is shown in Fig. 13-8 for a fixed initial state.

The characteristics of the ideal cycle as described above do not alone determine the best pressure and temperature for an actual plant because the characteristics of the actual apparatus, and especially cost considerations, influence the final result. For example, the use of superheated steam benefits the actual plant more than it does the ideal cycle, giving increased turbine efficiency and reduced mainte-

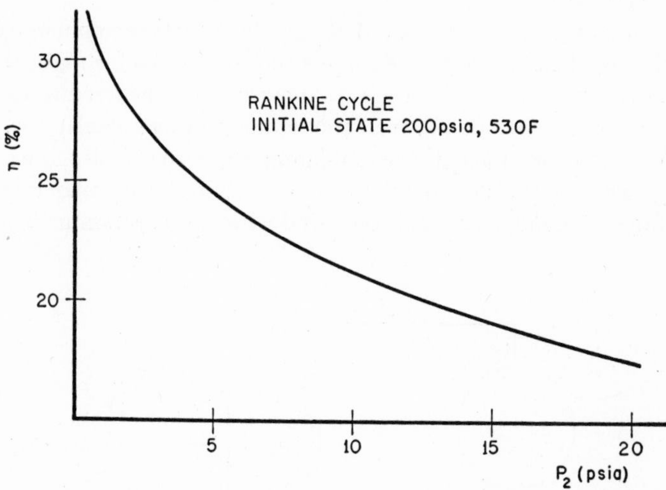

Fig. 13-8. Efficiency of Rankine cycle vs. exhaust pressure for cycles using a constant initial state.

nance. On the other hand the use of very low exhaust pressure is not as beneficial in the actual plant as it might appear in the ideal cycle, because of the cost of apparatus and the work required to maintain the low pressure.

13-4 Actual vapor cycle processes. The processes of an actual cycle differ from those of the ideal cycle; in analysis the difference is accounted for by conventional methods as described in the following sections, so that by the use of experimental factors the actual cycle characteristics may be predicted from the ideal cycle computations.

By definition the thermal efficiency of an actual cycle is

$$\eta_{th} = \frac{W_{\text{net}}}{Q_1}$$

where the work and heat quantities are the measured values for the actual cycle; these will differ from the corresponding quantities for the ideal cycle. In the actual cycle conditions might be as indicated in Fig. 13-9. The machinery and pipe lines are not frictionless and not adiabatic. The heat supplied and heat rejected are not the sole heat transfers, consequently the net work is not equal to their difference. The work produced by the turbine is less than the work that would have been produced by a reversible turbine and the work input to the pump is greater than that required by a reversible pump. (There may be several pumps; W_P is the total work for all of them.)

The various processes of the actual cycle are considered separately in developing relations between the actual and the ideal cycle. The actual turbine or engine, the actual pump, the actual boiler, the actual condenser and the actual pipe lines are taken, each under its actual imposed conditions, and compared with ideal apparatus working

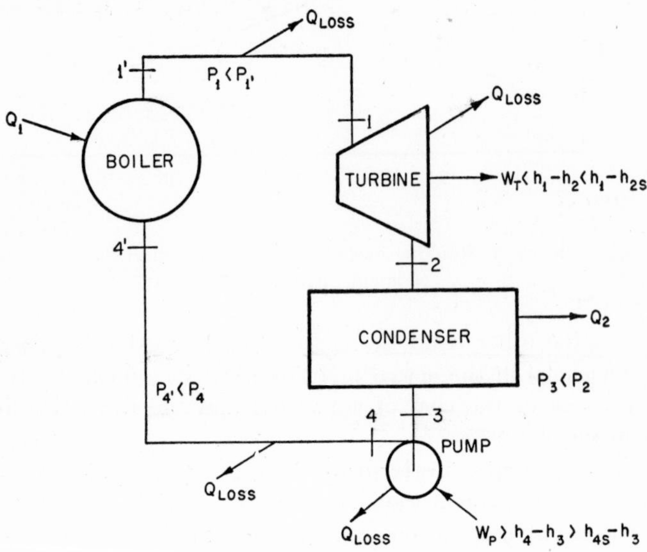

Fig. 13-9. Steam plant flow diagram showing losses.

under the same imposed conditions in each case. In this chapter particular consideration will be given to actual engines and pumps. It should also be noted that in boilers and condensers there is some pressure drop between inlet and outlet; moreover in the condenser the fluid is often cooled below the saturated liquid state. Although these effects are studied here they cannot be ignored in the analysis of an actual plant.

13-5 The actual engine or turbine. The work output of the actual turbine is compared to the output of a reversible adiabatic turbine by means of the *engine efficiency* η_T, defined as follows:

$$\eta_T = \frac{W_T}{h_1 - h_{2S}} \tag{7}$$

where W_T is the actual turbine work, h_1 is the actual enthalpy of the steam entering the turbine, and h_{2S} is the enthalpy at a state having the actual exhaust pressure and lying on an isentropic path through the state 1.

The actual steam rate, w_T, is given by

$$w_T = \frac{2545}{W_T} \frac{\text{lb}}{\text{hphr}} \tag{8}$$

The ideal steam rate, w_I, is given by

$$w_I = \frac{2545}{h_1 - h_{2S}} \frac{\text{lb}}{\text{hphr}} \tag{9}$$

Then from (7), (8) and (9)

$$\eta_T = \frac{w_I}{w_T} \tag{10}$$

Finally

$$\eta_T = \frac{2545}{w_T(h_1 - h_{2S})} \tag{11}$$

Thus the engine efficiency can be determined experimentally by measuring the turbine steam rate, the initial state of the steam, and the exhaust pressure.

The actual engine or turbine process is certainly not reversible and it may not be adiabatic. Applying the steady flow energy equation to the turbine in Fig. 13-9, assuming negligible elevation change and velocity change,

$$h_1 + Q = h_2 + W_x$$

or

$$W_x = W_T = h_1 - h_2 + Q = h_1 - h_2 - Q_{\text{loss}} \tag{12}$$

If a plot is made on the T–s plane for the process of expansion in the turbine it may follow any of a number of paths as shown in Fig. 13-10. For the reversible adiabatic process the path will be 1–2_S. For an ordinary real turbine the heat loss is small and the turbine work may be obtained from (12) as $h_1 - h_2$ by setting Q equal to zero. But for the ideal turbine the work is equal to $h_1 - h_{2S}$; then since the actual turbine produces less work than the reversible adiabatic turbine, it follows that for an adiabatic turbine h_2 must be greater than h_{2S}. The actual expansion might end at 2 in Fig. 13-10 for a turbine of high efficiency, or at 2_a for a turbine of low efficiency. If the heat loss is not negligible, as in a small turbine or engine, the value of h_2 need not always be greater than h_{2S}. As Eq. (12) shows, the value of h_2 may be decreased at will by increasing the heat loss. Thus it is possible for the end state of the expansion to be at point 2_b in Fig. 13-10, if the heat loss is exceptionally large. It is also possible to imagine a case in which the heat loss and the engine efficiency would be so related that the entropy would remain constant during

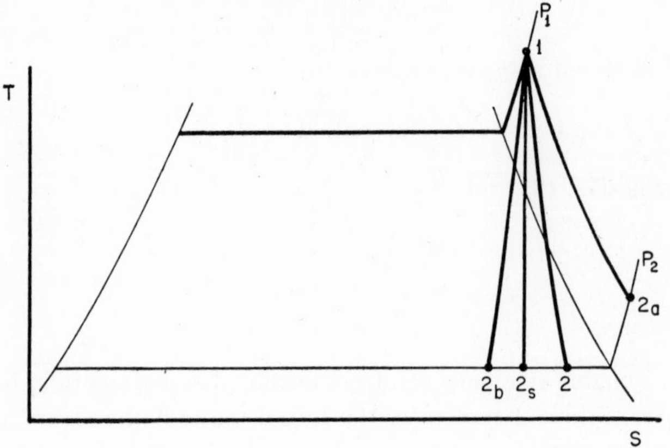

Fig. 13-10. Actual exhaust states.

the expansion process; such an expansion would be isentropic but it would not be either reversible or adiabatic. The above reasoning is consistent with the Second Law; in an actual *adiabatic* process the entropy must increase, but in a process involving heat transfer *out* there is no restriction on the direction of entropy change.

Example 2. A turbine operating under the same steam conditions as given for the cycle of Example 1 has a measured steam rate of 8 lb/hphr. Find the engine efficiency of the turbine and the state of the exhaust steam.

Solution: From Example 1, $h_1 = 1399.2$ Btu/lb, $h_{2S} = 976$ Btu/lb. The actual turbine work per pound of steam is, from the definition of the steam rate,

$$W_T = \frac{2545}{w_T} = 318.1 \text{ Btu/lb}$$

then

$$\eta_T = \frac{W_T}{h_1 - h_{2S}} = \frac{318.1}{423} = 0.751$$

The same result would be obtained from Eq. (10) taking the Rankine cycle steam rate from Example 1 and assuming pump work negligible, so that $w_I = w_R$.

$$\eta_T = \frac{w_I}{w_T} = \frac{6.01}{8.00} = 0.751$$

The state of the exhaust steam can be obtained only by applying the steady flow equation, (12). In this equation W_T and h_1 are known, h_2 and Q_{loss} are unknown. Lacking information we may assume Q_{loss} negligible and obtain

$$h_2 = h_1 - W_T = 1399.2 - 318.1 = 1081 \text{ Btu/lb}$$

If an estimate or a measurement of heat loss could be made, a more accurate result would be possible.

Except for very small turbines, heat loss from turbines is generally negligible. Heat losses from engines are relatively larger than from turbines, but may still be small in the energy balance. The kinetic energy of the steam leaving a turbine may in some cases be appreciably larger than the kinetic energy approaching the turbine; in such cases Eq. (12) needs a kinetic energy term.

13-6 The actual pump. The actual pump is compared to the ideal pump by means of the *pump efficiency*, η_P, defined as follows:

$$\eta_P = \frac{h_{4S} - h_3}{W_P} \tag{13}$$

where h_3 is the actual enthalpy entering the pump, h_{4S} is the enthalpy at a state having the actual discharge pressure and lying on the isentropic path through state 3, and W_p is the actual work supplied to the pump. By the steady flow equation, neglecting velocity and elevation,

$$W_P = h_4 - h_3 + Q_{\text{loss}} \tag{14}$$

The actual h_4 is found in the same manner as the actual h_2 in the case of the turbine.

In discussing vapor cycles the pump work is often neglected but it can be of great importance in some cases. In any event the designer of a plant must know what actual pump work to expect if he is to provide a suitable power supply for the pump.

13-7 More efficient cycles. The simple Rankine cycle has limited efficiency, depending upon the properties of the working substance, as shown in Sec. 13-3. Moreover the characteristics of available apparatus set limits upon the pressures and temperatures that can be used, thereby preventing full exploitation of the cycle possibilities. However, the limits of efficiency may be extended by modifications of the cycle or by using different working fluids. Some cycles in which this is done are: the *reheat* cycle, used in large-capacity central station plants; the *regenerative* cycle, now used in every steam power plant of any significance; the *by-product power* cycle, long used in industrial plants and now being adopted, where possible, by many public utilities; and the mercury-steam *binary* cycle, which has had limited commercial use.

13-8 The reheat cycle. It is a general deduction from the Second Law that the efficiency of a heat engine cycle may be increased by supplying heat at higher temperature. For a cycle like the Rankine cycle, in which heat is supplied through a range of temperatures, the efficiency will be increased if a greater proportion of the heat is supplied at higher temperatures. One way of doing this is by using the reheat cycle, Fig. 13-11.

In the reheat cycle the expansion of steam from the initial state is carried out in two (or possibly more) steps. The first step is from the initial state to approximately the saturation line, process 1–2 in Fig. 13-11. Following this partial expansion the steam is resuperheated at constant pressure, process 2–3, and the remaining expansion, process 3–4, is carried out. The resuperheating is usually done in heating tubes incorporated in the boiler structure to absorb heat from the combustion products; occasionally reheating by steam is used. For the cycle of Fig. 13-11

$$Q_1 = h_1 - h_6 + h_3 - h_2$$
$$W_T = h_1 - h_2 + h_3 - h_4$$
$$W_P = h_6 - h_5$$
$$\eta = \frac{W_T - W_P}{Q_1} = \frac{h_1 - h_2 + h_3 - h_4 - (h_6 - h_5)}{h_1 - h_6 + h_3 - h_2}$$

If the pump work is neglected,

$$\text{approximate } \eta = \frac{h_1 - h_2 + h_3 - h_4}{h_1 - h_5 + h_3 - h_2}$$

It may be observed that in practice the reheat cycle is used with high pressures where the pump work begins to be appreciable.

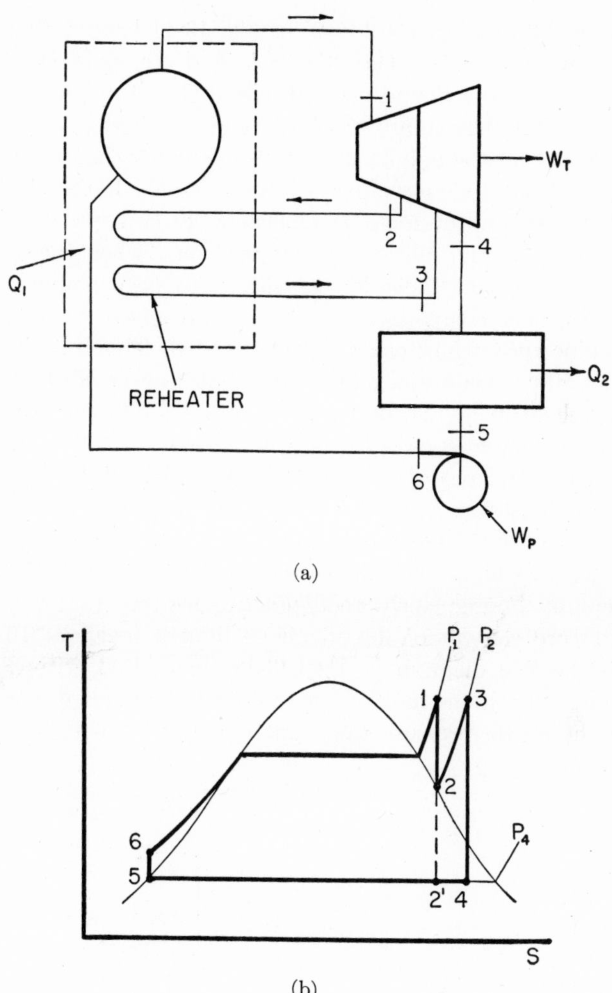

Fig. 13-11. Reheat vapor cycle: (a) flow diagram; (b) state plot.

The steam rate of the reheat cycle of Fig. 13-11 is

$$w = \frac{2545}{W_{\text{net}}} = \frac{2545}{h_1 - h_2 + h_3 - h_4 - (h_6 - h_5)} \text{ lb/hphr}$$

The reheat cycle may be evaluated relative to either of two basic Rankine cycles, depending upon the point of view. An obvious assumption is that the area 2–3–4–2′ in Fig. 13-11 has been added to the basic cycle 1–2′–5–6. On this basis a decrease in steam rate (increase in work per pound of steam) is apparent on the T–s plot; whether the efficiency improves depends upon whether the average temperature of heat supply in process 2–3 is higher than the average temperature of heat supply in process 6–1. Usually in the ideal cycle there is a small improvement in efficiency. From a practical point of view, however, the reheat cycle is used only when the Rankine cycle 1–2′–5–6 is unattainable for reasons to be explained below. Hence the proper comparison is between the reheat cycle and the best Rankine cycle which could have been used. This Rankine cycle would have the same initial temperature but a lower initial pressure than the reheat cycle. To understand this situation some information on the characteristics of turbines is needed.

The engine efficiency of a turbine is reduced by the presence of moisture in the steam; but what is worse, the presence of more than about 10 or 12 percent moisture results in excessive erosion of turbine blading. This reduces the efficiency still further and necessitates costly repairs. Consequently good practice requires that the maximum moisture content of the steam be limited to about 10 or 12 percent. The T–s diagram in Fig. 13-12 shows that with a given exhaust pressure, if the moisture is limited to 10 percent, there is a limit on the steam pressure that can be employed with any given

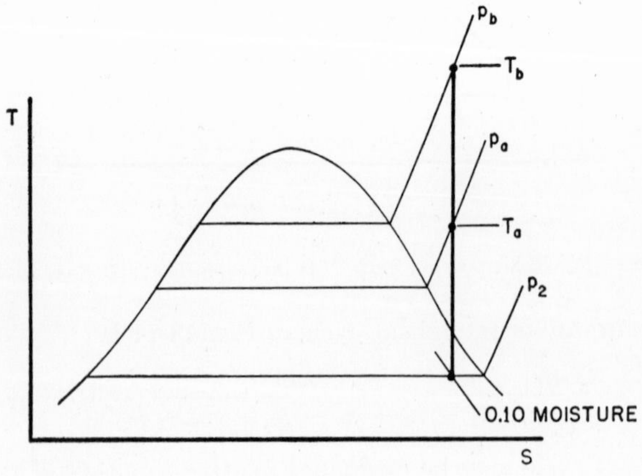

Fig. 13-12.

initial temperature. If T_a is the maximum steam temperature permissible with the available materials of construction then p_a is the maximum permissible pressure for an isentropic turbine. (In the real turbine the expansion will not be isentropic, so the limiting pressure will be somewhat higher, but a limit will still exist.) If better high-temperature materials become available, the limits may increase to T_b and p_b, Fig. 13-12. A table of reasonable steam pressures at given temperatures, for condensing cycles in modern practice, is as follows:

t_1	p_1
750°F	400 psi
825	600
900	850
950	1250
1050	1500

For comparison, reheat cycles with temperature 750°F have been used with initial pressure 1450 psi; and with temperature 950°F reheat cycles have been used with 2400 psi initial pressure.

Example 3. A reheat cycle is to operate with turbines of 85 percent engine efficiency, but otherwise with idealized processes. The initial pressure is 2000 psia, reheat pressure 400 psia, and exhaust pressure 0.5 psia. The maximum temperature of 1000°F is to be used for both initial and reheat temperature. Find the efficiency and steam rate of this cycle; also of a Rankine cycle working between 2000 psia, 1000°F, and 0.5 psia; also of a Rankine cycle working between 1400 psia, 1000°F, and 0.5 psia, this being taken as the cycle of maximum permissible pressure without reheat. Use turbines of 85 percent engine efficiency in the Rankine cycles.

Solution: The cycles to be computed are illustrated in Fig. 13-13. The properties at the various states are given below. For the reheat cycle:

$$h_1 = 1474.5 \text{ Btu/lb} \qquad \text{Table 3}$$

$$s_1 = 1.5603 \text{ Btu/lb °R}$$

$$h_{2S} = 1277.5 \text{ Btu/lb} \qquad \text{Mollier chart}$$

$$h_2 = h_1 - 0.85(h_1 - h_{2S}) = 1307 \text{ Btu/lb}$$

$$h_3 = 1522.4 \text{ Btu/lb} \qquad \text{Table 3}$$

$$s_3 = 1.7623 \text{ Btu/lb °R}$$

$$h_{4S} = 948 \text{ Btu/lb} \qquad \text{Mollier chart}$$

$$h_4 = h_3 - 0.85(h_3 - h_{4S}) = 1033 \text{ Btu/lb}$$

$$h_5 = 47.6 \text{ Btu/lb} \qquad \text{Table 2}$$

$$h_6 = 53.5 \text{ Btu/lb} \qquad \text{Fig. 3, } K \text{ and } K.$$

For the first Rankine cycle:

$$h_{7S} = 840 \text{ Btu/lb} \qquad \text{Mollier chart}$$
$$h_7 = h_1 - 0.85(h_1 - h_{7S}) = 935 \text{ Btu/lb}$$

For the second Rankine cycle:

$$h_8 = 1493.2 \text{ Btu/lb} \qquad \text{Table 3}$$
$$s_8 = 1.6093 \text{ Btu/lb °R}$$
$$h_{9S} = 866 \text{ Btu/lb} \qquad \text{Mollier chart}$$
$$h_9 = h_8 - 0.85(h_8 - h_{9S}) = 960 \text{ Btu/lb}$$
$$h_{11} = 51.5 \text{ Btu/lb} \qquad \text{Fig. 3, } K \text{ and } K.$$

The efficiencies and steam rates computed from the above data by equations given in this and the preceding chapter are tabulated below.

	η	w	*Moisture content of exhaust steam*
Reheat cycle.	0.401	3.9 lb/hphr	6 percent
First Rankine cycle.	0.375	4.75	15.3
Second Rankine cycle. . .	0.366	4.81	12.9

The improvement in efficiency by using the reheat cycle, compared to the Rankine cycles is for the first case

$$\frac{0.401 - 0.375}{0.375} = 7 \text{ percent}$$

and for the second case

$$\frac{0.401 - 0.366}{0.366} = 9.6 \text{ percent}$$

The first Rankine cycle, however, would not be used; therefore the reheat cycle is approximately 10 percent more efficient than the Rankine cycle it replaces and its steam rate is 18 percent less.

The actual reheat cycle will lose a percentage point or two from these figures because of pressure drop in the reheating piping system; the actual Rankine cycle will have slightly lower turbine efficiency because of higher moisture. To determine in any given case whether the reheat cycle should be used, the net final gain in efficiency must be balanced against the additional installation and operating costs of the reheat plant.

Actual reheat cycles are always built upon the regenerative cycle described in the following section, rather than upon the simple Ran-

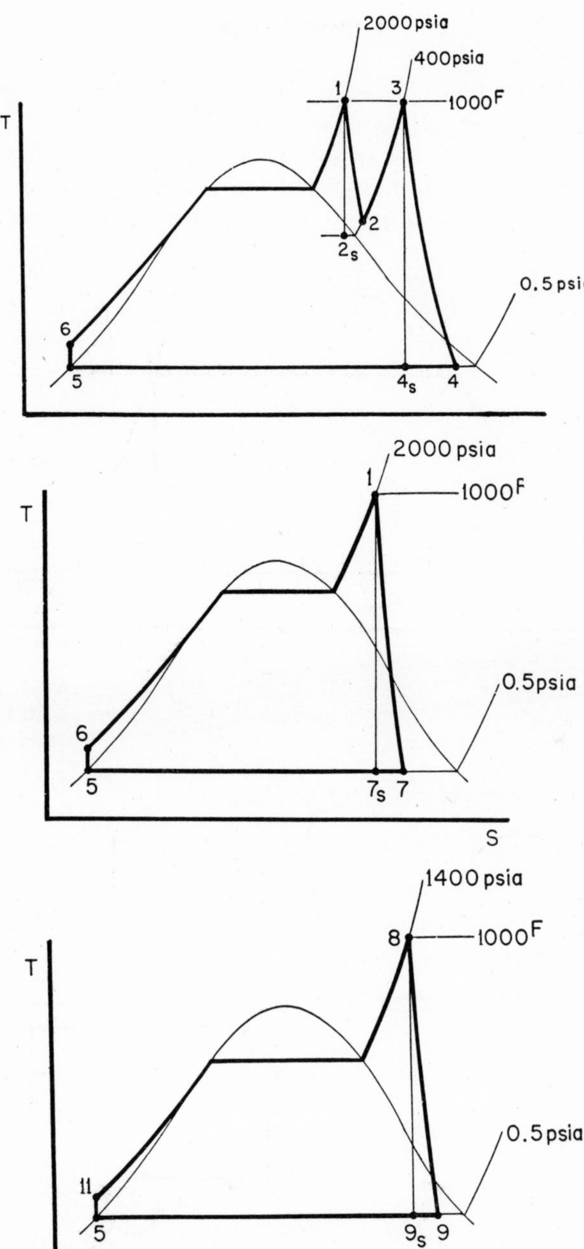

2000 psia
400 psia
1
3
1000F
T
2
2s
0.5 psia
6
5
4s
4
2000 psia
1
1000F
T
0.5psia
6
5
7s
7
S
1400 psia
8
1000F
T
0.5psia
11
5
9s
9
S

Fig. 13-13.

kine cycle. This does not alter the conclusions from the above example, regarding the advantages of reheating.

13-9 The regenerative vapor cycle. It is possible to increase the average temperature of heat supply to a vapor cycle not only by increasing the amount of heat supplied at high temperatures, but also

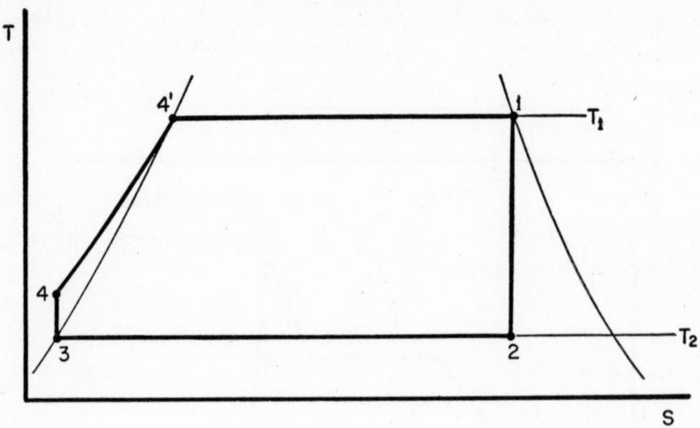

Fig. 13-14.

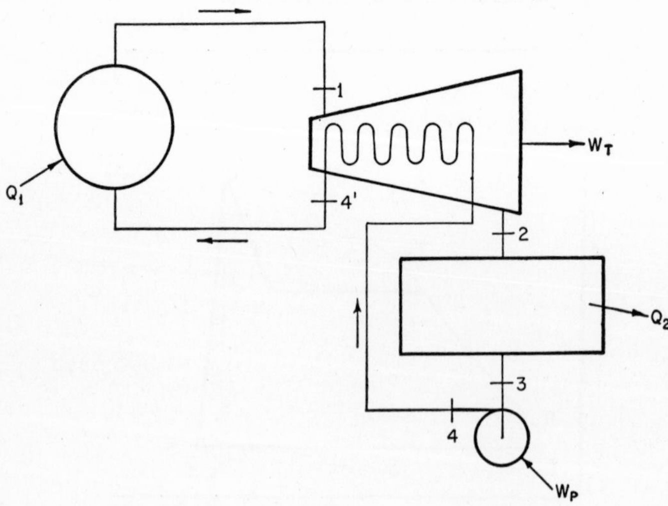

Fig. 13-15. Regenerative cycle; basic scheme.

by *decreasing* the amount of heat supplied at low temperatures. In the saturated steam Rankine cycle of Fig. 13-14 a considerable amount of heat is supplied in the process 4–4′ to water at a temperature lower than T_1, the maximum temperature of the cycle. For maximum efficiency all heat should be supplied at T_1, but this requires that means be found of raising the temperature of the liquid without taking heat from a source external to the system. This is done by a cycle in which heat transfer between two parts of the thermodynamic system takes the place of *externally* supplied heat at low temperatures. Such a cycle is called a regenerative cycle.

The purpose of the ideal regenerative cycle can be accomplished only if means can be found, *by reversible processes*, to make available at all temperatures between T_4 and T_4' energy which was received by the system as heat at T_1 (or in this case T_4'). In the ideal Rankine cycle the fluid expanding through the turbine passes reversibly through all the temperatures from T_1 to T_2; therefore it forms a suitable source of heat for the process 4–4′. Imagine a turbine, Fig. 13-15, containing a heat exchanger consisting of tubes so arranged that the boiler feed water can pass through them on its way from the pump to the boiler. The water passes in the direction opposite to the steam flow, absorbing heat from the steam and thereby rising in temperature to the initial steam temperature by reversible heat transfer. In such a case the ideal expansion process in the turbine is not a reversible adiabatic process but a reversible process in which heat is transferred. In Fig. 13-16 the resulting cycle is plotted on the T–s plane; the quantity of heat Q_x is transferred *from* the steam *to* the liquid water.

Assuming reversible heat transfer, for any small step in the process of heating the water,

$$\Delta T \text{ (water)} = -\Delta T \text{ (steam)}$$

and

$$\Delta s \text{ (water)} = -\Delta s \text{ (steam)}$$

Then the slopes of lines 1–2′ and 4′–3 on the T–s plot will be identical at every temperature and the lines must be identical in contour. Now for the complete cycle all *external* heat is supplied at T_1 and all heat is rejected at T_2. Then

$$Q_1 = h_1 - h_4' = T_1(s_1 - s_4')$$

$$Q_2 = h_2' - h_3 = T_2(s_2' - s_3)$$

But by the similarity of the paths 1–2′ and 4′–3

$$s_4' - s_3 = s_1 - s_2'$$

or

$$s_1 - s_4' = s_2' - s_3$$

Then

$$\frac{Q_1}{Q_2} = \frac{T_1}{T_2}$$

and the efficiency of the regenerative cycle is equal to the efficiency of the Carnot cycle. The two cycles are compared graphically in Fig. 13-17.

The ideal regenerative cycle using saturated steam will be more efficient than the Rankine cycle but will have a larger steam rate; the transfer of heat in the expansion process results in a loss of work, represented in Fig. 13-16 by the area 1–2–2′. The pump work of the ideal regenerative cycle is, however, the same as that of the Rankine cycle and is much smaller than that of the Carnot cycle.

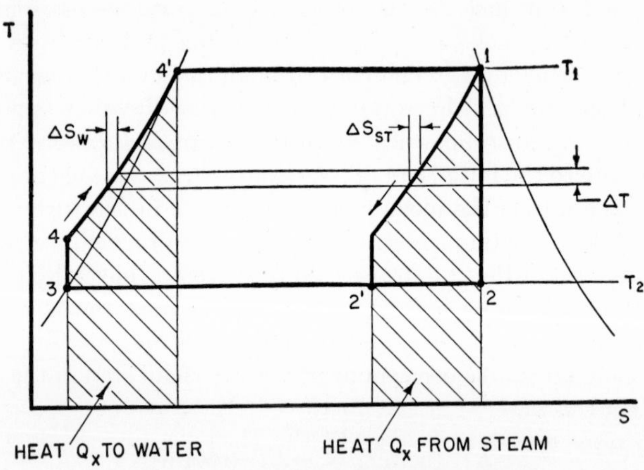

Fig. 13-16. Regenerative feed water heating.

The ideal cycle described above cannot be used for the following reasons: (a) reversible heat transfer is unobtainable in finite time, (b) the heat exchanger in the turbine is mechanically impractical, (c) the process described would increase the moisture content of the steam in the turbine, an undesirable result for reasons explained under the reheat cycle. However, it is possible to arrange an *extraction* cycle

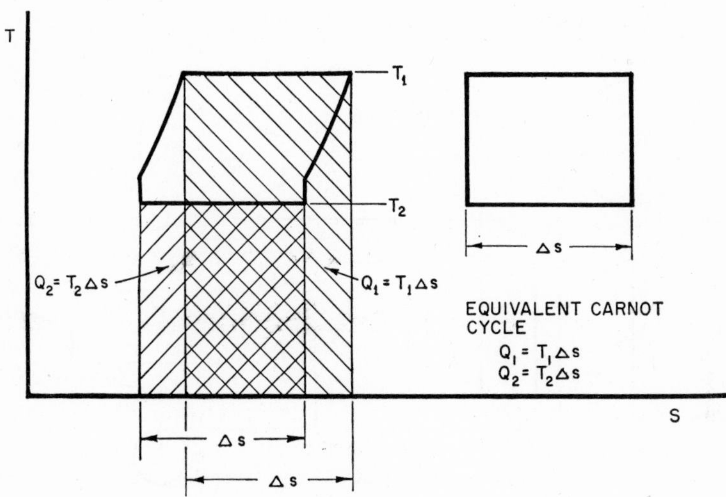

Fig. 13-17. Comparison of regenerative cycle with Carnot cycle.

which approximates the regenerative process in a practical way. The staged regenerative process, which is universally used in modern vapor cycles, differs from the ideal process as follows: (a) the boiler feed water is heated in a number of steps or stages, using temperature differences that are finite but much smaller than in the water-heating process of the Rankine cycle, (b) instead of condensing a portion of the steam within the turbine and permitting the moisture to flow on through the machine, the required amount of steam is *extracted* from the turbine and completely condensed in a separate feed water heater.

A flow diagram of a regenerative cycle using two stages of feed water heating is shown in Fig. 13-18, with the corresponding temperature-entropy diagram in Fig. 13-19. The total steam flow to the turbine, w_1, expands to state 2 where steam is extracted at the rate w_2; this leaves w_1–w_2 as the flow rate from 2 to 3, where w_3 is extracted. The remaining steam flows at rate w_4 to the condenser. From the condenser the water is pumped to the pressure of state 3 and then mixes in a heater with the steam extracted at 3. From the heater at state b a stream consisting of w_4 plus w_3 is pumped to the next heater where the extracted steam w_2 rejoins the boiler feed water; the total flow is then pumped back to the boiler.

In the staged heating process described above the heating of the

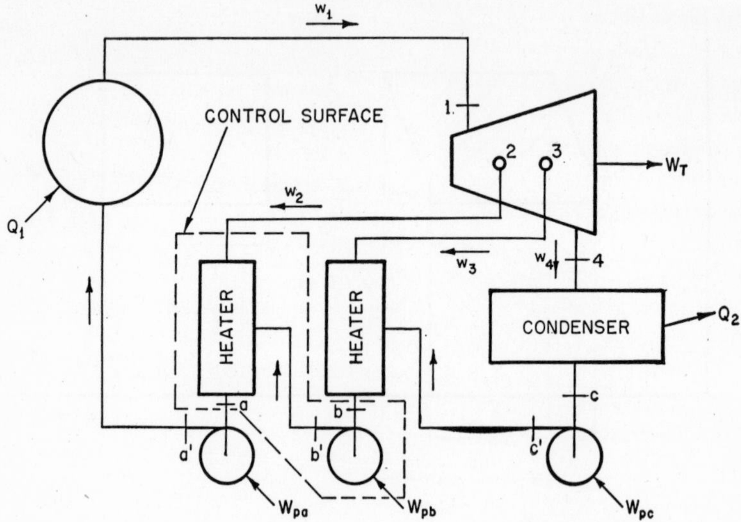

Fig. 13-18. Practical regenerative cycle flow diagram.

feed water is not reversible but in each stage the temperature difference, for example $T_b - T_c$, is only one-third of the difference $T_1 - T_4$ between the boiler feed water in the Rankine cycle and the saturated water in the boiler, which is the source of heat for the feed water in

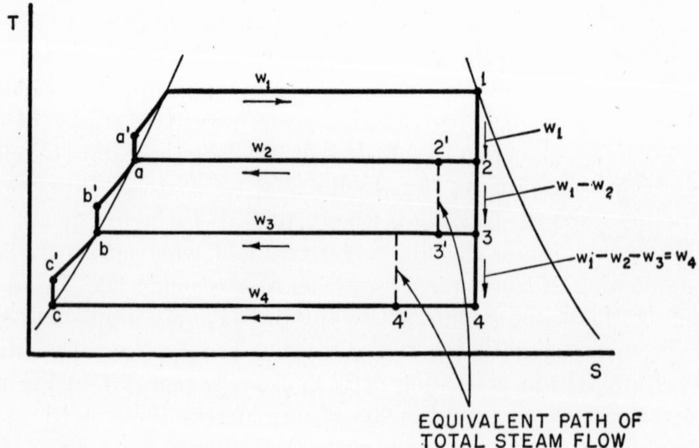

Fig. 13-19. Practical regenerative cycle.

the Rankine cycle. If more stages of heating were used the temperature difference could be reduced further and the heating could be made more nearly reversible. The relation to the reversible cycle may be seen in Fig. 13-19; the path 1–2–3–4 represents the states of a decreasing mass of fluid. If the extracted steam had been condensed *within* the turbine, and flowed through to the condenser, the states of the total mass of fluid would be represented by the dotted path 1–2′–3′–4′. It is seen that the stepped cycle 1–2′–3′–4′ c–c'–b–b'–a–a' approximates the ideal regenerative cycle of Fig. 13-16, and that a greater number of stages would give a closer approximation.

The work and heat quantities and the efficiency of a staged regenerative cycle are easily written in terms of the flow rates and enthalpies. For the cycle of Figs. 13-18 and 13-19 the following are found:

Turbine work

$$W_T = w_1(h_1 - h_2) + (w_1 - w_2)(h_2 - h_3) + (w_1 - w_2 - w_3)(h_3 - h_4)$$

Heat supplied $$Q_1 = w_1(h_1 - h'_a)$$

Cycle efficiency $$\eta = \frac{W_T - \Sigma W_P}{Q_1}$$

where ΣW_P is the sum of all pump work quantities.

To obtain numerical values, the enthalpies and flow rates must be known. In an idealized cycle the temperature levels in the heating of the feed water are usually chosen arbitrarily to divide the total temperature range between condenser and boiler into equal steps. Then, if the turbine process is reversible and adiabatic, the diagram of Fig. 13-19 is completely defined and all the necessary enthalpies can be found.* The flow rates may then be found by applying the steady flow equations as follows. Set up a control surface around the highest pressure heater as shown in Fig. 13-18, and write the equations for energy and mass, assuming heat transfer through the control surface, kinetic energy change, and potential energy change are negligible.

Then $$w_2h_2 + w_bh_b = w_ah_a + w_bW_{Pb}$$

and $$w_2 + w_b = w_a$$

* With a real turbine the detailed path of the expansion process must be known to locate the extraction points such as 2 and 3, Fig. 13-19.

Now
$$w_a = w_1$$
so
$$w_b = w_1 - w_2$$

Then combining and rearranging

$$w_2h_2 + (w_1 - w_2)(h_b + W_{Pb}) = w_1h_a$$
$$w_1(h_a - h_b - W_{Pb}) = w_2(h_2 - h_b - W_{Pb})$$
$$\frac{w_2}{w_1} = \frac{h_a - h_b - W_{Pb}}{h_2 - h_b - W_{Pb}}$$

For the next heater, by similar reasoning,

$$\frac{w_3}{w_1 - w_2} = \frac{h_b - h_c - W_{Pc}}{h_3 - h_c - W_{Pc}}$$

The procedure can be repeated for any number of heaters, the equation giving for each extraction point the fraction of the steam approaching that point which is there extracted. Starting with any given *throttle flow*, w_1, the flow in all branches of the system can thus be determined.

When superheated steam is used in a regenerative cycle the cycle cannot, even in the ideal case, approach the Carnot cycle as a limit because only an infinitesimal amount of heat is supplied at T_1; nevertheless the elimination of the external heat supply to the feed water is still advantageous.

As illustrated in Fig. 13-20, an extraction point occurring in the superheat region is located not by the temperature of the feed water

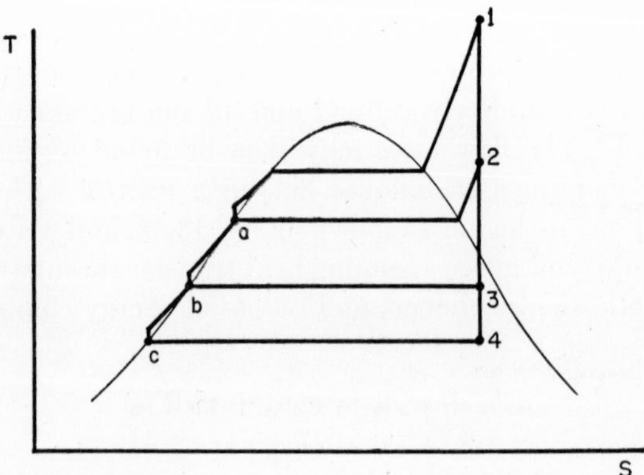

Fig. 13-20. Regenerative cycle with superheated steam.

heater, T_a, but by the saturation pressure p_2 corresponding to this temperature. Since T_2 is greater than T_a, the heating process suffers some added irreversibility compared to a process receiving saturated steam at T_a. This effect can be minimized by suitable heater design.

13-10 By-product power and heat. The concept of the efficiency of a heat engine cycle attaches ultimate value only to energy in the form of work. However, in many places there is use for heat as well as work, and then the value of a heat supply is not measured solely by the work that can be obtained from it. Steam is used for

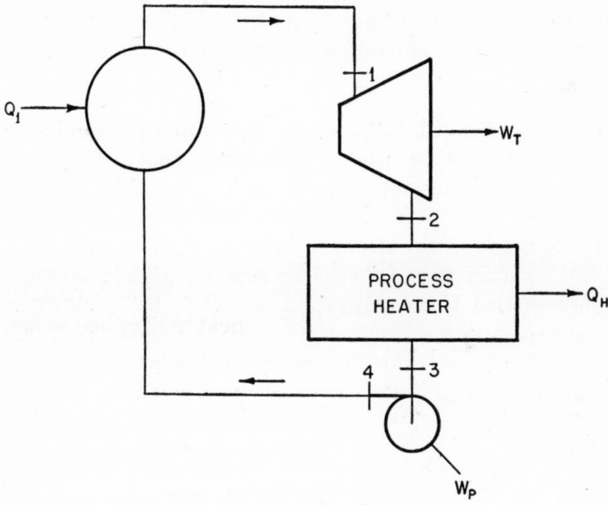

Fig. 13-21. By-product power scheme.

heating in many industrial processes as well as for comfort heating of buildings.

The steam necessary for a heating system can be generated especially for that purpose; in this case the steam system is considered to have performed perfectly if the heat transferred for a useful purpose is equal to the heat received by the steam at the same temperature. It is also possible to generate steam at a higher pressure than p_2, expand it through a turbine to obtain work, and exhaust from the turbine at p_2 to the process heating plant. Such an arrangement is pictured in Fig. 13-21. It is apparent that the process heater takes the place of a condenser for the power cycle but the rejected heat has become useful energy rather than waste energy. Assuming that there is use for both the net work and the heat rejected, all the heat supplied can serve a useful purpose. In a real plant there would be some losses to the surroundings, but there is no inherently wasted energy as in the case of a plant existing to produce work only. The arrangement is called a by-product power cycle on the assumption that the process heat is the basic need and the power is produced incidentally. In this case the cost of power is only the *additional* cost over the cost of the process heat alone. Since part of the installation and labor cost is borne by the process heating, and since the additional heat supplied for power is utilized at a very high efficiency, the by-product power scheme may offer great economies when properly applied.

Example 4. A steam plant operates with initial pressure 250 psia and temperature 700°F, and exhausts to a heating system at 25 psia. The condensate from the heating system is returned to the boiler plant at 150°F, and the heating system utilizes for its intended purpose 90 percent of the energy transferred from the steam it receives. The turbine efficiency is 70 percent. (a) What fraction of the energy supplied to the steam plant serves a useful purpose? (b) If two separate steam plants had been set up to produce the same useful energy, one to generate heating steam at 25 psia and the other to generate power through a cycle working between 250 psia, 700°F, and 1 psia, what fraction of the energy supplied would have served a useful purpose?

Solution: Refer to Fig. 13-21.

(a) From the tables and mollier chart,

$$h_1 = 1371 \text{ Btu/lb}, \qquad h_{2S} = 1149, \qquad h_3 = 118$$

Then, neglecting pump work,

$$Q_1 = 1253 \text{ Btu/lb}, \qquad W = 156 \text{ Btu/lb}$$

The thermal efficiency *as a power plant* is

$$\eta = \frac{W}{Q_1} = 0.124$$

The net energy supplied to the heating system is

$$Q_H = h_2 - h_3$$

Assuming the turbine adiabatic

$$h_2 = h_1 - W$$

and

$$Q_H = h_1 - W - h_3 = 1096 \text{ Btu/lb}$$

Of this, 90 percent is usefully applied; then the useful energy is

$$W + 0.9Q_H = 1142 \text{ Btu/lb}$$

The fraction of the energy supplied that serves a useful purpose is then

$$\frac{1142}{1253} = 0.91$$

It is seen that this is greater than the individual efficiency of either the power plant or the heating plant.

(b) If two separate plants were used, the plant for heating steam would be supplied with Q_H or 1096 Btu/lb of heating steam. The power plant, under the given conditions, would have a thermal efficiency of about 23 percent. Thus, to produce the work of 156 Btu as in part (a) would require a heat input of 156/0.23 or 680 Btu. The total input per pound of heating steam for both heating and power, would be

$$1096 + 680 = 1776 \text{ Btu}$$

The useful energy, as in part (a), is 1142 Btu; then the fraction of the energy supplied which appears as useful energy is

$$\frac{1142}{1776} = 0.65$$

To obtain the same total output under method (b) it would be necessary to burn $0.91/0.65 = 1.4$ times as much fuel as under method (a); moreover method (b) requires an additional boiler and a condensing plant which are not needed under method (a).

This example shows clearly the basic thermodynamic advantage of the by-product power scheme. Whether it is *economical* in any particular application depends upon the cost of fuel, the cost of purchased power, the existence of simultaneous demands for power and process heat, and other factors which cannot be considered here.

13-11 Two-fluid or binary vapor cycles. It will have become evident by now that the efficiency of a vapor cycle is limited by the characteristics of the working substance. The complications of regenerative cycles and reheat cycles could be avoided if it were possible to specify at will the property relations of the working fluid and to obtain the specified fluid. Some of the characteristics that would be specified in a made-to-order fluid are listed below, with the reasons therefor:

(a) A high critical temperature so that the heat of vaporization could be supplied at the maximum temperature of the cycle, which would be the maximum temperature for which apparatus could be built.

(b) A large enthalpy of vaporization and a small specific heat of

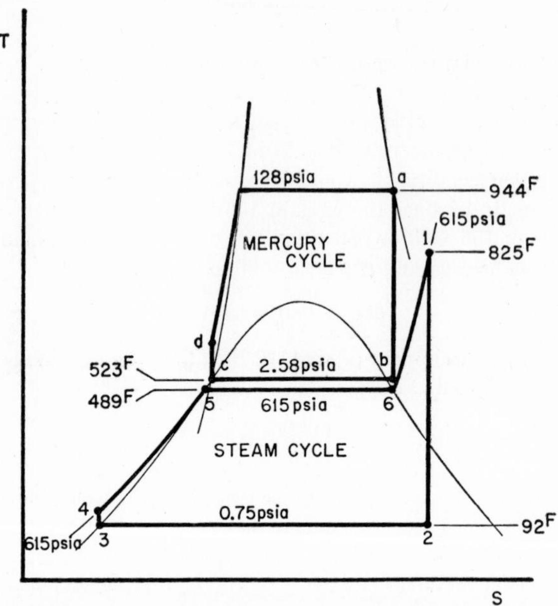

Fig. 13-22. Mercury-steam cycle.

the liquid so that relatively little of the heat supplied would be used to raise the liquid temperature to the boiling point.

(c) Saturation pressures, at the maximum and minimum temperatures of the cycle, within an order of magnitude of the atmospheric pressure (in the range from 0.1 atm to 10 atm) so that stress problems and leakage problems would be minimized.

(d) Freezing point below room temperature to facilitate draining and filling the apparatus.

(e) Saturated vapor line close to the path of a turbine expansion process (slightly increasing entropy with decreasing pressure) so that excessive moisture would not appear during the expansion.

(f) Chemical stability and inertness toward materials of construction at all attainable temperatures so that neither the fluid nor the apparatus would deteriorate in use.

(g) The fluid should be non-toxic and low in cost.

The only fluid at present that is able to satisfy reasonably well all the above conditions is water, though it is somewhat deficient in

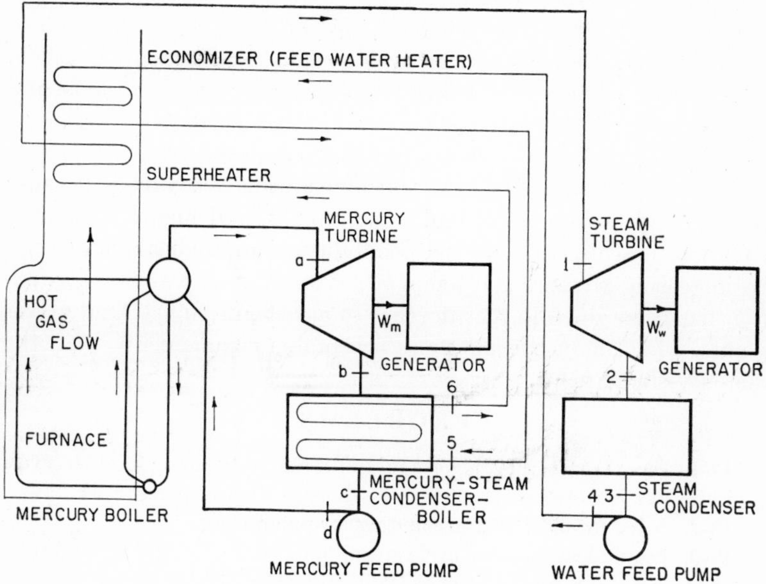

Fig. 13-23. Mercury-steam plant flow diagram.

categories (a), (c), and (e). The critical temperature of water (705°F) is not as high as the maximum temperature of modern steam cycles (1100°F), and the saturation pressures, even at moderate temperatures, are high (2200 psia at 650°F). A more suitable fluid at high temperatures is mercury, which has a critical temperature above 2700°F and a saturation pressure of only 180 psia at 1000°F. However, mercury is unsuitable at low temperatures because its saturation pressure becomes exceedingly low and it would be impractical to maintain in a condenser the low pressure needed to condense mercury at atmospheric temperatures.

These circumstances have led to the development of a binary (two-fluid) cycle in which mercury is used in a Rankine type cycle for the high-temperature range. The heat rejected at relatively high temperature from the mercury cycle is used to evaporate water in a regenerative steam cycle which operates to the usual low-temperature limit for a steam cycle. The diagrams, Figures 13-22, 13-23 show how this method is used, the cycle conditions being based upon the operating conditions of the Schiller Station of the Public Service Company of New Hampshire, which went into operation in 1950.

It is evident from the T–s plot, Fig. 13-22, that the addition of the mercury cycle to the steam cycle results in a marked increase in the average temperature of heat supply to the plant as a whole and consequently the efficiency is increased.

The mercury-steam cycle has been in commercial use for a quarter of a century but it has never attained wide acceptance because (a) mercury is expensive, limited in supply, and highly toxic, and (b) there has always been the possibility of improving steam cycles by increasing pressure and temperature, and by reheat. Some attention has been given in recent years to gas-steam binary fluid power plants, which may become more acceptable commercially than the mercury-steam plant.

PROBLEMS

13-1. Find the thermal efficiency and steam rate for a Rankine cycle operating between 150 psia, 500°F, and 15 psia. Neglect pump work.

13-2. Solve Problem 13-1 accounting for pump work.

13-3. In an ideal Rankine cycle steam flows to the turbine at 1200 psia, 900°F, and the exhaust pressure is 1.5 in. Hg. The turbine power output is 10,000 kw. Find: (a) the heat supplied, Btu/hr; (b) the steam flow rate, lb/hr; (c) the heat rejected, Btu/hr; (d) the power input to the pump, kw.

13-4. In a certain steam plant the pressure and temperature at the turbine inlet are 200 psia and 450°F; the exhaust pressure is 15 psia, and the condensate is cooled to 140°F before it goes from the condenser to the boiler feed pump. The boiler efficiency is 82 percent (i.e. the heat transferred to the steam is 82 percent of the heating value of the fuel consumed). Neglecting pump work, find: (a) the plant heat rate, Btu of fuel heating value per horsepower-hour of work output; (b) the heat rejected in the condenser, Btu/hphr; (c) the results under (a) and (b) if the plant operated on an ideal Rankine cycle based upon the turbine inlet and exhaust conditions.

13-5. Solve Problem 13-2 for a cycle using an engine and a pump each of 65 percent efficiency.

13-6. It is proposed to change the steam conditions of Problem 13-5 to 250 psia, 700°F, keeping machine efficiencies constant. The cost of heat energy for the cycle is $0.50 per million Btu. If the plant operates at 500 hp for 3500 hours per year what would be the reduction in annual cost of heat energy?

13-7. It is proposed to change the exhaust pressure of Problem 13-5 to 1.5 in. Hg_{abs}, keeping machine efficiencies constant. If energy cost and operating schedule are as in Problem 13-6, what would be the reduction in annual cost of heat energy?

13-8. In a steam power plant the turbine inlet and exhaust conditions and the power output are the same as in Problem 13-3, but the turbine has an efficiency of 75 percent. If the other processes of the cycle are in accordance with the ideal Rankine cycle, find the four quantities asked for in Problem 13-3.

13-9. A small steam turbine operating with steam supply at 180 psia and 400°F exhausts to a condenser at 2 psia; at 50 kw load the steam rate is 30 lb/kwhr. The exhaust steam is found by test to contain 5 percent moisture. Find the rate of heat transfer (Btu/hr) from the turbine.

13-10. In a steam power plant steam is generated at 160 psia, saturated. Steam enters the engine at 150 psia saturated, and the exhaust is to the atmosphere at 15 psia. The feed water is taken from a lake at 65°F, and before entering the boiler it is heated to 180°F in a feed water heater; the source of heat is the exhaust steam on its way to the atmosphere. The steam rate of the plant is 32 lb/hphr. The boiler efficiency is 70 percent (i.e. 70 percent of the fuel heating value is transferred to the steam).

(a) Sketch a flow diagram for the process. (b) Find the thermal efficiency of the plant (work output/heating value of the fuel consumed). (c) Find the efficiency of an ideal Rankine cycle based upon the steam and exhaust conditions at the actual engine. (d) Find the difference between the heat rates of the ideal cycle and the actual plant. (e) Show approximately how much the heat rate could be reduced by substituting an idealized component for each of the following: (1) the engine; (2) the feed water heater; (3) the boiler. In each case only one component is to be idealized.

13-11. An ideal steam power cycle has a turbine inlet steam state of 600 psia, 725°F; the steam expands reversibly and adiabatically in a turbine to 120 psia after which it is reheated to 725°F, 120 psia, and expands reversibly and adiabatically to the exhaust pressure of 1.5 in. Hg_{abs}. Find: (a) the steam rate, lb/kwhr; (b) the cycle efficiency.

13-12. For comparison with the reheat cycle of Problem 13-11, find the turbine inlet pressure of a Rankine cycle which has a turbine inlet temperature of 725°F, an exhaust pressure of 1.5 in. Hg_{abs}, and an exhaust moisture content of 12 percent (use the h–s chart). For this cycle find: (a) the steam rate, lb/kwhr; (b) the cycle efficiency.

13-13. In a plant to which the basic conditions of Problem 13-11 apply, the turbine efficiency for each part of the expansion is 75 percent; there is a 10 psi pressure drop in the reheater (steam comes out at 110 psia, 725°F); the feed pump efficiency is 60 percent. Heat losses from the machines are negligible. Find: (a) the steam rate, lb/kwhr; (b) the plant cycle efficiency.

13-14. In a certain steam power plant the original installation operated between 400 psia, 750°F, and 1.5 in. Hg_{abs}. In order to increase the power output a new high-pressure boiler and turbine were installed; the new turbine receives steam at 2000 psia, 950°F, and exhausts at 420 psia; the exhaust steam is reheated and supplied to the old turbines at 400 psia, 750°F. (a) With a turbine efficiency of 80 percent, how much power (kw) is delivered, per lb/hr of steam, by the new turbine? (b) If the old turbine has 75 percent efficiency how much power does it deliver per lb/hr of steam? (c) What is the total turbine output, new and old, when the new turbine exhausts enough steam for the old turbine to produce its rated output of 70,000 kw? (d) Under conditions in (c) what fraction of the total heat supplied to the steam is transferred in the reheater? (e) Neglecting pump work, compare the thermal efficiencies of the old and new cycles.

13-15. In a single-heater ideal regenerative steam cycle the steam enters the turbine at 400 psia, 760°F, and the exhaust pressure is 2 psia. The feed water heater is a direct-contact type which operates at 50 psia. Find: (a) the efficiency and steam rate of the cycle; (b) the percentage improvement in efficiency, compared to the Rankine cycle; (c) the percentage increase in steam rate, compared to the Rankine cycle.

13-16. In a cycle operating under the conditions of Problem 13-15 the efficiency of the portion of the turbine before the extraction opening is 65 percent, and of the remaining portion is 70 percent. Both feed water pumps operate at 60 percent efficiency. There are no external heat transfers except in the boiler and in the condenser; there are no pressure losses due to friction in piping or apparatus. Find the efficiency and the steam rate of the cycle.

13-17. An industrial plant requires 20,000 lb/hr of heating steam at 25 psia, and 1000 kw of power from a turbine. Consider an ideal cycle in which steam goes to the turbine at 200 psia, 600°F, and the exhaust steam goes to the heating system at 25 psia. (a) Find the state of the exhaust steam. (b) Will the exhaust steam flow be sufficient for the heating requirement?

13-18. Solve Problem 13-17 with a turbine efficiency of 70 percent.

13-19. For the conditions of Problem 13-18 all boiler feed water is taken from a lake at 60°F. Exhaust steam which is not used in the heating system is discharged to the atmosphere; heating system condensate is thrown away. Compare the heat required per hour to make steam in two cases: (a) operating as in Problem 13-18; (b) using the same steam conditions and turbine efficiency as in Problem 13-18, with 15 psia exhaust pressure, to generate power; and using a separate boiler operating at 25 psia to supply saturated heating steam.

13-20. A power plant design is to be based upon the following conditions: power output 2500 kw; heating steam, 60,000 lb/hr at 30 psia, approximately saturated vapor, is to be extracted from the turbine; the turbine inlet pressure and temperature are respectively 175 psia and 500°F; the exhaust pressure is 2 psia; the turbine efficiency is 75 percent for each part (before and after the extraction opening). (a) How much steam must be supplied to the turbine? (b) The extracted steam is returned to the boiler as liquid at 200°F; the exhaust steam as liquid at 120°F. How much heat must be supplied per hour?

13-21. Solve Problem 13-20 if sufficient additional steam is extracted at 30 psia to heat all the feed water to 215°F in a direct-contact heater. What percentage saving in heat supplied results from this change?

13-22. An inventor has proposed using Freon-12 as a working fluid for a vapor power cycle. Discuss the probable advantages and disadvantages of this fluid, considering efficiency, size and strength of equipment, and any other points on which you can comment.

REFERENCES

Barnard, W. N., F. O. Ellenwood, and C. F. Hirshfeld, *Heat-Power Engineering*, Parts I and II. New York: Wiley, 1935.

Gaffert, G. A., *Steam Power Stations*. New York: McGraw-Hill, 1952.

Skrotzki, B. G. A. and W. A. Vopat, *Applied Energy Conversion*. New York: McGraw-Hill, 1945.

Morse, F. T., *Applied Energy*. New York: Van Nostrand, 1947.

Chapter 14

COMBUSTION PROCESSES: FIRST LAW ANALYSIS

Combustion processes are important in heat engineering because they are the means of obtaining the high temperatures necessary for efficient power plants. A combustion process is essentially the chemical reaction of combining a fuel with oxygen. Such processes are subject to the laws of thermodynamics, but before these laws can be applied it is necessary to know something of the properties of the systems involved and of the nature of the processes of combustion. The discussion of the application of the First Law to combustion processes is therefore preceded by an outline of the elementary chemistry of combustion.

14-1 Fuels. Fuels used in engineering practice may be solid, liquid, or gaseous; they may range from pure substances to heterogeneous mixtures of combustible materials with rock or other minerals. In all ordinary cases the essential chemical elements in the fuel are carbon and hydrogen. The other elements present may be considered impurities; they often have a profound effect upon the burning process, but assuming that the fuel does burn, they have only a minor effect upon the thermodynamic analysis of the process.

The *ultimate analysis* of a fuel is given in terms of the mass fractions of carbon, C; hydrogen, H; oxygen, O; sulfur, S; nitrogen, N; ash, A; and moisture, M. Except for ash and moisture the analysis is in terms of the chemical elements, with no indication of the compounds in which they may be combined. *Moisture* represents water which is simply mixed with the fuel; there may also be water chemically combined as an intrinsic part of the fuel structure and which is accounted for by part of the H and O of the analysis. *Ash* is the solid residue after all combustible and volatile components of the fuel have been removed by burning. The ultimate analysis is used in making the material balance for a combustion process.

For data on typical fuels see Appendix Table A-2.

14-2 Combustion reactions—material balance. Combustion is the oxidation of a fuel at such a rate as to cause an appreciable temperature rise (of several hundred degrees) in the substances involved in the process. The details of combustion reactions may be exceedingly complex, and perhaps not fully understood, but the overall results can be described by simple chemical equations. The basic equations describe the reaction between fuel and oxygen, and indicate the amount of oxygen necessary to burn the fuel. Since the usual source of oxygen is atmospheric air it is desirable also to set up equations for the air required (*theoretical air*).

Assumptions. Certain assumptions will be made as to the properties of the substances involved in the reactions.

Dry air is assumed to have the composition 21 percent oxygen and 79 percent nitrogen by volume. (For convenience the argon and other inert constituents, which make up about one percent of dry air, are taken as nitrogen.) The fractions by mass are then, oxygen 0.232, and nitrogen 0.768.

The molecular weights of the elements involved in combustion may be taken as follows: O_2, 32; N_2, 28; H_2, 2; C, 12; S, 32. Then the equivalent molecular weight of air becomes

$$(0.21)(32) + (0.79)(28) = 28.8$$

which is often rounded off to 29. The molecular weights of combustion products become: CO_2, 44; CO, 28; H_2O, 18; SO_2, 64.

It is assumed that all gases involved are perfect gases.

Combustion in Oxygen. The oxygen required to burn the elements, carbon, hydrogen, and sulfur, will be determined. The equation for the combustion of carbon to carbon dioxide (complete combustion) is

$$C + O_2 \rightarrow CO_2$$

This equation is on a mol basis; that is, it states that one mol (molecular weight) of carbon combines with one mol of oxygen to yield one mol of carbon dioxide. By substituting the molecular weights of the substances, a mass balance may be written:

$$12 \text{ lb C} + 32 \text{ lb } O_2 \rightarrow 44 \text{ lb } CO_2$$

This may be reduced to the basis of one pound of fuel as follows:

$$1 \text{ lb} + 2.67 \text{ lb} \rightarrow 3.67 \text{ lb}$$

In the same way, for the burning of hydrogen,

$$\begin{aligned} &H_2 + \tfrac{1}{2}\,O_2 &&\rightarrow H_2O \\ &1\text{ mol} + \tfrac{1}{2}\text{ mol} &&\rightarrow 1\text{ mol} \\ &2\text{ lb} + 16\text{ lb} &&\rightarrow 18\text{ lb} \\ &1\text{ lb} + 8\text{ lb} &&\rightarrow 9\text{ lb} \end{aligned}$$

For sulfur

$$\begin{aligned} &S + O_2 &&\rightarrow SO_2 \\ &1\text{ mol} + 1\text{ mol} &&\rightarrow 1\text{ mol} \\ &1\text{ lb} + 1\text{ lb} &&\rightarrow 2\text{ lb} \end{aligned}$$

Carbon does not always burn to carbon dioxide; it may burn to carbon monoxide, a result sometimes called incomplete combustion. For the combustion of carbon to carbon monoxide the material balance is:

$$\begin{aligned} &C + \tfrac{1}{2}\,O_2 &&\rightarrow CO \\ &1\text{ lb} + 1.33\text{ lb} &&\rightarrow 2.33\text{ lb} \end{aligned}$$

If the carbon monoxide is then burned, the balance is

$$\begin{aligned} &CO + \tfrac{1}{2}\,O_2 &&\rightarrow CO_2 \\ &1\text{ lb} + 0.57\text{ lb} &&\rightarrow 1.57\text{ lb} \end{aligned}$$

Observe that one pound of carbon will require the same amount of oxygen to burn to carbon dioxide whether it burns directly to carbon dioxide or indirectly by burning first to carbon monoxide; 2.33 pounds of CO, which result from burning one pound of carbon, will require 1.33 lb of oxygen and will yield 3.67 lb of CO_2.

In all the reactions the mass of products must equal the mass of reactants and the mass of each individual element must be the same on both sides of the equation.

Combustion in Air. When the oxygen is supplied from air the nitrogen of the air may be added to both sides of the equation since it passes through the reaction unchanged. As shown in Sec. 12-3, the fractions by volume in a perfect gas mixture are equal to the fractions by mols. Therefore, since air is 21 percent oxygen and 79 percent nitrogen by volume, each mol of oxygen is accompanied by 79/21 mols of nitrogen. Taking the combustion of carbon to carbon dioxide as an example,

$$\begin{aligned} &C + O_2 + \tfrac{79}{21}\,N_2 &&\rightarrow CO_2 + \tfrac{79}{21}\,N_2 \\ &12\text{ lb} + 32\text{ lb} + (\tfrac{79}{21})\,28\text{ lb} &&\rightarrow 44\text{ lb} + (\tfrac{79}{21})\,28\text{ lb} \\ &1\text{ lb} + 2.67\text{ lb} + 8.78\text{ lb} &&\rightarrow 3.67\text{ lb} + 8.78\text{ lb} \end{aligned}$$

Then each pound of carbon requires 2.67 + 8.78 or 11.45 lb of air. This is called the *theoretical air* for burning carbon to carbon dioxide. The theoretical air for several other reactions may be found in Appendix Table A-2. If the ultimate analysis of a fuel is known, the theoretical air required to burn it may be calculated from the data in the table.

Example 1. Find the theoretical air for combustion of octane (C_8H_{18}) to CO_2 and H_2O.

Solution: Each C requires 1 O_2, and each H_2 requires $\frac{1}{2}$ O_2; then 8 C and 9 H_2 require $12\frac{1}{2}$ mols of O_2, accompanied by $(12\frac{1}{2})(\frac{79}{21})$ mols of N_2.

$$C_8H_{18} + 12\tfrac{1}{2}\,O_2 + (12\tfrac{1}{2})(\tfrac{79}{21})N_2 \rightarrow 8\,CO_2 + 9\,H_2O + (12\tfrac{1}{2})(\tfrac{79}{21})N_2$$
$$114\text{ lb} + 400\text{ lb} + 1315\text{ lb} \rightarrow 352\text{ lb} + 162\text{ lb} + 1315\text{ lb}$$
$$1\text{ lb} + 3.51\text{ lb} + 11.6\text{ lb} \rightarrow 3.09\text{ lb} + 1.42\text{ lb} + 11.6\text{ lb}$$

The theoretical air is 3.51 lb + 11.6 lb = 15.1 lb air per lb C_8H_{18}.

Example 2. Find the theoretical air for a fuel of ultimate analysis C, H, O, S, N, *A*, and *M*.

Solution: The fuel contains three combustible elements, carbon, hydrogen, and sulfur, each of which requires a certain amount of air as indicated in Table A-2. However, there is some oxygen already in the fuel, which need not be supplied from the air. It is assumed for convenience that all of this oxygen will eventually be combined with hydrogen; then the amount of hydrogen which is finally combined with *oxygen from the fuel* is, per pound of fuel, O/8, where O is the fraction by mass of oxygen in the fuel. The remaining hydrogen $\left(H - \frac{O}{8}\right)$, called *free hydrogen*, requires oxygen from the air. The theoretical air per pound of fuel is then given by

$$\text{theo. air} = 11.5\,C + 34.2\left(H - \frac{O}{8}\right) + 4.31\,S$$

where the letters are the fuel analysis fractions, and the numerical factors are the air requirements per pound of the respective elements.

Example 3. Fuel oil containing 86 percent carbon and 14 percent hydrogen by mass is to be burned with 10 percent *excess air* (110 percent theoretical air).

(a) If the air is dry, and combustion is complete, what will be the composition by mass of the products? (b) If the pressure is 15 psia what is the dew point of the products? (c) If the products are cooled to 100°F at 15 psia how much liquid water will condense per pound of fuel burned?

Solution: (a) To burn 1 lb of carbon with theoretical air,

$$1 \text{ lb C} + 2.67 \text{ lb } O_2 + 8.78 \text{ lb } N_2 \rightarrow 3.67 \text{ lb } CO_2 + 8.78 \text{ lb } N_2$$

With 110 percent theoretical air

$$1 \text{ lb C} + 2.94 \text{ lb } O_2 + 9.66 \text{ lb } N_2 \rightarrow 3.67 \text{ lb } CO_2 + 0.267 \text{ lb } O_2 + 9.66 \text{ lb } N_2$$

One pound of fuel contains 0.86 lb of carbon; the products of combustion of this carbon with 110 percent theoretical air will be

$$0.86\,(3.67 \text{ lb } CO_2 + 0.267 \text{ lb } O_2 + 9.66 \text{ lb } N_2)$$

or

$$3.16 \text{ lb } CO_2 + 0.23 \text{ lb } O_2 + 8.30 \text{ lb } N_2$$

By similar procedure, burning 0.14 lb of hydrogen with 110 percent theoretical air is found to give products as follows:

$$1.26 \text{ lb } H_2O + 0.112 \text{ lb } O_2 + 4.09 \text{ lb } N_2$$

Then for one pound of fuel the products will be

$$3.16 \text{ lb } CO_2 + 1.26 \text{ lb } H_2O + 0.34 \text{ lb } O_2 + 12.39 \text{ lb } N_2$$

or 17.15 pounds of products per pound of fuel. In fractions by mass the composition is 0.184 CO_2, 0.074 H_2O, 0.020 O_2, 0.722 N_2.

(b) The dew point of the products is the saturation temperature corresponding to the partial pressure of the water vapor (Sec. 12-6). The partial pressure of the water vapor is given by

$$\frac{p_w}{p} = x_w$$

where p_w is the partial pressure of the water, p is the total pressure of the gas, and x_w is the mol fraction of water vapor.

$$x_w = \frac{\text{mols water vapor per pound of mixture}}{\text{mols mixture per pound of mixture}}$$

$$= \frac{\dfrac{0.074}{18}}{\dfrac{0.184}{44} + \dfrac{0.074}{18} + \dfrac{0.020}{32} + \dfrac{0.722}{28}} = \frac{0.00411}{0.0347} = 0.118$$

Then

$$p_w = (0.118)(15) = 1.77 \text{ psia}$$

From the steam tables the saturation temperature for this pressure is 121.6°F, which is the dew point required.

(c) If the products are cooled to 100°F the partial pressure of the water vapor will be the saturation pressure for 100°F; this is 0.9492 psia, from the tables. Then neglecting the volume of any liquid water present, the mol fraction of water *vapor* in the gas is

$$x_w = \frac{p_w}{p} = \frac{0.9492}{15} = 0.0631$$

Let y be the pounds of water *vapor* per pound of mixture at 100°F; then the mols of water vapor per pound of mixture is $y/18$. Similarly the mols of each component will be the mass of that component divided by its molecular weight; then the total mols of all gaseous components of the mixture, per pound of mixture, is

$$\overset{CO_2}{\frac{0.184}{44}} + \overset{H_2O}{\frac{y}{18}} + \overset{O_2}{\frac{0.020}{32}} + \overset{N_2}{\frac{0.722}{28}} = 0.0346 + \frac{y}{18}$$

The mol fraction of water vapor may then be written

$$x_w = \frac{y/18}{0.0346 + (y/18)}$$

Setting equal the two expressions for x_w,

$$y = 0.0421 \, \frac{\text{lb water vapor}}{\text{lb mixture}}$$

The liquid water is the difference between the total water and the vapor; $0.074 - 0.042 = 0.032$ lb of liquid water per pound of products mixture. Since there are 17.15 lb of products per pound of fuel, the liquid water formed per pound of fuel is $17.15(0.032) = 0.53$ lb.

14-3 Products analysis. In an actual combustion process if no more than the theoretical air is supplied, the reaction will not be complete; the products will contain some carbon monoxide, unburned hydrocarbons or hydrogen. This is wasteful of fuel since, to obtain a given amount of useful energy from a combustion process, more fuel must be burned if the reactions are incomplete. Complete combustion of fuel can be obtained only by supplying *excess air*; the necessary excess depends upon the fuel and the apparatus used for combustion, and may vary in practice from a few percent to 50 percent or more of the theoretical air. It is undesirable to supply more air than necessary because the excess air, being taken in at a low temperature and discharged at a high temperature, represents a direct waste of energy.*

In order to determine the completeness of combustion and the amount of excess air actually used in a combustion process the actual products analysis is used. In Fig. 14-1 a general steady flow combustion process is represented diagrammatically. The products appear as flue gas and, in some cases, solid refuse. The flue gas composition may often be obtained by means of an Orsat gas analysis apparatus, in which the components of the gas are selectively absorbed by chemical solutions. The usual Orsat analysis gives the composition of the gas as fractions by volume of CO_2, CO, O_2, and N_2. The analysis is on a dry gas basis, moisture being accounted for by other methods. Small amounts of SO_2 are measured as CO_2 and are ignored in the calculations.

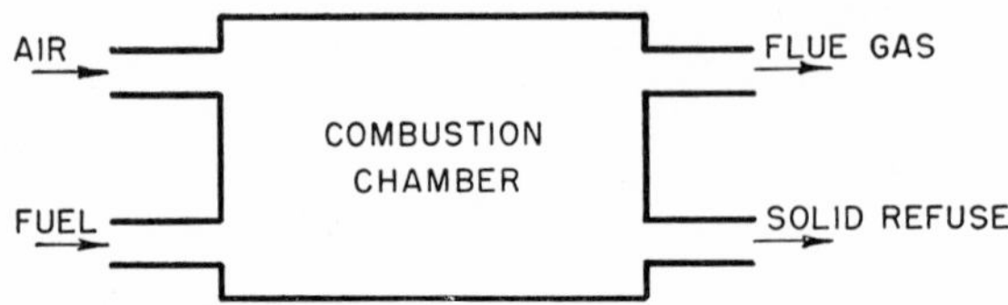

Fig. 14-1. Combustion process; steady flow diagram.

When solid refuse is appreciable as in coal burning furnaces, it is analyzed by "igniting" a sample, i.e. heating it above the ignition temperature in the presence of air for a long period. The loss of weight of the sample is measured, and is taken to be carbon in the refuse, while the residue is the ash in the refuse.

14-4 Energy equation for a chemical reaction. When a system undergoes a process involving a chemical reaction the First Law applies as for any other process. From the viewpoint of thermodynamics a chemical reaction is a change of phase in a system analogous to the physical phase change of vaporization. When one pound of carbon combines with 2.67 pounds of oxygen to form 3.67 pounds of carbon dioxide, the material involved is still the same system even

* It is sometimes necessary to use large amounts of excess air in order to limit the temperature of the products to a desired value; this happens in gas turbine power plants. On the other hand it is sometimes desirable to use less than theoretical air; this is done in internal combustion engines to obtain maximum power output (but not maximum efficiency).

though it has become a different chemical substance. It would be possible to construct tables of properties for such a system so that the internal energy change could be taken from the tables for a process starting with the reactants at any given pressure and temperature and ending with the products at any given pressure and temperature. This would be analogous to the tables for liquid and for superheated vapor in the steam tables. In general, however, such highly organized data are not available, and a step by step procedure is used as described below.

Consider a constant-volume reaction in the absence of motion; the First Law for this process may be written

$$Q = E_2 - E_1 \tag{1}$$

where E_2 is the internal energy of the products and E_1 is the internal energy of the reactants.* It is clear, however, that values of E_1 and E_2 taken from any arbitrary tables of data for the reactants and products would not necessarily satisfy the equation because the state of zero E might have been chosen arbitrarily for each substance and $E_2 - E_1$ could have any value whatsoever. In order to be used directly in this equation, the data for products and reactants must have a common base, just as the liquid and vapor tables for water are referred to the same base state. The relation between data on arbitrary bases and on a common base is illustrated in Fig. 14-2. At (a) is a plot of E vs. t at constant volume with the zero of E taken arbitrarily at zero t for both phases. For all processes which involve the products or the reactants *individually* this plot would be correct, but for any process involving a change from one phase to the other only one of the curves has an arbitrary zero point; the other must be shifted vertically to account for the property relationship between the two phases. The proper relative position of the two curves can be determined by an experiment in which the heat quantity of Eq. (1) is measured. As shown at (b) in Fig. 14-2, if states 1 and 2, and ΔE_{12} can be measured, the relative positions of the curves can be fixed.

In practice the experimental determination of the relation between the internal energies of products and reactants is carried out so that the final temperature is close to the initial temperature. The

* Under the conditions here imposed E is the E_U of Sec. 5-5, but for simplicity the subscript U will be omitted.

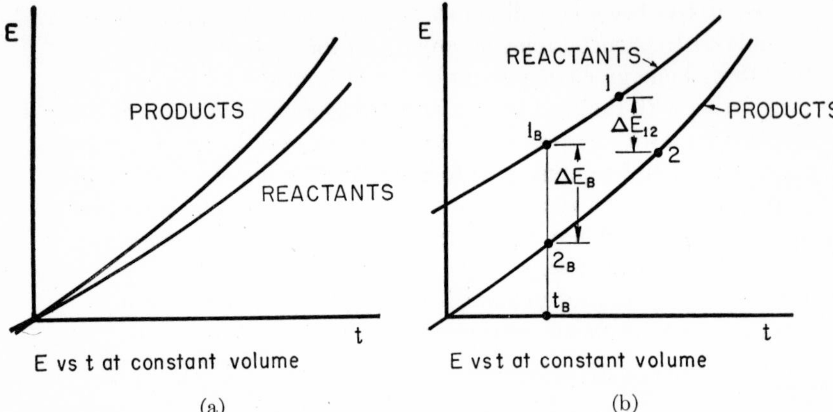

Fig. 14-2. *E–t* diagrams for a constant-volume process involving a chemical reaction.

internal energy change ΔE_B at a fixed base temperature t_B, Fig. 14-2, is called the *internal energy of reaction at* t_B. The determination is made by carrying out the reaction in a *bomb calorimeter*, Fig. 14-3, and measuring the heat transferred out.*

Given the internal energy of reaction at some fixed temperature, and the property relations for the reactants and products individually, it is easy to calculate the change of internal energy between *any* two states, 1 and 2, because the internal energy change is independent of the path. Taking a path composed of three steps, $1\text{–}1_B$, $1_B\text{–}2_B$, $2_B\text{–}2$, it follows that

$$E_2 - E_1 = (E_2 - E_{2B}) + (E_{2B} - E_{1B}) + (E_{1B} - E_1) \qquad (2)$$

$(E_2 - E_{2B})$ can be found from the properties of the products, $(E_{2B} - E_{1B})$ is the internal energy of reaction, and $(E_{1B} - E_1)$ can be found from the properties of the reactants.

Example 4. The internal energy of reaction for burning carbon to carbon dioxide at 68°F is −14,087 Btu/lb of carbon. Find the heat transferred when a system composed of 1 lb of carbon and 4 lb of oxygen at 300°F burns at

* In a constant-temperature process the internal energy of substances that do not react chemically is almost constant, assuming no phase changes occur. Therefore the internal energy of reaction depends only upon the reacting substances in the system, and is independent of the amount of non-reacting substances, such as excess air, in the system.

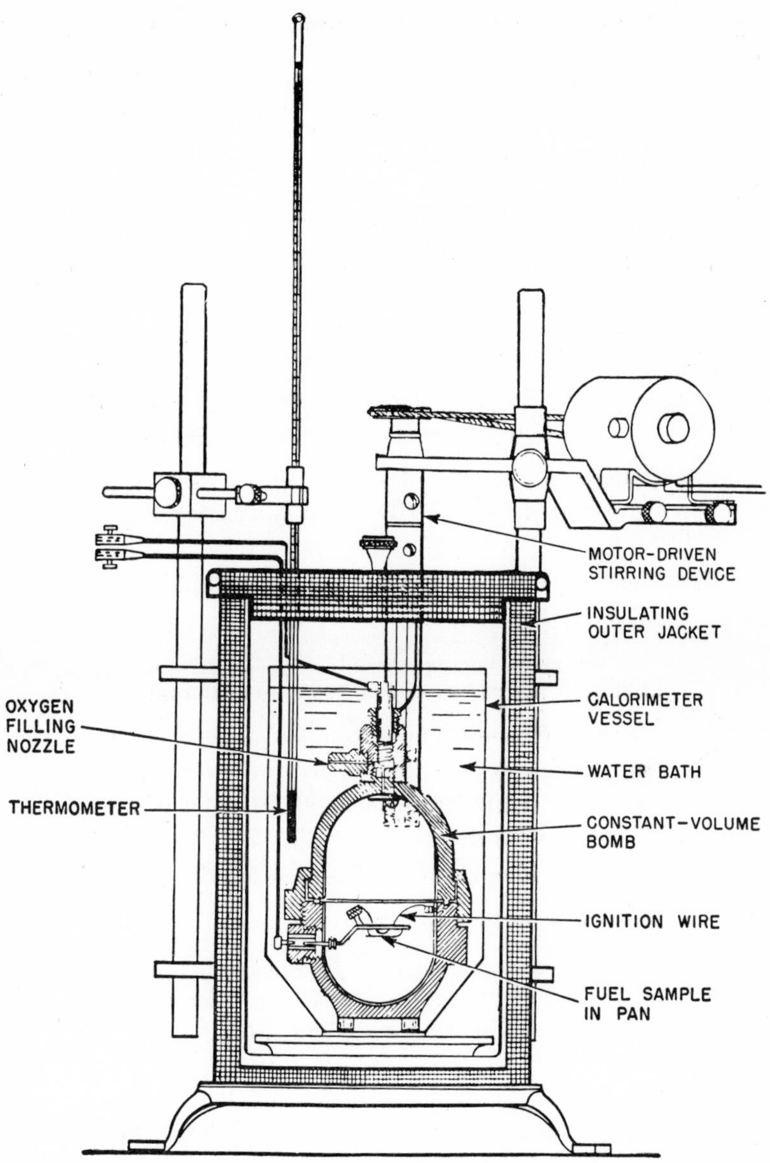

Fig. 14-3. Bomb calorimeter. A weighed sample of fuel is burned completely in oxygen and the small temperature rise of the water and the bomb is measured. From the measured temperature rise and the known heat capacity of the apparatus the internal energy of combustion may be calculated. (Courtesy Emerson Apparatus Company.)

constant volume to a mixture of CO_2 and O_2 at a final temperature of 1000°F. The specific heat of carbon is 0.17 Btu/lb F.

Solution: For a constant-volume process with no work the energy equation is

$$Q = E_2 - E_1 = (E_2 - E_{2B}) + (E_{2B} - E_{1B}) + (E_{1B} - E_1)$$

The material balance gives for the burning of carbon with oxygen as specified

$$1 \text{ lb C} + 4 \text{ lb } O_2 \rightarrow 3.67 \text{ lb } CO_2 + 1.33 \text{ lb } O_2$$

Taking the separate terms of the energy equation,

$$\begin{aligned} E_2 - E_{2B} &= (U_2 - U_{2B}) \quad \text{for} \quad O_2 + (U_2 - U_{2B}) \quad \text{for} \quad CO_2 \\ &= m_{O_2}(c_v)_{O_2}(t_2 - t_B) + m_{CO_2}(c_v)_{CO_2}(t_2 - t_B) \end{aligned}$$

From the material balance $m_{O_2} = 1.33$ lb and $m_{CO_2} = 3.67$ lb; taking the specific heats as $(c_v)_{O_2} = 0.155$ Btu/lb °F and $(c_v)_{CO_2} = 0.165$ Btu/lb °F, it follows that

$$\begin{aligned} E_2 - E_{2B} &= (1.33)(0.155)(1000 - 68) + (3.67)(0.165)(1000 - 68) \\ &= 755 \text{ Btu} \end{aligned}$$

$E_{2B} - E_{1B}$ is the given internal energy of reaction, −14,087 Btu/lb of carbon, multiplied by the pounds of carbon, or −14,087 Btu.

$$\begin{aligned} E_{1B} - E_1 &= (U_{1B} - U_1) \text{ for carbon} + (U_{1B} - U_1) \text{ for } O_2 \\ &= m_C(c)_C(t_B - t_1) + m_{O_2}(c_v)_{O_2}(t_B - t_1) \\ &= (1)(0.170)(68 - 300) + (4)(0.155)(68 - 300) \\ &= -183 \text{ Btu} \end{aligned}$$

Then

$$Q = 755 - 14{,}087 - 183 = -13{,}515 \text{ Btu}$$

This is a heat transfer *from* the system.

The relations between enthalpy of reactants and enthalpy of products are similar to the internal energy relations. The enthalpy change for a process involving a reaction may be written

$$\mathrm{H}_2 - \mathrm{H}_1 = (\mathrm{H}_2 - \mathrm{H}_{2B}) + (\mathrm{H}_{2B} - \mathrm{H}_{1B}) + (\mathrm{H}_{1B} - \mathrm{H}_1) \qquad (3)$$

where the terms are analogous to those of Eq. (2). The enthalpy relation is illustrated in Fig. 14-4.

The *enthalpy of reaction at* t_B $(\mathrm{H}_{2B} - \mathrm{H}_{1B})$ may, in principle, be found by carrying out a steady flow reaction at t_B, making shaft work, kinetic energy change, and potential energy change negligible, or by carrying out a constant-pressure reaction in a stationary system in which all work is $p\ dv$ work. In either case a measurement of the heat transfer will give the enthalpy change:

$$Q = \mathrm{H}_2 - \mathrm{H}_1 \qquad (4)$$

In most cases in practice the enthalpy of reaction is not measured directly but is computed from the measured internal energy of reaction. Gases are an exception; their enthalpies of reaction are often measured by a steady flow calorimeter process. The computation of enthalpy of reaction from internal energy of reaction is based upon the definition of enthalpy, Eq. (5-13), and the assumption that the gaseous reactants and products follow the perfect gas laws, while the solids and liquids involved have negligible volume. From (5-13) (noting that the E of this chapter is the E_U of Chap. 5), for any reaction,

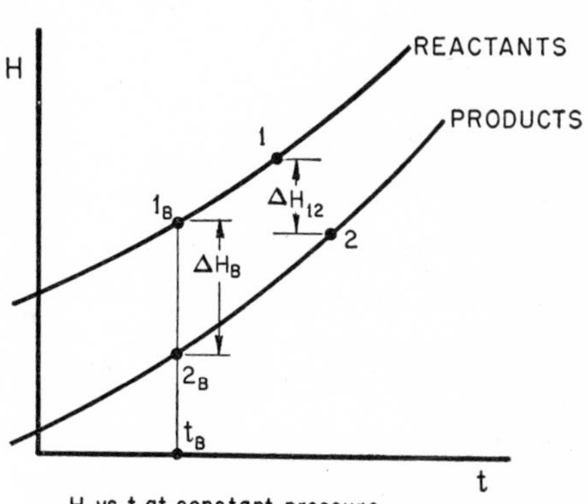

Fig. 14-4. H–t diagram for a constant-pressure process involving a chemical reaction.

$$\mathrm{H}_2 - \mathrm{H}_1 = E_2 - E_1 + p_2V_2 - p_1V_1 \qquad (5)$$

Let states 1 and 2 be the initial and final states of a reaction at t_B and at some constant pressure p; then

$$\mathrm{H}_2 - \mathrm{H}_1 = \Delta H_B$$

Now since the products follow the perfect gas law (or are solid or liquid), the internal energy of the products depends only upon temperature. E_2 is therefore the same at the end of the given constant-pressure reaction as it would have been at the end of a constant-volume reaction at the same temperature, and

$$E_2 - E_1 = \Delta E_B$$

This is true even though state 2 is not the state 2 of the constant-volume reaction by which ΔE_B was determined. To find a value for $(p_2V_2 - p_1V_1)$ observe that the perfect gas equation on the mol basis is independent of the chemical composition of the gas; thus for both the gaseous products and the gaseous reactants

$$pV = n\overline{R}T$$

At the base temperature T_B, if only the gaseous components have appreciable volume,

$$(p_2V_2 - p_1V_1)_B = \overline{R}T_B(n_2 - n_1)$$

where n_1 is the number of mols of *gaseous* reactants, and n_2 is the number of mols of *gaseous* products. Substituting in (5),

$$\Delta H_B = \Delta E_B + \overline{R}T_B(n_2 - n_1) \qquad (6)$$

The use of Eq. (6) is illustrated in Example 5 in the next section.

The terms *internal energy of reaction* and *enthalpy of reaction* have general application to reactions of any kind, but for special kinds of reaction special names are often used. *Enthalpy of formation* and *enthalpy of combustion* are used for the case of formation of a compound from its elements, and combustion of a fuel, respectively. The terms *constant-pressure heat of reaction* and *constant-pressure heating value* are used to signify the *negative* of the enthalpy of reaction. For example, the same information is conveyed by any one of the following statements:

(a) $C + O_2 \rightarrow CO_2$; $\Delta H = -94{,}052$ cal/gm formula weight.

(b) $C + O_2 \rightarrow CO_2$; constant-pressure heat of reaction is 94,052 cal/gm formula weight.

(c) The enthalpy of formation of CO_2 is $-94{,}052$ cal/gm mol.

(d) The constant-pressure heat of combustion (or heating value) of carbon, burned to CO_2, is 14,087 Btu/lb.

The units in the statements above are related as follows: 1 gm formula weight of carbon in the equations above is 12.010 gm, or 0.02648 lb; this is the mass of 1 gm mol, since the formula contains 1 mol of carbon. Since 94,052 cal/gm = 169,180 Btu/lb, 94,052 cal/gm formula weight = 14,087 Btu/lb.

Values of heat of combustion for some important substances are given in Table A-2.

14-5 Heating values of fuels. The heat of combustion or heating value of a fuel is generally given for a reaction starting at 1 atm and 20°C or 25°C. In tables of scientific data for reactions between definite chemical substances the temperature and pressure are always specified, but in engineering data on fuel samples these details are not usually given. Since heating values do not change much with pressure, nor with temperature in the range of room temperatures, the heating value is assumed to apply to any normal room condition.

It is necessary, however, to distinguish between different heating values which may be obtained for a single fuel, depending upon the conditions of the reaction. For a fuel containing no hydrogen there may be at least two heating values, the constant-pressure heating value ($-\Delta H_B$), and the constant-volume heating value ($-\Delta E_B$). These may sometimes be equal, as in the case of carbon. For the reaction $C + O_2 \rightarrow CO_2$ there is one mol of *gas* on each side of the equation; hence by Eq. (6) $\Delta H_B = \Delta E_B$ and the constant-volume heating value is equal to the constant-pressure heating value. For the combustion of carbon monoxide to carbon dioxide the two heating values are different.

Example 5. Given the reaction $2\,CO + O_2 \rightarrow 2\,CO_2$ and the constant-pressure heating value, 4,344 Btu/lb of CO, find the constant-volume heating value.

Solution: The mol basis, using Eq. (6), is most convenient. The two mols of CO contain 56 lb, so the heat of reaction is (56)(4,344) = 243,300 Btu/lb formula weight. There are 3 mols of gaseous reactants and 2 mols of gaseous products; then by Eq. (6), taking $R = 1.986$ Btu/lb mol °R, and taking T_B as 530°R,

$$\begin{aligned} \Delta E_B &= -243{,}300 - (1.986)(530)(2 - 3) \\ &= -243{,}300 + 1050 \\ &= -242{,}200 \text{ Btu/lb formula wt} \end{aligned}$$

Then the constant-volume heating value is 4,325 Btu/lb of CO.

For fuels containing hydrogen another variable is encountered in the fact that the water in the products may be either liquid or vapor; this may have an appreciable effect upon the final properties of the system. Suppose hydrogen is burned to water under such conditions that all the water vapor formed is condensed to liquid.* The chemist writes for this reaction at 77°F

$$H_2 + \tfrac{1}{2}\,O_2 \rightarrow H_2O(l); \quad \Delta H = -68{,}317 \text{ cal/gm formula wt}$$

* This is done in calorimetric practice by putting a drop of liquid water in the calorimeter bomb so that the space is already saturated with water vapor before the reaction, and any additional vapor formed must condense.

where the (l) signifies that the water is finally all liquid. The engineer computes from this that the heat of combustion of hydrogen to *liquid water* is 60,958 Btu/lb of hydrogen. This is called the *higher heating value* (HHV) because it includes the heat transferred out from the system as the result of the condensation of the H_2O, and is therefore larger than if the final state had been vapor. The difference is appreciable; each pound of hydrogen produces 9 pounds of water and at 77°F the enthalpy of vaporization is 1050.4 Btu/lb. Therefore if the reaction had ended with water *vapor* at 77°F, instead of liquid, the heat transferred out would have been reduced by 9 × 1050.4 or 9454 Btu/lb of hydrogen. The *lower heating value* (LHV) is given by

$$\text{LHV} = \text{HHV} - m_{\text{H}_2\text{O}} h_{fg} \tag{7}$$

where $m_{\text{H}_2\text{O}}$ is the mass of H_2O produced per pound of fuel, and h_{fg} is taken at the base temperature.* Then for this case

$$\text{LHV} = 60{,}958 - 9(1050.4) = 51{,}504 \text{ Btu/lb hydrogen}$$

It should be observed that the higher and lower heating values obtained above were both at constant pressure. It is possible also to determine higher and lower heating values at constant volume; for this case

$$(\text{LHV})_v = (\text{HHV})_v - m_{\text{H}_2\text{O}} u_{fg} \tag{8}$$

The higher heating value gives the maximum heat which could be transferred to constant-temperature surroundings from a given reaction. In practice, however, the heat of vaporization of the water in the products is not utilized, first, because serious corrosion problems usually arise if moisture is permitted to condense from flue gas and, second, because low temperature heat is of small thermodynamic value. Since in most actual processes the moisture will be vapor, it is convenient to use the lower heating value in computations involving the properties of the products. In experimental measurement of heating value, however, it is easier to insure complete condensation of the water than complete vaporization. Therefore the higher heating value is measured in the calorimeter, and the lower heating value is computed.

* In practice it is usual to take h_{fg} as some fixed arbitrary value such as 1050 Btu/lb.

The relation of higher and lower heating values to the efficiency of combustion processes is discussed in Sec. 14-7.

Some fuels may be supplied to a combustion process either as liquid or as gas; in such cases the heating value of the fuel is taken for liquid or gaseous fuel as the case may be. The two values will differ by the latent heat of vaporization of the fuel. For any ordinary fuel this is a smaller difference than that resulting from the latent heat of the water in the products. For example, Keenan and Kaye give for the enthalpy of combustion at 25°C of octane C_8H_{18} to gaseous products, −19,100 Btu/lb for liquid octance, and −19,256 Btu/lb for gaseous octane. These numbers (taken positive) are both lower heating values; the corresponding higher heating values are 1,490 Btu/lb greater, or 20,590 and 20,746 Btu/lb respectively.

Heating values for some typical fuels are given in Table A-2.

14-6 Temperature of products. The temperature of the products of combustion is important in the design and operation of combustion apparatus. In some cases the maximum possible temperature may be desired, but for power-plant purposes the problem is usually to limit the temperature of the products to some desirable range, neither too high nor too low. From the thermodynamic viewpoint high temperature is desirable, but in practice, temperature must be limited to avoid such effects as slag deposits in steam boilers or over-heating of metal parts of gas turbines.

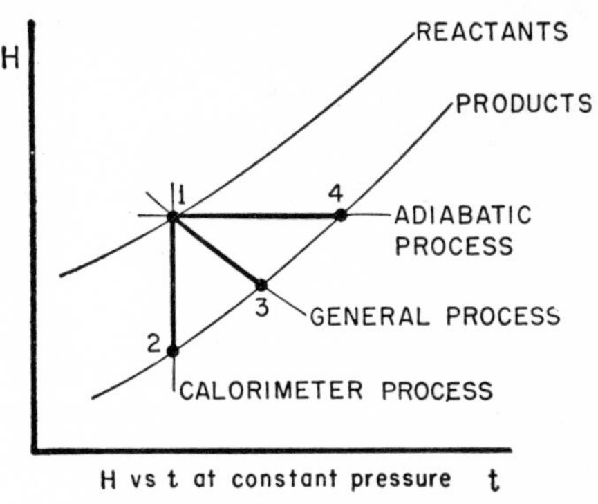

Fig. 14-5. Temperature change in a combustion process.

The temperature of the products of a combustion process in which there is no work other than $p\,dv$ work is a function of the reactions involved and the heat transferred. In Fig. 14-5 the temperature change and enthalpy change are plotted for a given reaction with three different amounts of heat transfer. For the process 1–2, the temperature is constant; this is the calorimeter type of process in which all the heat of combustion is transferred from the system. In the process 1–3, the heat transferred is less than the heat of combustion and there is a temperature rise in the system; this is the usual type of combustion process. In the process 1–4, there is no heat transfer and H_4 is equal to H_1; the temperature t_4 is called the temperature of adiabatic combustion. This temperature may be approached in uncooled combustion chambers for producing hot gas streams, such as the combustion chambers of gas turbine power plants or rockets.

The factors which influence the products temperature may be illustrated by an example.

Example 6. Assume carbon burns with air to carbon dioxide in a steady flow process, Fig. 14-6. If theoretical air is used, calculate the products temperature for adiabatic combustion, assuming the products have constant specific heats of room temperature magnitude.

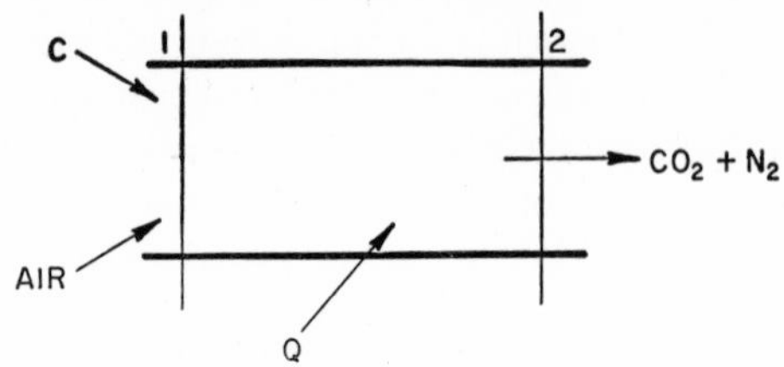

Fig. 14-6. Steady flow combustion process.

Solution:

$$C + O_2 + 3.76\ N_2 \rightarrow CO_2 + 3.76\ N_2$$

$$1\text{ lb} + 2.67\text{ lb} + 8.78\text{ lb} \rightarrow 3.67\text{ lb} + 8.78\text{ lb}$$

$$Q = H_2 - H_1 = H_2 - H_{2B} + H_{2B} - H_{1B} + H_{1B} - H_1$$

Assume that $t_1 = t_B = 77°F$; then $H_{1B} - H_1$ is zero and $H_{2B} - H_{1B}$ is $-14{,}087$ Btu/lb of carbon.

Then

$$Q = 0 = H_2 - H_{2B} - 14{,}087$$

$$H_2 - H_{2B} = 14{,}087$$

The enthalpy change of the products may also be written in terms of the specific heats and temperature change:

$$H_2 - H_{2B} = [(mc_p)_{CO_2} + (mc_p)_{N_2}](t_2 - t_B)$$

Then

$$t_2 - t_B = \frac{14{,}087}{(3.67)(0.196) + (8.78)(0.248)} = 4{,}900°F$$

Assuming t_B approximately 100°F, $t_2 = 5000°F$.

The temperature found above would never be reached in an actual process for several reasons.

Excess Air. In an actual process complete combustion will not be obtained without some excess air. This will increase the mass of products, thereby reducing the temperature rise for a given enthalpy of combustion.

Heat Transfer. In an actual process the heat transfer will not be zero.

Variation of Specific Heats. The specific heats of the gases are greater at higher temperatures, giving a smaller temperature rise than calculated above. This may be shown by taking average specific heat values for the conditions of Example 6 from Fig. A-4, p. 391; taking for carbon dioxide $c_p = 0.300$, and for nitrogen 0.285,

$$t_2 - t_B = \frac{14{,}087}{(3.67)(0.300) + 8.78(0.285)} = 3{,}910°\text{F}$$

Then

$$t_2 = 4000°\text{F}$$

Chemical Equilibrium or *Dissociation.* The combustion reaction does not go to completion at high temperatures. The composition of the products tends to approach a certain equilibrium state which depends upon the relative quantities of the various chemical elements present and upon the temperature and pressure. At low temperatures the equilibrium condition for carbon and oxygen, if sufficient oxygen is supplied, is carbon dioxide. At high temperatures the equilibrium state is a mixture of carbon dioxide, carbon monoxide, and oxygen. The higher the temperature (and the lower the pressure) the greater is the fraction of carbon monoxide for equilibrium. If sufficient oxygen is supplied to form carbon dioxide, there will be negligible carbon monoxide at equilibrium for temperatures below 3500°F. At 5000°F and 1 atm there will be approximately as many mols of carbon monoxide as of carbon dioxide. The heat of reaction for carbon burned to carbon monoxide is only about 28 percent of that for carbon burned to carbon dioxide; hence for temperatures above 3500°F the temperature actually reached in the combustion of carbon will be lower than that calculated on the assumption of complete combustion. At temperatures above 4500°F the combustion of hydrogen begins to be appreciably incomplete.

The term *dissociation* is often used in connection with equilibrium problems, but it has a rather restricted implication which may be misleading.

Products of combustion that are dissociated at a high temperature may recombine as the temperature drops, provided sufficient time is available before the temperature becomes too low for the maintenance of combustion. The time needed depends upon the amount of excess oxygen present and upon the degree of mixing of the gases, as well as upon the pressure, since all these factors influence the probability of the molecules of oxygen meeting the molecules of carbon monoxide, for example.

In steam boiler furnaces it is usually practical to provide sufficient time and oxygen to obtain substantially complete combustion before the gases leave the furnace and enter the relatively cold tube banks of the boiler. Provided combustion is finally complete, the existence of dissociation in the hot zones of a boiler furnace does not reduce the available energy from the combustion, since the energy is later to be transferred as heat to relatively low-temperature steam. The thermodynamic advantage of the high temperatures reached in the furnace would be wasted by the irreversible heat transfer, regardless of dissociation. In an internal combustion engine, however, some work would be lost through dissociation even if combustion were finally complete, because the reduction of the maximum temperature of the gases would reduce the work available from their expansion.

14-7 Efficiency of processes involving combustion. The efficiency of processes involving combustion may be defined in many different ways, depending upon the purposes of the processes and the arbitrary definitions of input and output.

Considering the combustion process alone, without regard to heat or work effects, an efficiency may be based upon the fraction of complete combustion obtained.

Example 7. Suppose one pound of carbon burns at constant pressure so that 0.9 pound goes into CO_2, 0.05 lb goes into CO, and 0.05 lb emerges as unburned carbon (these fractions are determined by a products analysis); find the efficiency of the combustion process.

Solution: For complete combustion of the pound of carbon the heat of reaction is 14,087 Btu; for combustion of 0.9 lb of carbon to CO_2 the heat of reaction is 0.9(14,087) or 12,678 Btu; for combustion of 0.05 lb of carbon to CO the heat of reaction is 0.05(3952) or 198 Btu. Then the efficiency on the basis here considered is $(12{,}678 + 198)/14{,}087$ or 0.91.

In reciprocating internal combustion engines the purpose of combustion is the production of work by a process which is not easily broken down into separate parts; therefore it is convenient to define an efficiency for the process as a whole. In this case the numerator of the efficiency ratio may be taken as the work output of the engine per unit of fuel supplied, while the denominator may be taken as the heat of reaction for complete combustion of the fuel. For example, if an engine consumes 0.5 lb of fuel per horsepower-hour of work output, and the heating value of the fuel is 19,000 Btu/lb, the efficiency is 2545/(0.5)(19,000), or 0.257.

In steam boiler furnaces the purpose of the combustion process is to supply heat to water; the efficiency of the overall process in a steam generator is defined as the ratio of the heat transferred to the water divided by the heating value of the fuel.

Example 8. A steam generator produces 10,000 lb/hr of steam at 150 psia, saturated vapor. In addition there is drawn off from the boiler 500 lb/hr of saturated liquid at 150 psia as "blow down" (this water carries with it the dissolved solids which would otherwise accumulate and precipitate as scale in the boiler). The feed water is supplied to the boiler at 210°F. Oil fuel having a heating value of 19,500 Btu per pound is burned in the furnace at the rate of 700 lb/hr. Find the efficiency of the steam generator, as defined above.

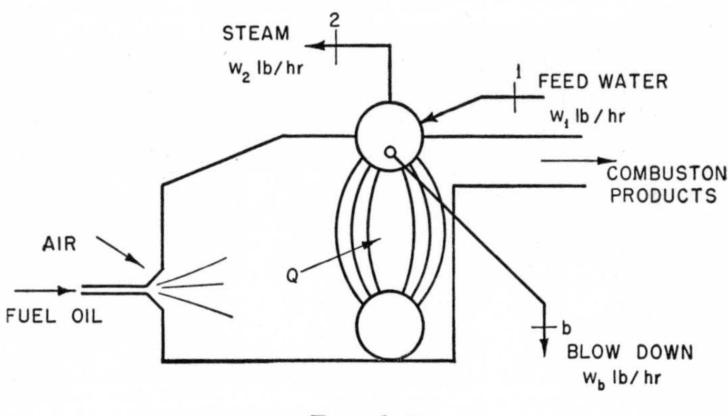

Example 8.

Solution: The heat transferred to the water is found by applying the steady flow energy equation and mass equation; referring to the sketch,

$$w_1 = w_2 + w_b$$
$$w_1h_1 + Q = w_2h_2 + w_bh_b$$

Substituting values from the data and the steam tables

$$Q = 10{,}240{,}000 \text{ Btu/hr}$$

The heating value of the fuel burned, on an hourly basis, is

$$(700)(19{,}500) = 13{,}650{,}000 \text{ Btu/hr}$$
$$\text{efficiency} = 1024/1365 = 0.75$$

In the discussion above no consideration was given to the question of which of the various heating values should be used as a basis for input in efficiency computations. This choice is arbitrary, since an efficiency is simply a comparison to an arbitrary standard. Since most technical combustion processes are associated with operations which are essentially of the steady flow type, it is customary to use constant-pressure heating values, which represent enthalpy changes. This applies even to engines in which the combustion process proper is a constant-volume process, since from the external viewpoint such engines operate on the steady flow basis. Another choice remains, in the case of fuels containing hydrogen, of using either the higher or lower heating value; thus two different efficiency values may be obtained for the same process.

In steam power plants in the United States the higher heating value is used as a basis for efficiency because it is the experimentally measured heating value and it gives the lower value of efficiency. Internal combustion plant efficiencies and European steam plant efficiencies are often based on lower heating values, which give larger numbers for the same performance. Because of such variations in practice, quoted efficiency values should not be compared indiscriminately.

PROBLEMS*

14-1. (a) Find the "theoretical" quantity of oxygen for burning 1 lb of acetylene, C_2H_2. (b) Find the analysis by mass of the products of complete combustion of acetylene with 125 percent theoretical oxygen.

14-2. A gaseous fuel consists of a mixture of methane, CH_4, and carbon monoxide, CO, in equal parts by volume. Find the theoretical air, lb per lb of fuel, and the analysis by mass of the products of complete combustion of the fuel with 110 percent theoretical air.

14-3. A power plant burns coal having an analysis as follows: C, 71.04; H, 5.28; O, 6.72; S, 4.96; N, 1.34; *A*, 10.66; in percent by mass. The combustion equipment will operate efficiently with 20 percent excess air. For this case find the pounds of air per pound of fuel, and the analysis by mass of the

* Heats of reaction at 25°C, 1 atm, are given in Appendix Table A-2.

flue gas, assuming combustion is complete and everything but the ash appears in the flue gas.

14-4. For the combustion process of Problem 14-3 find the saturation temperature of the water vapor in the products for a total pressure of 15 psia. (The actual dew point of the flue gas will be higher than determined here; because of sulfur compounds in the gas the condensate will not be pure water but will be acid, which has a higher saturation temperature.)

14-5. Carbon monoxide is burned completely with theoretical oxygen in steady flow. The reactants are supplied at 400°F, 20 psia, and the products leave at 600°F, 15 psia. Velocities are negligible and there is no shaft work. Assuming all the gases are perfect gases find the heat transfer, Btu/lb of CO, for the process.

14-6. Hydrogen is burned with 50 percent excess air in steady flow at a pressure of 1 atm. Dry air and hydrogen are supplied at 77°F; the products leave at 77°F. Velocities are negligible and there is no shaft work. Find: (a) the pounds of *liquid* water in the products per pound of hydrogen; (b) the heat transfer per pound of hydrogen.

14-7. For the fuel of Problem 14-2 find the constant-pressure heat of combustion, Btu/lb of fuel, at 77°F, (a) if the water in the products is liquid, and (b) if the water in the products is vapor.

14-8. Hydrogen peroxide, H_2O_2, is to be used as a source of oxygen for a power plant which is to be independent of an air supply. The following reaction will take place:

$$C + 2\,H_2O_2\ (\text{liq}) \rightarrow 2\,H_2O + CO_2$$

Enthalpies of formation, calories per gram formula weight, at 68°F:

$$H_2O_2\ (\text{liq})\ -44{,}500;\ H_2O\ (\text{liq})\ -68{,}320;\ CO_2\ (\text{gas})\ -94{,}050.$$

Find: (a) the pounds of hydrogen peroxide needed per pound of carbon; (b) the enthalpy of reaction, Btu per lb of carbon, at 68°F, for the reaction, if the water in the products is all liquid.

14-9. The coal of Problem 14-3 has a higher heating value of 13,025 Btu/lb. Find its lower heating value.

14-10. According to US Bureau of Standards *Circular C*461, methane, CH_4, has a constant-pressure higher heating value at 25°C of 23,861 Btu/lb. Assuming that all gases involved are perfect gases, and that liquid water has negligible volume, find for methane at 25°C: (a) the constant-pressure lower heating value; (b) the constant-volume higher heating value; (c) the constant-volume lower heating value.

14-11. The enthalpy of combustion of octane vapor, C_8H_{18}, at 25°C, is −20,747 Btu/lb when all the water in the products is liquid. The enthalpy of vaporization of octane at 25°C is 156 Btu/lb. In a certain process octane, supplied as liquid at 25°C, is burned completely with 400 percent theoretical air supplied at 25°C. There is no heat transfer, but work is transferred out. If the gaseous products of combustion are discharged at 500°F, how much work is done per pound of octane burned?

14-12. A certain fuel oil is 84 percent carbon and 16 percent hydrogen by mass, and the constant-pressure heating value (to gaseous products) is 19,000 Btu/lb at 70°F. The oil is burned in an adiabatic gas-turbine combustion chamber with sufficient air so that the products temperature is 1450°F. Air and oil are both supplied at 70°F. Using average specific heats from the plots in the Appendix, find the air-fuel ratio required if combustion is complete.

14-13. The fuel of Problem 14-12 is burned completely with 125 percent theoretical air, and liquid water at 70°F is then injected into the combustion products so that the final temperature of the mixture, after the water evaporates, is 1450°F. How much water must be injected per pound of fuel?

14-14. A diesel engine uses 29 lb of fuel per hour when the brake output is 75 hp. If the heating value of the fuel is 19,700 Btu/lb what is the brake thermal efficiency of the engine?

REFERENCES

Barnard, W. N., F. O. Ellenwood, and C. F. Hirshfeld, *Heat Power Engineering*, Vol. II. New York: Wiley, 1935.

de Lorenzi, O., ed., *Combustion Engineering*. New York: Combustion Engineering-Superheater, Inc., 1948.

Griswold, J., *Fuels, Furnaces and Combustion*. New York: McGraw-Hill, 1946.

Chapter 15

GAS CYCLES

As pointed out in the introduction to Chap. 13, no cyclic heat engines using gas for working fluid have yet attained great importance in practice. The study of gas cycles is important, however, for two reasons: (1) considerable insight into the characteristics of non-cyclic gas power plants of the greatest practical importance (internal combustion engines and turbines) can be obtained from the study of gas cycles; (2) as a result of improved materials and designs the prospects for the practical use of cyclic gas power plants seem better now than at any time in the past.

15-1 Hot-air engines. The earliest gas engines were cyclic hot-air engines in which a fixed mass of air in the engine was alternately heated by a fire and cooled by circulating water. This was accomplished by keeping one part of the engine in contact with the fire, and another part in contact with the water, and transferring the air periodically from one place to the other. It was found, however, that the transfer of heat to the gas raised two serious obstacles: (1) it was difficult to transfer heat to a gas at high rates, so the engine speeds were limited, and consequently the power output was small for a given size of engine; (2) the metal through which the heat was transferred deteriorated rapidly since its temperature had to be much higher than the maximum gas temperature. If the metal temperature were kept low, not only would the efficiency be reduced but the rate of heat transfer would also be reduced, thereby reducing the capacity (power output) as well. Improvements in heat-resisting metals and in the design of heat exchangers have in recent years made possible greatly improved hot-air engines. Since hot-air engines can use cheap fuels, they may eventually find a field of use where the cost of internal combustion engine fuels is excessive.

15-2 Air-standard cycles. Historically, the obstacles to the effective use of the hot-air engine were overcome by producing high temperature through a combustion process directly in the working fluid (internal combustion). In this way the gas can reach a high temperature while the metal of the cylinder can be cooled by external air or water circulation in order to avoid rapid deterioration. Moreover, it is easy to raise the gas temperature very rapidly so that the engine can operate at high speed and have a large power output capacity for its size.

When the internal combustion process is examined, it is seen to differ from the process of heat supply in a heat engine cycle because there is a permanent change in the working fluid during combustion. Therefore the fluid does not pass through a cycle and the internal combustion engine is not a thermodynamically cyclic heat engine. There is, in fact, no heat supplied to the engine, so it cannot be a heat engine in the sense of Sec. 7-2. Nevertheless a heat engine cycle can be constructed to correspond approximately to the operation of an internal combustion engine by substitution of heat transfer processes for some of the actual engine processes. Such cycles, called *air-standard cycles*, are useful in the elementary study of internal combustion engines. More detailed study requires analysis of combustion processes.

15-3 The Otto cycle (constant-volume cycle). One of the most commonly encountered types of internal combustion engine, used in automobiles and aircraft, is the *spark-ignition* gasoline engine, for which the air-standard cycle is the *Otto* cycle. The sequence of processes in the elementary operation of such an engine is given below, with reference to the sketch and indicator diagram, Fig. 15-1.

Process 1–2, *intake*. With the inlet valve open the piston moves to increase the cylinder volume, taking in fuel-air mixture at constant pressure.

Process 2–3, *compression*. With the valves closed the piston returns, compressing the combustible mixture to the minimum cylinder volume.

Process 3–4, *combustion*. At the minimum cylinder volume a spark ignites the mixture, which burns with resulting increase in temperature and pressure.

Process 4–5, *expansion*. The piston moves out to a maximum cylinder volume, the average pressure being greater than in the compression process.

Process 5–6, *blow-down*. At the maximum cylinder volume the

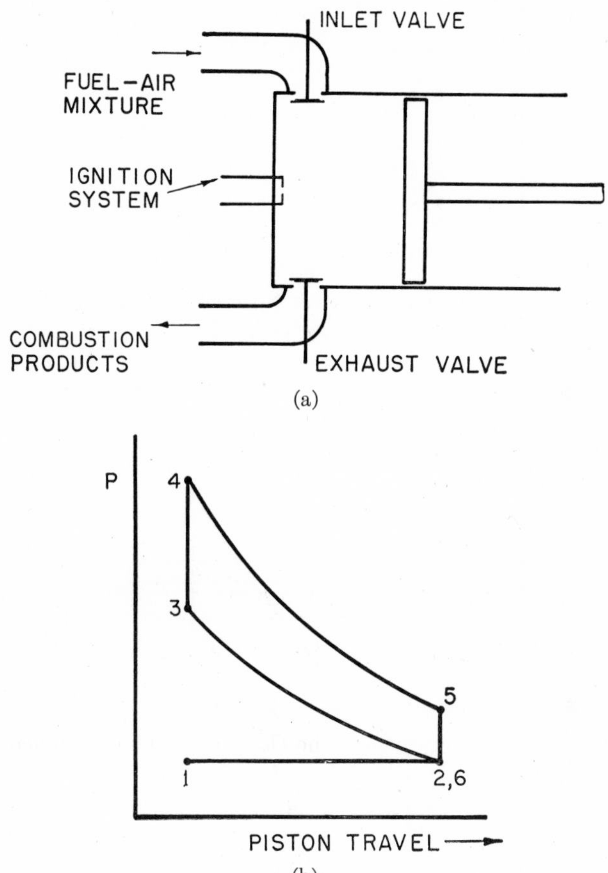

Fig. 15-1. (a) Spark ignition engine. (b) Indicator diagram.

exhaust valve opens and fluid leaves the cylinder until the pressure drops to the initial pressure, p_1.

Process 6–1, *exhaust*. With the exhaust valve open, the piston moves to expel the combustion products from the cylinder at constant pressure.

While the series of processes described above constitutes a mechanical cycle, it is not a thermodynamic cycle. The designation *four-stroke cycle*, commonly applied to such a sequence, refers to the mechanical action.

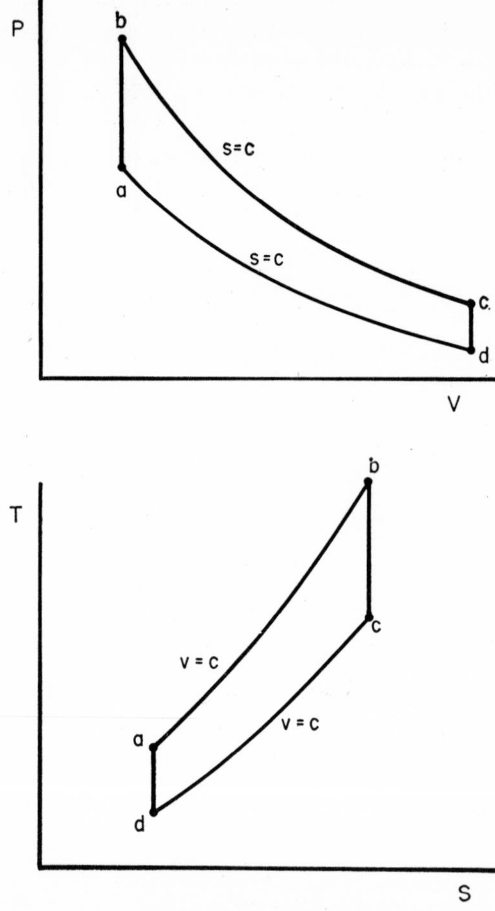

Fig. 15-2. Otto cycle.

Figure 15-2 shows the air-standard cycle corresponding to the engine of Fig. 15-1 (the Otto cycle). It consists of the following processes, carried out with a perfect gas system composed of air:

Process *a–b*, heat supplied at constant volume, reversibly.

Process *b–c*, reversible adiabatic expansion.

Process *c–d*, heat rejected at constant volume, reversibly.

Process *d–a*, reversible adiabatic compression.

Heat transfer processes have been substituted for the combustion and blow-down processes of the engine; the intake and exhaust processes of the engine are not needed in the cycle, for they would merely cancel each other. Note the distinction between the indicator diagram of Fig. 15-1 and the state diagram of Fig. 15-2; the former refers to pressure and volume in a cylinder containing varying masses of gas, while the latter refers to the properties of a fixed mass of air.

The efficiency analysis of the air-standard Otto cycle for a system of mass m follows.

Heat supplied $\quad Q_1 = Q_{a-b} = mc_v(T_b - T_a) \quad (1)$

Heat rejected $\quad Q_2 = -Q_{c-d} = mc_v(T_c - T_d) \quad (2)$

Efficiency $\quad \eta = \dfrac{W_{\text{net}}}{Q_1} = 1 - \dfrac{Q_2}{Q_1} = 1 - \dfrac{mc_v(T_c - T_d)}{mc_v(T_b - T_a)}$

$$\eta = 1 - \frac{T_c - T_d}{T_b - T_a} \tag{3}$$

This can be simplified still further by the relations among the temperatures. Since d–a and b–c are reversible adiabatic processes it follows that

$$\Delta s_{a-b} = \Delta s_{d-c}$$

or

$$mc_v \ln \frac{T_b}{T_a} = mc_v \ln \frac{T_c}{T_d}$$

or

$$\frac{T_b}{T_a} = \frac{T_c}{T_d}$$

Then

$$\frac{T_b - T_a}{T_a} = \frac{T_c - T_d}{T_d}$$

or, from (3),

$$\eta = 1 - \frac{T_d}{T_a} \tag{4}$$

Since d–a is a reversible adiabatic process, (4) can be rewritten

$$\eta = 1 - \left(\frac{V_a}{V_d}\right)^{k-1} \tag{5}$$

For the cycle of Fig. 15-2 or the engine of Fig. 15-1 the compression ratio r_k is defined as the ratio of the volume at the beginning of compression to the volume at the end of compression;

$$r_k = \frac{V_d}{V_a} \tag{6}$$

Then

$$\eta = 1 - \frac{1}{r_k^{k-1}} \tag{7}$$

Equation (7) shows that efficiency increases with increasing compression ratio. The equation applies strictly only to an ideal cycle using a perfect gas, but it serves to indicate within limits the trend of efficiency variation in actual engines. As an example, Fig. 15-3 shows the relation between air-standard efficiency and the brake thermal efficiency of a certain engine when the compression ratio was changed, other conditions being constant. Brake thermal efficiency is the ratio of the work output of the engine to the heating value of the fuel it burns.

A somewhat closer approach to actual engine conditions may be

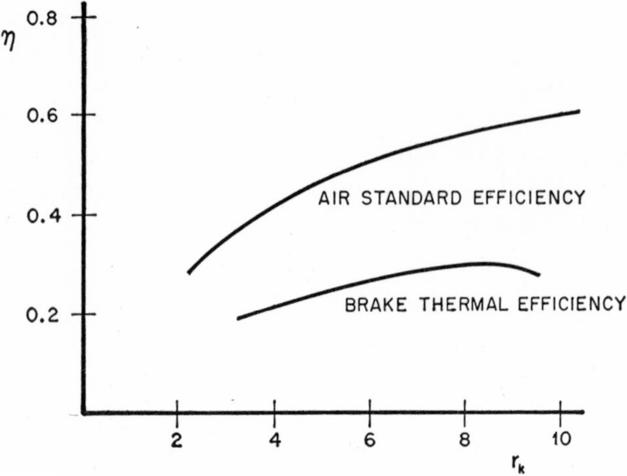

Fig. 15-3. Otto cycle efficiency.

made by taking account of the variation of the specific heat of air at high temperature, using the *variable specific-heat* Otto cycle. In this case the heat transferred in a reversible constant-volume process is given by the change of internal energy, which is most conveniently obtained from the Gas Tables; then referring to Fig. 15-2,

$$\eta = 1 - \frac{Q_2}{Q_1} = 1 - \frac{u_c - u_d}{u_b - u_a} \tag{8}$$

In the case of the air-standard cycle the efficiency is independent of the temperature level at which the cycle operates, and of the rate of working, but this is not true for the variable specific-heat cycle. In the latter case efficiency is a function of compression ratio, temperature at beginning of compression, and heat supplied per unit mass of working fluid.

In general, as pointed out above, the efficiency of an actual engine working on the Otto cycle increases with increased compression ratio; however, there is a limit to the compression ratio that can be used in any particular case in practice. In a given engine under given operating conditions, the compression ratio may not be increased beyond a certain limit without encountering *detonation*, a noisy and destructive combustion phenomenon ("knocking"). The avoidance of detonation requires different values of compression ratio depending upon the fuel, the engine design, and the operating conditions, but it fixes some limit on the efficiency of all ordinary gasoline engines.

15-4 The Diesel cycle (constant-pressure cycle). One method of avoiding the limitation on compression ratio in the spark-ignition engine is by compressing air alone, and injecting the fuel into the cylinder only when combustion is desired. By this method the air can be raised to a temperature higher than the ignition temperature of the fuel so that no spark is necessary to start combustion. The fuel can begin to burn spontaneously, soon after it is sprayed into the hot air, and the rate of burning can be controlled to some extent by the rate of injection of the fuel. An engine operating in this way is called a compression-ignition engine.

The classical air-standard cycle for the compression-ignition engine is the Diesel cycle, Fig. 15-4. In this cycle the processes are:

Process *a–b*, heat supplied at constant pressure, reversibly.

Process *b–c*, reversible adiabatic expansion.

Process *c–d*, heat rejected at constant volume, reversibly.

Process *d–a*, reversible adiabatic compression.

The efficiency analysis of the cycle for a system of mass m follows.

Heat supplied $\quad Q_1 = Q_{a-b} = mc_p(T_b - T_a)$

Heat rejected $\quad Q_2 = -Q_{c-d} = mc_v(T_c - T_d)$

Efficiency $$\eta = \frac{W_{\text{net}}}{Q_1} = 1 - \frac{Q_2}{Q_1} = 1 - \frac{mc_v(T_c - T_d)}{mc_p(T_b - T_a)}$$

$$\eta = 1 - \frac{T_c - T_d}{k(T_b - T_a)} \tag{9}$$

The efficiency of the air-standard Diesel cycle may be expressed in terms of volume ratios but not so simply as in the case of the Otto

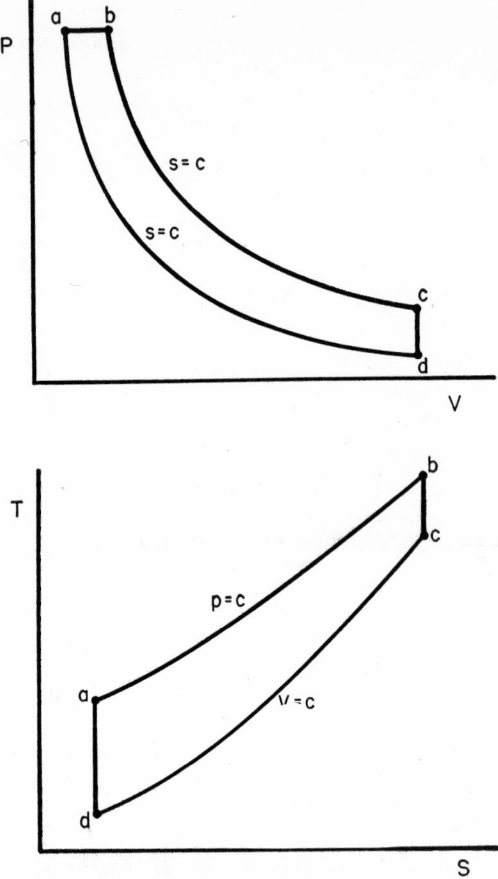

Fig. 15-4. Diesel cycle.

cycle. The efficiency of the Diesel cycle depends upon the amount of heat supplied as well as upon the compression ratio, for as the *T–s* diagram shows, the ratio of heat supplied to heat rejected decreases as the line *b–c* moves toward greater entropy. Thus the efficiency may be expressed in terms of any two of the three ratios below.

$$\text{Compression ratio, } r_k, = \frac{V_d}{V_a}$$

$$\text{Expansion ratio, } r_e, \quad = \frac{V_c}{V_b}$$

$$\text{Cut-off ratio, } r_c, \quad = \frac{V_b}{V_a}$$

Probably the most enlightening form of efficiency expression is in terms of r_k and r_c, as follows:

$$\eta = 1 - \frac{1}{r_k^{k-1}} \frac{r_c^k - 1}{k(r_c - 1)} \tag{10}$$

This equation gives a family of curves on the plane of efficiency vs. compression ratio, Fig. 15-5. The curve for unity cut-off ratio

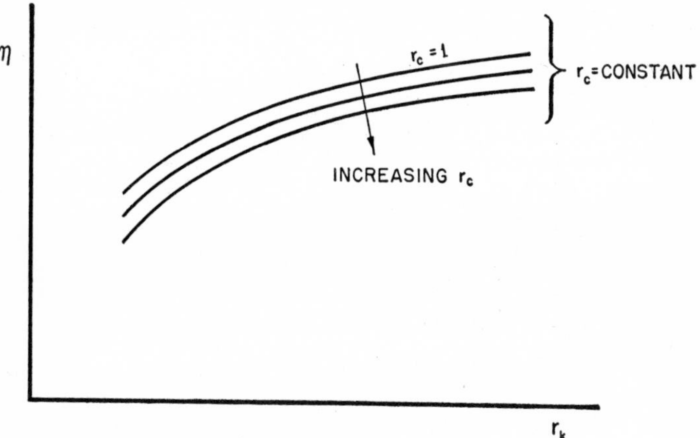

Fig. 15-5. Diesel cycle efficiency.

(zero heat supply) is the same as the Otto cycle curve, Fig. 15-3; with increasing cut-off ratio the curves become lower. Thus, for the same compression ratio at zero load, the Diesel cycle is as efficient as the Otto cycle, but it is less efficient at normal loads. However, engines using the Diesel cycle usually operate at higher compression ratios than engines using the Otto cycle, so the resulting efficiencies under operating conditions are not necessarily much different for the two types.

The variable specific-heat Diesel cycle efficiency may be seen by inspection of the cycle diagram to be

$$\eta = 1 - \frac{u_c - u_d}{h_b - h} \tag{11}$$

15-5 The mixed cycle (limited pressure cycle). The operation of modern compression-ignition engines is better represented by a mixed cycle than by the Diesel cycle. In the mixed cycle the heat is supplied partly in a constant-volume process and partly in a constant-pressure process. In the actual engine, fuel injection is started before the end of the compression stroke so that a portion of the combustion is completed while the piston is reversing its motion and not moving rapidly. This causes a pressure rise and combustion is completed at approximately constant pressure by the continuing

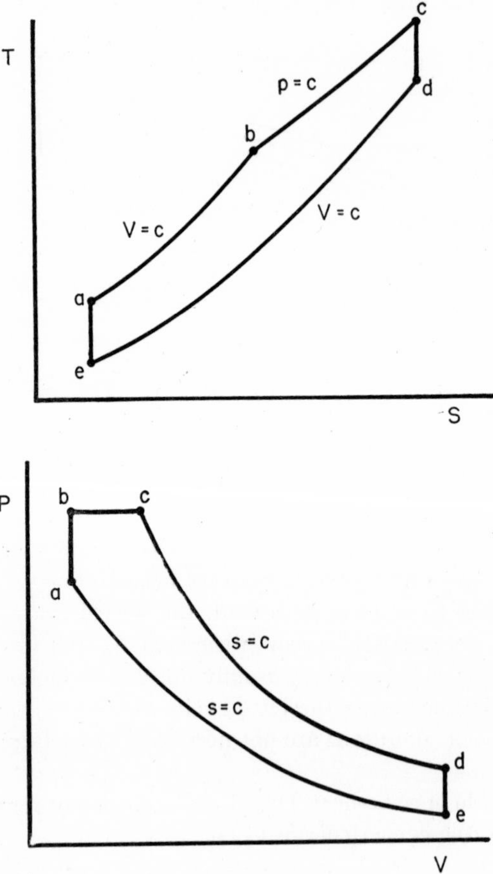

Fig. 15-6. Mixed cycle.

injection of fuel during the first part of the outward piston travel. The air-standard mixed cycle is shown in Fig. 15-6.

The efficiency of the air-standard mixed cycle is given by

$$\eta = 1 - \frac{Q_2}{Q_1} = 1 - \frac{mc_v(T_d - T_e)}{mc_v(T_b - T_a) + mc_p(T_c - T_b)}$$

$$= 1 - \frac{T_d - T_e}{T_b - T_a + k(T_c - T_b)} \tag{12}$$

By substitutions as in the preceding sections

$$\eta = 1 - \frac{1}{r_k^{k-1}} \frac{r_p r_c^k - 1}{r_p - 1 + k r_p (r_c - 1)} \tag{13}$$

where r_k and r_c are V_e/V_a and V_c/V_b respectively, and r_p is the constant-volume pressure ratio p_b/p_a. For any given compression ratio the efficiency of the mixed cycle is intermediate between that of the Otto cycle and that of the Diesel cycle, depending upon the ratio of the heat supplied at constant volume to that supplied at constant pressure.

15-6 The Brayton cycle. The Brayton cycle is an air-standard cycle composed of two reversible constant-pressure processes and two reversible adiabatic processes as shown in Fig. 15-7. A reciprocating engine operating on this cycle and patented in the 1870's by George B. Brayton, was the first successful gas engine built in the United States.

The efficiency of the Brayton cycle is given by

$$\eta = 1 - \frac{Q_2}{Q_1} = 1 - \frac{mc_p(T_c - T_d)}{mc_p(T_b - T_a)}$$

By the same reasoning as for the Otto cycle, Sec. 15-3, this reduces to

$$\eta = 1 - \frac{T_d}{T_a} = 1 - \left(\frac{p_d}{p_a}\right)^{(k-1)/k} = 1 - \left(\frac{V_a}{V_d}\right)^{k-1} = 1 - \frac{1}{r_k^{k-1}} \tag{14}$$

where r_k is the adiabatic compression ratio. Thus the efficiency of the Brayton cycle is the same as that of the Otto cycle for the same adiabatic compression ratio.

The Brayton cycle differs from the Otto cycle of the same adiabatic compression ratio and work output, in having a larger range of volume and smaller ranges of pressure and temperature. The larger volume of low-pressure gas cannot be handled efficiently in a reciprocating engine because of increased friction losses in the mechanism and in the exhaust process; also the Brayton engine is more bulky

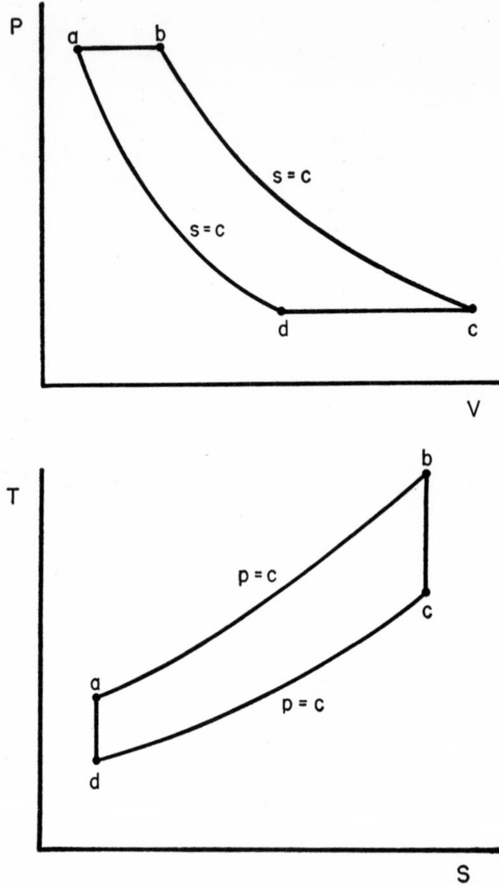

Fig. 15-7. Brayton cycle.

than the Otto engine for the same power capacity. Therefore the Brayton cycle cannot compete with the Otto cycle in the reciprocating engine field. For turbine machinery, however, the Brayton cycle is much better adapted than the Otto cycle because turbine machinery can handle large volumes of gas more efficiently than small volumes. Moreover, in steady flow machinery it is much easier to carry out heat transfer or combustion processes at constant pressure than at constant volume; also the turbine is not adapted to as high maximum gas temperatures and pressures as is the reciprocating engine. For these reasons the Brayton cycle is the basic air-standard cycle for all modern gas turbine plants, whether of the truly cyclic type or of the internal combustion type.

A flow diagram for a cyclic gas turbine plant using the Brayton cycle is shown in Fig. 15-8; this arrangement is popularly called a "closed cycle" gas turbine plant. In Fig. 15-8 the points a, b, c, and d correspond to the identically marked points in the state diagrams, Fig. 15-7. The compressor work W_C is taken from the turbine

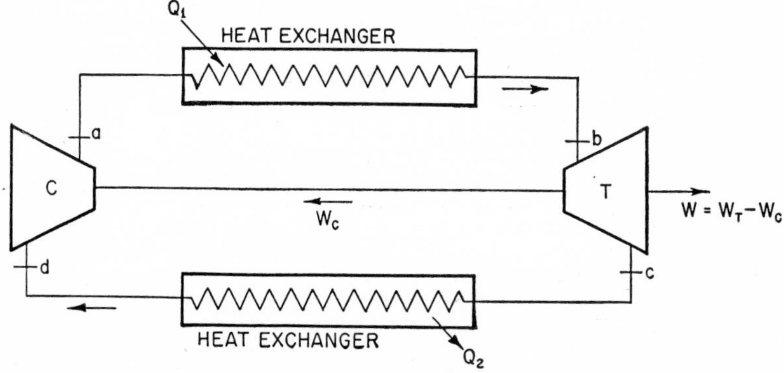

Fig. 15-8. Gas-turbine plant flow diagram.

directly to the compressor, leaving the net work W available for use. By the steady flow analysis

$$Q_1 = h_b - h_a$$
$$W_T = h_b - h_c$$
$$Q_2 = h_c - h_d$$
$$W_C = h_a - h_d$$
$$W_{\text{net}} = W_T - W_C$$
$$\eta = \frac{W_{\text{net}}}{Q_1} = \frac{h_b - h_c - h_a + h_d}{h_b - h_a}$$
$$\eta = 1 - \frac{h_c - h_d}{h_b - h_a} \tag{15}$$

The enthalpies may be obtained from the gas tables or by use of the perfect gas rules.

Example 1. Find the efficiency, and the work per pound of fluid circulated, for a Brayton cycle working between pressures of 15 psia and 75 psia, if the minimum temperature in the cycle is 550°R and the maximum temperature is 1700°R.

(a) **Gas law solution:**

From Fig. 15-7 the minimum temperature occurs at d and the maximum at b. The temperature ratios for the reversible adiabatic processes are given by

$$\frac{T_a}{T_d} = \left(\frac{p_a}{p_d}\right)^{(k-1)/k}, \qquad \frac{T_b}{T_c} = \left(\frac{p_b}{p_c}\right)^{(k-1)/k}$$

$$\frac{p_a}{p_d} = \frac{p_b}{p_c} = \frac{75}{15} = 5, \qquad k = 1.4$$

$$\frac{T_a}{T_d} = \frac{T_b}{T_c} = 1.58, \qquad T_a = (550)(1.58) = 870°\text{R}$$

$$T_c = \frac{1700}{1.58} = 1075°\text{R}$$

$$Q_1 = c_p(T_b - T_a) = (0.24)(1700 - 870) = 199 \text{ Btu/lb}$$

$$Q_2 = c_p(T_c - T_d) = (0.24)(1075 - 550) = 126 \text{ Btu/lb}$$

$$W_{\text{net}} = Q_1 - Q_2 = 73 \text{ Btu/lb}$$

$$\eta = \frac{W_{\text{net}}}{Q_1} = 0.367$$

Also $$\eta = 1 - \frac{T_d}{T_a} = 1 - \frac{700}{1110} = 1 - 0.63 = 0.37$$

(b) **Gas table solution:**

In the tables, for reversible adiabatic processes,

$$\frac{p_a}{p_d} = \frac{p_{ra}}{p_{rd}}, \qquad \frac{p_b}{p_c} = \frac{p_{rb}}{p_{rc}}$$

At 550°R, $p_{rd} = 1.4779$; $h_d = 131.46$ Btu/lb

At 1700°R, $p_{rb} = 90.95$; $h_b = 422.59$ Btu/lb, Table 1, Keenan and Kaye

Then $p_{ra} = 5(1.478) = 7.390$; $T_a = 868°$R; $h_a = 208.41$

$p_{rc} = 90.95/5 = 18.19$; $T_c = 1113°$R; $h_c = 269.27$

$$Q_1 = h_b - h_a = 214.18 \text{ Btu/lb}$$

$$Q_2 = h_c - h_d = 137.81 \text{ Btu/lb}$$

$$W_{\text{net}} = Q_1 - Q_2 = 76.37 \text{ Btu/lb}$$

$$\eta = \frac{W_{\text{net}}}{Q_1} = 0.357$$

The efficiency of the Brayton cycle may sometimes be increased by the use of a *regenerator* or *heat exchanger* to transfer heat from the turbine exhaust gas to the gas leaving the compressor, thereby reducing the heat supplied from an external source. Such a cycle is illustrated in Fig. 15-9. In this case the gas leaving the turbine at *c* has higher temperature than the gas leaving the compressor at *a*. Consequently if the two streams are passed through a heat exchanger, the temperature of the gas leaving the compressor can be raised by heat transfer from the turbine exhaust gas. The maximum temperature to which the cold gas can be heated is the temperature of the

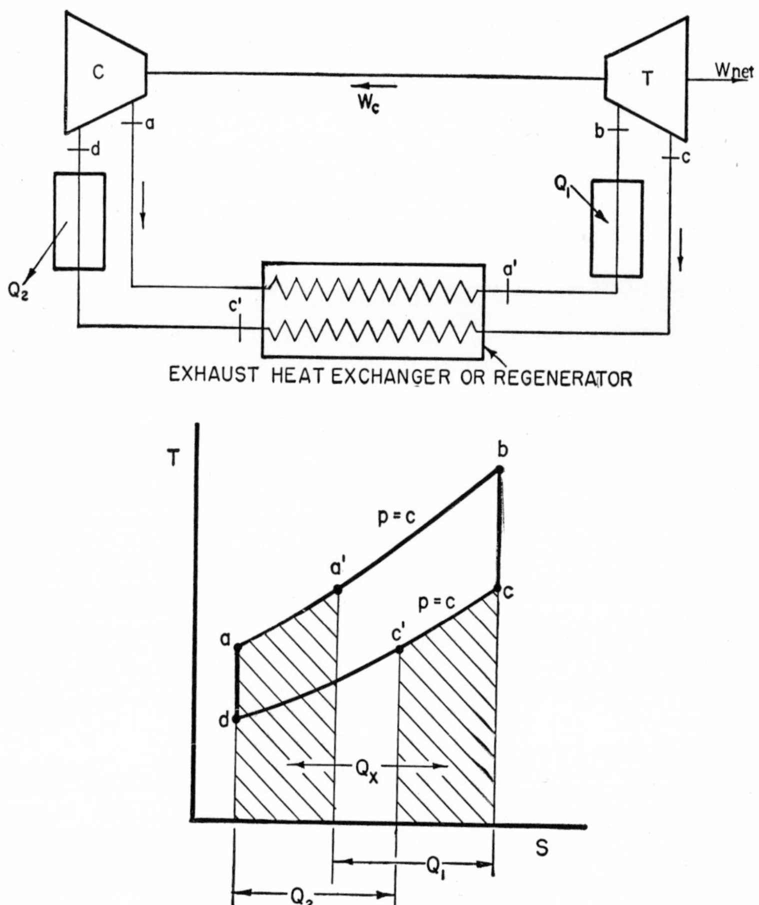

Fig. 15-9. Gas-turbine plant with regenerator.

hot gas entering the exchanger; this could be done only with an infinitely large heat exchanger operating reversibly. In a real case the temperature at a', Fig. 15-9, cannot reach the temperature at c; the fraction of the maximum possible temperature rise actually obtained is called the *effectiveness* (sometimes the *efficiency*) of the regenerator. For the case illustrated,

$$\text{effectiveness} = \frac{t'_a - t_a}{t_c - t_a} \tag{16}$$

As indicated in Fig. 15-9, when the regenerator is used in the idealized cycle, the heat supplied and the heat rejected are each reduced by the same amount Q_x hence the work of the cycle is unchanged but the efficiency is increased. The heat and work quantities are as follows:

$$Q_1 = h_b - h'_a, \qquad Q_2 = h'_c - h_d$$
$$W_T = h_b - h_c, \qquad W_C = h_a - h_d$$
$$W_{\text{net}} = W_T - W_C = Q_1 - Q_2$$

The effect of the regenerator upon the cycle is best shown by an example.

Example 2. Repeat Example 1 for a cycle using a regenerator of 75 percent effectiveness.

Solution: The states at a, b, c, and d, Fig. 15-9, will be the same as in Example 1. The state at a' is determined by Eq. (16).

$$0.75 = \frac{T'_a - 870}{1075 - 870}, \qquad T'_a = 1024°\text{R}$$

T'_c is determined by the energy balance for the regenerator.

$$c_p(T'_a - T_a) = c_p(T_c - T'_c)$$
$$1024 - 870 = 1075 - T'_c$$
$$T'_c = 921°\text{R}$$
$$Q_1 = c_p(T_b - T'_a) = 162 \text{ Btu/lb}$$
$$Q_2 = c_p(T'_c - T_d) = 89 \text{ Btu/lb}$$
$$W_{\text{net}} = 73 \text{ Btu/lb}$$
$$\eta = \frac{W_{\text{net}}}{Q_1} = 0.45$$

Compare this with 0.367 for the same cycle without the regenerator.

In practice the regenerator is costly, heavy, and bulky, and causes pressure losses which reduce the efficiency of the cycle. These factors have to be balanced against the ideal efficiency gain to decide whether it is worthwhile to use the regenerator.

If the turbine and compressor of the Brayton cycle are real (irreversible) machines the states a and c will have to be located by use of the machine efficiencies together with the isentropic end states a_s and c_s, Fig. 15-10. In the simple Brayton cycle the use of real machines will reduce the heat supplied by the amount $h_a - h_{as}$, and will increase the heat rejected by $h_c - h_{cs}$. Thus the net work of the cycle will be reduced by the sum, $h_a - h_{as} + h_c - h_{cs}$. The effects of using machines of 80 percent efficiency in Examples 1 and 2, above, are tabulated below, using air table solutions.

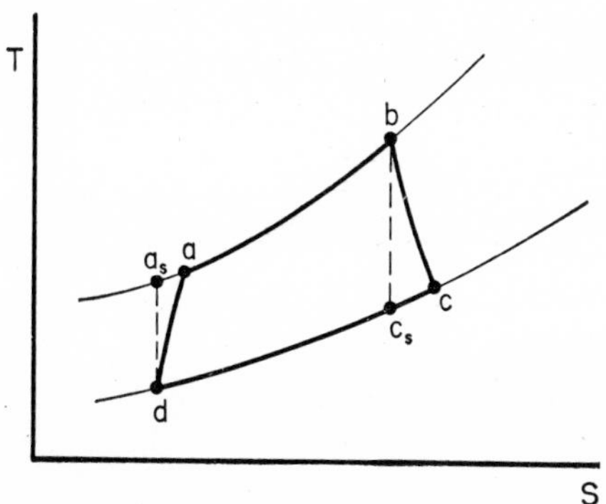

Fig. 15-10. Brayton cycle with real machines.

	Example 1		*Example 2*	
Machine Efficiency	1.00	0.80	1.00	0.80
W_T	153	123	153	123
W_C	77	96	77	96
W_{net}	76	27	76	27
Q_1	214	195	169	142
η	0.36	0.13	0.45	0.19

Example 1 had no regenerator, Example 2, a regenerator of 75 percent effectiveness.

The evident sensitivity of the cycle efficiency to machine efficiency occurs because of the large back work or compressor work. Because of its large back work, the gas turbine plant must use machines having a total power rating of five to ten times the net cycle power rating, and the efficiency of these machines must be very high to obtain a satisfactory plant efficiency. In fact the cycle efficiency may go to zero with machines efficiencies of 60 to 70 percent for cycle conditions that give good reversible efficiencies. With a vapor cycle a machine efficiency of 60 to 70 percent would give a cycle efficiency of almost 60 to 70 percent of the reversible cycle efficiency. These facts, and the problem of temperature, discussed in the following section, show why the gas turbine plant has been so far behind the steam plant in its development.

15-7 Efficiency and capacity characteristics of the Brayton cycle. Although the efficiency of the Brayton cycle may be expressed in terms of pressure ratio, Eq. (14), there are other influences that require consideration if the characteristics of the cycle in its practical application are to be understood. The following discussion refers only to design characteristics.

Referring to Fig. 15-10 if T_d, the lowest temperature of the cycle, is the temperature of the surroundings, the maximum efficiency attainable will be limited by T_b, the highest temperature of the cycle. In a turbine plant T_b is limited by the characteristics of metals available for burner and turbine construction. In aircraft turbines of short life the maximum temperature may be of the order of 1800°F, while for long-life power plants the limit is currently from 1400° to 1500°F. These limits may be extended in the future by improved metals or methods of cooling, but there will always be a definite limit for any particular combination of design, materials, and life expectancy. The reason turbine plants are much more sharply limited in this respect than reciprocating engines is that the hot parts of the steady flow machine are constantly exposed to hot gases, with no intervals of contact with a cool gas.

Assuming that the temperature limits for an ideal cycle are fixed as shown in Fig. 15-11, the cycle shape will change as the pressure ratio changes. Cycles of low pressure ratio (*a–b–c–d*) have low efficiency, since the average temperature of heat supply is little greater than the average temperature of heat rejection; moreover the work capacity is small, as shown by the area enclosed in the diagram. At the lower limit of pressure ratio (unity) both work and efficiency become zero.

As the pressure ratio is increased, the efficiency steadily increases because the average temperature of heat supply steadily approaches

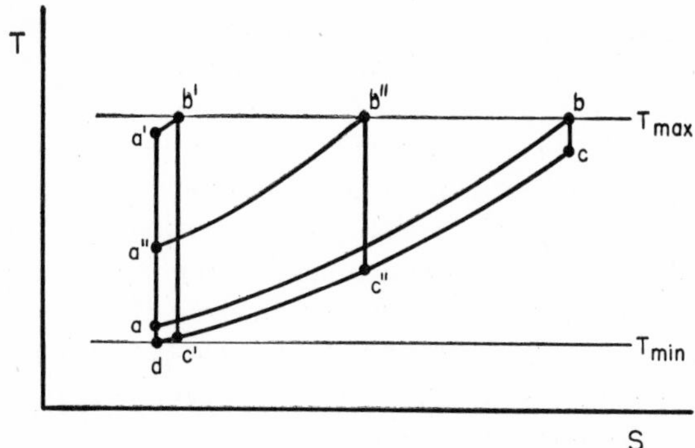

Fig. 15-11. Brayton cycles of various pressure ratios.

T_{max} while the average temperature of heat rejection steadily approaches T_{min}. In the limit, when the compression process ends at T_{max}, the Carnot efficiency is reached. The work capacity, however, passes through a maximum as the pressure ratio increases, and when the Carnot efficiency is reached the work output is zero. This can happen because the heat supplied approaches zero simultaneously with the work. Considering cycles a'–b'–c'–d and a''–b''–c''–d, Fig. 15-11, it is obvious that to obtain reasonable work capacity some reduction of efficiency must be accepted.

If the cycle uses real (irreversible) machines, the efficiency will

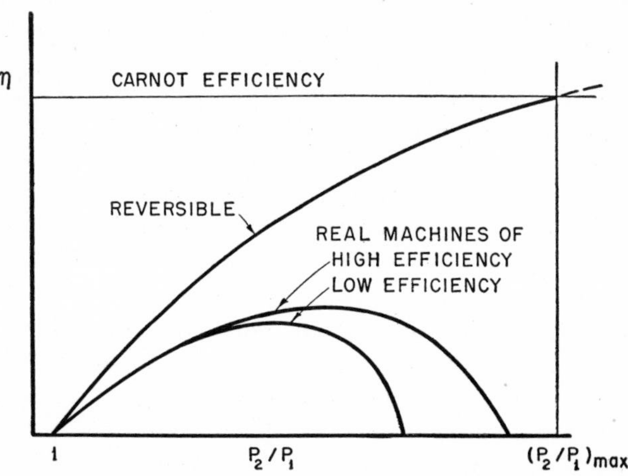

Fig. 15-12. Effect of pressure ratio on efficiency; simple Brayton cycle.

not increase continuously to the maximum pressure ratio, because some energy is used to supply machine losses. The net work output will then go to zero before the heat supplied reaches zero. Figure 15-12 shows how the efficiency will vary with pressure ratio at fixed maximum temperature for the cases of reversible machines and irreversible machines. In the latter case, operation at maximum efficiency will not necessarily involve much loss of capacity.

15-8 More complex gas turbine cycles. The efficiency or the work capacity of a gas turbine plant may often be increased by using *staged compression with intercooling*, or by using staged heat supply, called *reheat*. The first scheme is illustrated in Fig. 15-13. It is clear that the staged compression with intercooling increases the work of

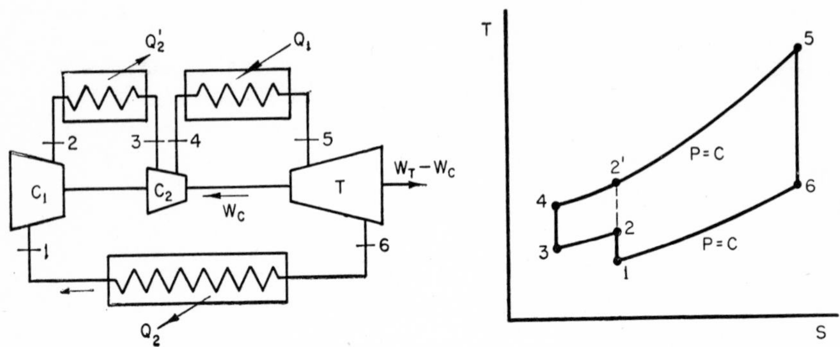

Fig. 15-13. Cycle with compressor intercooling.

the cycle by the area 2–3–4–2′, but the additional work involves a corresponding additional heat supply from 4 to 2′. Since this heat is supplied at a lower temperature than any heat supplied in the basic cycle, the intercooling results in a reduced efficiency for the cycle. However, when a regenerator is used, intercooling can give a net gain in efficiency of the reversible cycle because the additional low-temperature heat supply comes from further cooling of the exhaust gas rather than from an external source. With a real compressor intercooling may so reduce the back work as to result in improved cycle efficiency even without a regenerator.*

A cycle using reheat is shown in Fig. 15-14. Here it can be seen that there is a gain in work, area 4–5–6–4′, which is obtained from

* See Chap. 17 for details on gas compression with intercooling.

high-temperature heat; however, the small Brayton cycle 4–5–6–4′ added to the basic cycle has a smaller pressure ratio than the basic cycle, and therefore a smaller efficiency. Thus as in the case of intercooling, the simple cycle efficiency is reduced by reheating. With a regenerator the cycle efficiency may be improved by reheating, because the higher temperature exhaust permits greater heat recovery and smaller external heat supply for the basic part of the cycle. With real machines the increased work may give increased efficiency through reduction in the relative back work.

Although the simple flow diagrams presented here show only a single shaft with all compressors and turbines on that shaft, it is often desirable for operating convenience and part-load efficiency to divide

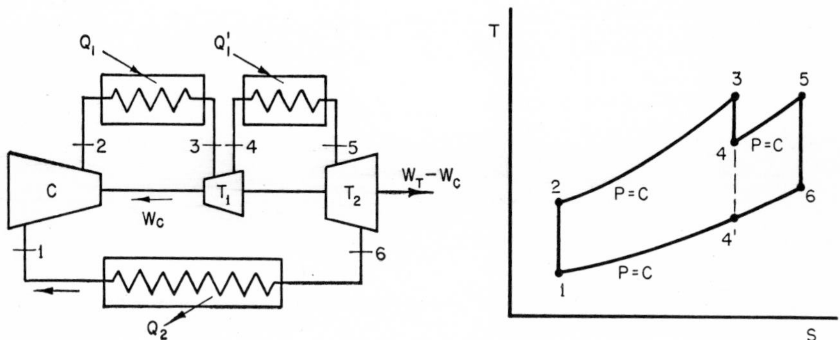

Fig. 15-14. Reheat gas cycle.

the machinery between two shafts. For example, there may be a turbine driving the compressor on one shaft, and a separate turbine driving the load on another shaft. The detailed study of such arrangements belongs in the field of plant design rather than in that of thermodynamics.

15-9 The combustion gas turbine plant. In the preceding sections cyclic gas-turbine plants were discussed. Their general characteristics apply to non-cyclic internal combustion plants such as shown in Fig. 15-15, but the analysis of the combustion-plant process differs in detail from that of the cycle because of the introduction of the combustion process in place of a heat exchanger. The fact that the plant process is "open," that it takes in air from atmosphere and discharges to atmosphere, would not in itself affect the analysis.

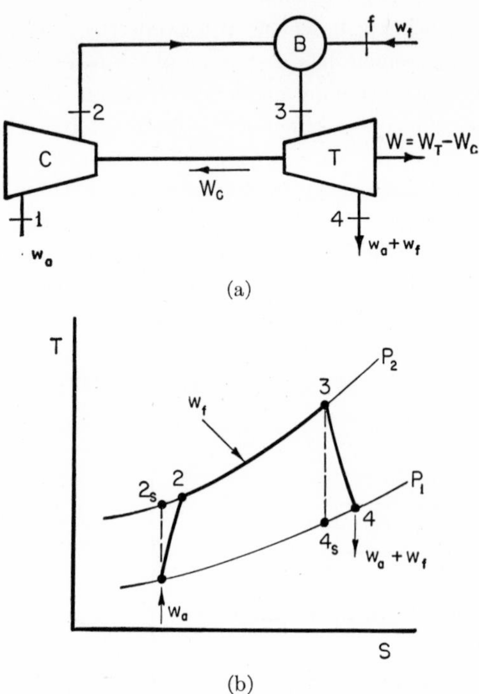

Fig. 15-15. Internal combustion turbine plant. (a) Flow diagram; (b) state plot.

In the simple plant of Fig. 15-15 air is taken into the compressor C from atmosphere, state 1, at flow rate w_a lb/sec. After compression it goes to the burner B where it burns with fuel supplied at flow rate w_f lb/sec. The products stream then expands through the turbine T which exhausts to atmosphere. The burner outlet temperature T_3 is controlled to suit the turbine requirements by regulating the fuel-air ratio. As brought out in Chap. 14, adiabatic combustion of a chemically correct mixture of an ordinary fuel will give a temperature of the order of 3500°F, so only a fraction of theoretical fuel is needed to reach permissible turbine inlet temperatures.

The energy analysis for the simple plant is as follows, based upon the steady flow energy equation.

Compressor. Assuming an adiabatic machine of efficiency η_c,

$$W_c = w_a(h_2 - h_1) = \frac{w_a}{\eta_c}(h_{2s} - h_1)$$

Burner. For an adiabatic combustion process

$$Q = 0 = (w_a + w_f)(h_3 - h_{3B}) + w_a(h_{2B} - h_2) + w_f(h_{fB} - h_f) + w_f(\Delta h_B)$$

or

$$\frac{w_f}{w_a} = \frac{h_{3B} - h_3 + h_2 - h_{2B}}{h_3 - h_{3B} + h_{fB} - h_f + (\Delta h_B)}$$

where the enthalpy differences are defined as follows:

$(h_3 - h_{3B})$ = enthalpy change of products between T_B and T_3

$(h_{2B} - h_2)$ = enthalpy change of air between T_2 and T_B

$(h_{fB} - h_f)$ = enthalpy change of fuel between its inlet state f and its state at T_B; this may sometimes include a latent heat.

(Δh_B) = enthalpy of combustion of fuel at T_B, Btu/lb

Turbine. Assuming an adiabatic turbine of efficiency η_T,

$$W_T = (w_a + w_f)(h_3 - h_4) = (w_a + w_f)\eta_T(h_3 - h_{4s})$$

The same analysis applies if the pressure drops in a real piping system and a real burner are taken into account as indicated in Fig. 15-16.

The properties of working fluids for gas turbine plants may be obtained with good accuracy from the gas tables. For product mixtures the products tables may be suitable in some cases, but often the fuel-air ratio is so low that the air tables are suitable, and in any practical case the air tables will give a good approximation.

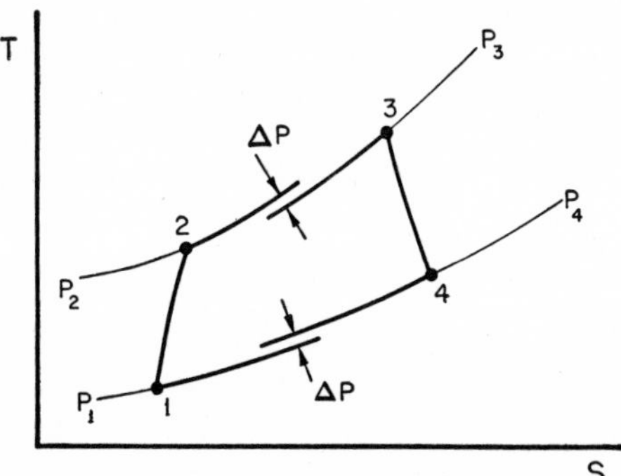

Fig. 15-16. State plot showing pressure losses.

Example 3. Find the efficiency, air rate (lb/hphr), and back work ratio W_c/W_{net} for a gas turbine plant of the following description: flow diagram Fig. 15-15; inlet air temperature 60°F; inlet pressure, $p_1 = p_4 = 15$ psia; pressure ratio $p_2/p_1 = 6$; machine efficiencies of compressor and turbine both 80 percent; fuel, $C_{16}H_{30}$; lower heating value 18,500 Btu/lb; turbine inlet temperature 1450°F; heat losses and pressure losses negligible.

Solution: Use air tables; for the compressor at 60°F = 520°R,

$$h_1 = 124.27; \quad p_{r1} = 1.2147$$

$$p_{r2} = \frac{p_2}{p_1} p_{r1} = 6.073$$

At this value of p_r, $h_{2s} = 197.5$

$$h_2 = h_1 + \frac{h_{2s} - h_1}{\eta_c} = 215.8 \text{ Btu/lb}$$

For the combustion process,

$$(\Delta h_B) = -18{,}500 \text{ Btu/lb}$$

Assuming that the fuel is supplied at T_B, and that T_B is 520°R,

$$h_{fB} - h_f = 0$$

For the air supply, since $T_B = 520°\text{R}$,

$$h_{2B} - h_2 = 124.3 - 215.8 = -91.5$$

Assuming the products properties are the same as those of air, at

$$T_3 = 1910°\text{R}, \quad h_3 = 479.85, \quad p_{r3} = 144.53$$

$$h_3 - h_{3B} = 479.85 - 124.3 = 355.5$$

$$\frac{w_f}{w_a} = \frac{-355.5 + 91.5}{355.5 + 0 - 18{,}500} = 0.0146 \text{ lb fuel/lb air}$$

This is about 470 percent theoretical air, so the products differ little from air.

For the turbine

$$p_{r4} = \frac{p_4}{p_3} p_{r3} = 28.91$$

at this value of p_r, $h_{4s} = 306.9$

$$h_4 = h_3 - \eta_T(h_3 - h_{4s}) = 341.5 \text{ Btu/lb}$$

$$W_T = (1 + 0.0146)(479.85 - 341.5) = 140.4 \text{ Btu/lb air}$$

$$W_c = (1)(215.8 - 124.3) = 91.5 \text{ Btu/lb air}$$

$$W_{\text{net}} = 48.9 \text{ Btu/lb air}$$

$$\eta = \frac{W_{\text{net}}}{(w_f/w_a)(\Delta h_B)} = \frac{48.9}{(0.0146)(18{,}500)} = 0.181$$

$$\text{air rate} = \frac{2545}{48.9} = 52.1 \text{ lb air/hphr}$$

$$\text{back work ratio } \frac{W_c}{W_{\text{net}}} = \frac{91.5}{48.9} = 1.87$$

Internal combustion plants are built with intercooling and reheat as described in Sec. 15-8 for cyclic plants; the analysis of such plants follows the same pattern as that of the simple plant. In the case of the reheat plant the total fuel burned may approach much closer to the theoretical fuel than in the simple plant, so the air tables may not be as suitable for the properties of the products of the reheat combustion as for the properties in the simple process.

Another type of plant, the jet propulsion plant for aircraft, is

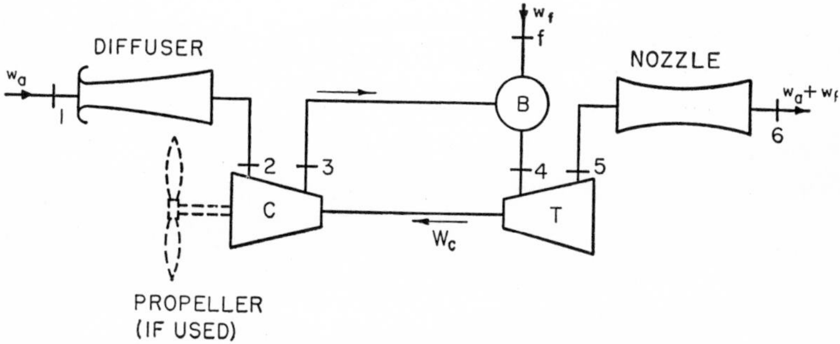

Fig. 15-17. Jet-propulsion plant flow diagram.

shown in the flow diagram, Fig. 15-17. Since the jet plant is ordinarily moving with respect to the atmosphere it is necessary to use a *diffuser* to bring the air from zero absolute velocity to approximately zero velocity relative to the plant. The air taken in through the diffuser passes through a compressor, combustion chamber, and turbine, as in the simple gas-turbine plant, and then expands through an exhaust nozzle or tail pipe to form the driving jet for the aircraft. In the jet plant the turbine exhaust pressure is higher than atmosphere, in order to provide the pressure required for the exhaust nozzle. There is usually no shaft work taken out, the compressor and auxiliaries absorbing all the turbine work. The propeller shown in Fig. 15-17 is absent in the jet engine, but is used with the so-called *turbo-prop* engine.

PROBLEMS

15-1. Plot the efficiency of the air-standard Otto cycle as a function of compression ratio for compression ratios from 4 to 16.

15-2. Find the efficiency of the variable specific-heat Otto cycle for compression ratio of 8, using air tables, and taking the maximum and minimum temperatures as 3500°R and 600°R, respectively. Compare with the results of Problem 15-1.

15-3. Tabulate the efficiencies at a compression ratio of 6 for Otto cycles using perfect gases having specific heat ratios of 1.3, 1.4, 1.67. What would be the advantages and disadvantages of helium as a working fluid for an Otto cycle?

15-4. In an air-standard Otto cycle the compression ratio is 7, and compression begins at 100°F, 14 psia; the maximum temperature of the cycle is 2000°F. Find: (a) the heat supplied per pound of air; (b) the work done per pound of air; (c) the cycle efficiency; (d) the temperature at the end of the isentropic expansion; (e) the maximum pressure of the cycle.

15-5. Solve Problem 15-4, if the maximum temperature is 3000°F.

15-6. Solve Problem 15-4, using air tables.

15-7. For an air-standard Diesel cycle with compression ratio of 12 plot the efficiency as a function of cut-off ratio for cut-off ratios from 1 to 4. Compare with the results of Problem 15-1.

15-8. In an air-standard Diesel cycle the compression ratio is 13, and compression begins at 15 psia, 140°F; the maximum temperature of the cycle is 2540°F. Find: (a) the heat supplied per pound of air; (b) the work done per pound of air; (c) the cycle efficiency; (d) the temperature at the end of the isentropic expansion; (e) the cut-off ratio; (f) the maximum pressure of the cycle.

15-9. Solve Problem 15-8, using air tables.

15-10. Derive Eq. (15-13) from Eq. (15-12).

15-11. An air-standard mixed, or limited-pressure, cycle has a compression ratio of 10, and compression begins at 15 psia, 140°F; the maximum pressure of the cycle is 600 psia, and the maximum temperature is 2540°F. Find: (a) the heat supplied at constant volume, per pound of air; (b) the heat supplied at constant pressure, per pound of air; (c) the work done per pound of air; (d) the cycle efficiency; (e) the temperature at the end of the constant-volume heating process; (f) the cut-off ratio.

15-12. Solve Problem 15-4 for an air-standard Brayton cycle having the same adiabatic compression process as the Otto cycle of Problem 15-4. Compare the maximum specific volumes and the maximum pressures for the Brayton and Otto cycles having the same compression process and the same maximum temperature.

15-13. In the Brayton cycle of Problem 15-12 it is proposed to use a regenerator to increase cycle efficiency. Investigate this proposal and give an opinion of its value; show your reasoning.

15-14. A gas turbine plant operates on the Brayton air cycle between a minimum temperature of 40°F and a maximum temperature of 1540°F. (a) Find the pressure ratio at which the cycle efficiency equals the Carnot efficiency. (b) Find the pressure ratio at which the work per pound of air is maximum. (Take your choice of an algebraic solution or a plot of computed values of work at various pressure ratios.) (c) Compare the efficiency at this pressure ratio with the Carnot efficiency for the given temperatures.

15-15. The plant of Problem 15-14 is changed to have compressor efficiency of 80 percent and turbine efficiency of 85 percent. Make a plot of efficiency, and work per pound of air, vs. pressure ratio, for pressure ratios from 5 to 10. Find: (a) the pressure ratio at which the work per pound of air is maximum; (b) the efficiency and the work per pound of air at this pressure ratio; (c) the pressure ratio at which the efficiency is maximum; (d) the efficiency and the work per pound of air at this pressure ratio. Compare results with Problem 15-14.

15-16. In a gas turbine plant working on the Brayton air cycle the air pressure and temperature before compression are respectively 15 psia and 80°F. The ratio of maximum pressure to minimum pressure is 6.25, and the temperature before expansion in the turbine is 1440°F. The turbine and compressor efficiencies are each 80 percent. Find: (a) the compressor shaft work per pound of air; (b) the turbine shaft work per pound of air; (c) the heat supplied per pound of air; (d) the cycle efficiency; (e) the turbine exhaust temperature.

15-17. Solve Problem 15-16 if a regenerator of 75 percent effectiveness is added to the plant.

15-18. Solve Problem 15-16 if the compression is divided into two steps, each of pressure ratio 2.5 and efficiency 80 percent, with intercooling to 80°F.

15-19. Solve Problem 15-18 if a regenerator of 75 percent effectiveness is added to the plant.

15-20. Solve Problem 15-16 if a reheat cycle is used; the turbine expansion is divided into two steps, each of pressure ratio 2.5 and efficiency 80 percent, with reheat to 1440°F.

15-21. Solve Problem 15-20 if a regenerator of 75 percent effectiveness is added to the plant.

15-22. Solve Problem 15-21 if the staged compression of Problem 15-18 is used in the plant.

15-23. In a gas turbine plant based upon the Brayton cycle the inlet air pressure and temperature are respectively 15 psia and 80°F. The pressure ratio is 6.25, and the temperature entering the turbine is 1440°F. The fuel is liquid octane, C_8H_{18}, at 80°F. The compressor and turbine efficiencies are each 80 percent; heat losses and pressure losses are negligible. Find: (a) the air-fuel ratio; (b) the back-work ratio, W_c/W_{net}; (c) the air rate, lb/hphr; (d) the specific fuel consumption, lb/hphr; (e) the thermal efficiency based on lower heating value. Compare results with Problem 15-16.

15-24. Solve Problem 15-23 if a regenerator of 75 percent effectiveness is added to the plant. Compare with results of Problem 15-17.

15-25. Solve Problem 15-23 if a reheat cycle is used; the turbine expansion is divided into two steps, each of pressure ratio 2.5 and efficiency 80 percent, with reheat to 1440°F. Compare with results of Problem 15-20.

15-26. Solve Problem 15-25 if a regenerator of 75 percent effectiveness is added to the plant. Compare with results of Problem 15-21.

15-27. A gas turbine plant operates under the same conditions as in Problem 15-23 except that the air flow is limited to 150 percent theoretical air for the fuel burned, and liquid water at 80°F is injected into the combustion chamber where it evaporates. The mixture of gases and water vapor enters the turbine at 1440°F, the amount of water injected being regulated to obtain this temperature. Find all the results called for in Problem 15-23, and also the ratio of water injected to fuel burned. Compare results with Problem 15-23.

REFERENCES

Taylor, C. F. and E. S. Taylor, *The Internal Combustion Engine*. Scranton: International Textbook Company.

Lichty, L. C., *Internal Combustion Engines*. New York: McGraw-Hill.

Durham, F. P., *Aircraft Jet Powerplants*. New York: Prentice-Hall, 1951.

Keenan, J. H. and J. Kaye, "Calculated Efficiencies of Jet Power Plants." *J. Aero. Sci.*, Vol. 14, August 1947, p. 437.

Keller, C., "The Escher Wyss-AK Closed-Cycle Turbine." *Trans. ASME*, Vol. 68, Nov. 1946, p. 791.

Chapter 16

FLUID FLOW: NOZZLES AND TURBINES

In Chap. 6 equations were developed for applying the First Law to processes involving fluids in motion. In this chapter these equations will be applied to certain important processes in which flow of the fluid is the principal factor. Such processes include the production of fluid jets to drive turbine wheels and the use of flow nozzles and orifices in fluid metering. The subject matter of this chapter is also treated in books on Fluid Mechanics, being at least as much a part of that field as of thermodynamics.

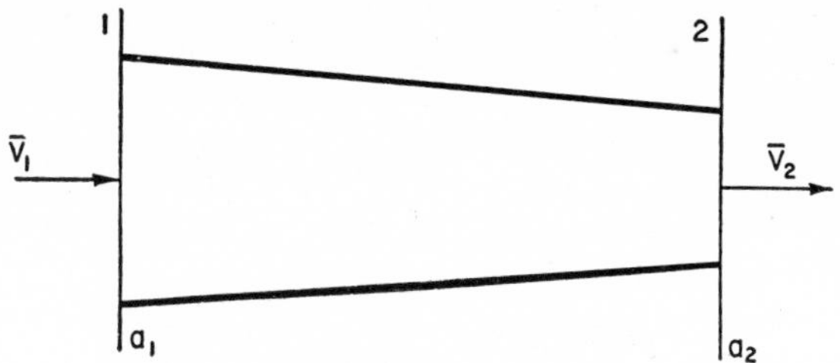

Fig. 16-1. Elementary nozzle.

16-1 Nozzles—mass and energy relations. A nozzle is a duct or passage in which the kinetic energy of a flowing fluid increases as a result of a drop in pressure along the stream. The following analysis of nozzles is based upon the assumption of a steady flow process with uniform properties across the stream at every cross-section (see Sec. 6-3). This is called a *one-dimensional* analysis. It is further assumed that potential energy change and heat transfer are negligible, and it is noted that there is no shaft work involved. Then for any such nozzle, Fig. 16-1, having inlet section 1 and outlet section 2 the steady flow equations for mass and energy will be

$$w = \frac{a_1 \overline{V}_1}{v_1} = \frac{a_2 \overline{V}_2}{v_2} \tag{1}$$

$$h_1 + \frac{\overline{V}_1^2}{2g_0} = h_2 + \frac{\overline{V}_2^2}{2g_0} \tag{2}$$

It is convenient to introduce here the concept of *stagnation enthalpy* which is the enthalpy that would be reached if the stream were brought to rest under the conditions assumed above. Using the subscript zero for stagnation conditions

$$\overline{V}_0 = 0$$

hence
$$h_0 = h_1 + \frac{\overline{V}_1^2}{2g_0} = h_2 + \frac{\overline{V}_2^2}{2g_0} \tag{3}$$

If the nozzle is supplied with fluid from a very large tank the enthalpy in the tank will approach h_0 as the cross-section of the tank approaches infinite area. Stagnation enthalpy is changed only by work transfer or heat transfer.*

It should be observed that while (3) defines the stagnation enthalpy it does not fix a definite stagnation *state*. Stagnation proper-

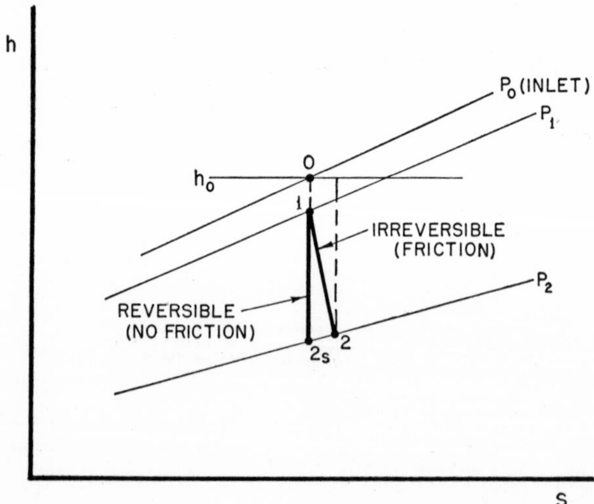

Fig. 16-2. Nozzle process state paths.

* Stagnation enthalpy is analogous to *total head* in hydraulics. When appropriate, a gravity potential energy term, relative to some fixed datum, may be included in the stagnation enthalpy as it is in the total head.

ties other than enthalpy are taken for an isentropic stagnation state. The isentropic stagnation state corresponding to any section of the nozzle is the state fixed by h_0 and the entropy at the section in question. Thus at the inlet stagnation state $s_0 = s_1$ and at the exit stagnation state $s_0 = s_2$, but in both cases h_0 is given by (3). Hence there may be many stagnation pressures or temperatures for a given adiabatic nozzle process, but only one stagnation enthalpy. This is illustrated in Fig. 16-2. (For a perfect gas, since h is a function of t only, there is only one stagnation temperature.)

Stagnation pressure is often called *total* or *impact* pressure, while pressure independent of velocity is called *static* pressure. Total pressure is measured by a tube with open end turned directly against the flow so the stream is brought to rest at the opening. Static pressure would be measured by a gage moving along with the stream at the stream velocity. When the flow is parallel to a long straight wall the static pressure may be measured through a small hole in the wall.

A practical device for measuring impact pressure and static pressure in duct flow is the pitot-static tube, Fig. 16-3.

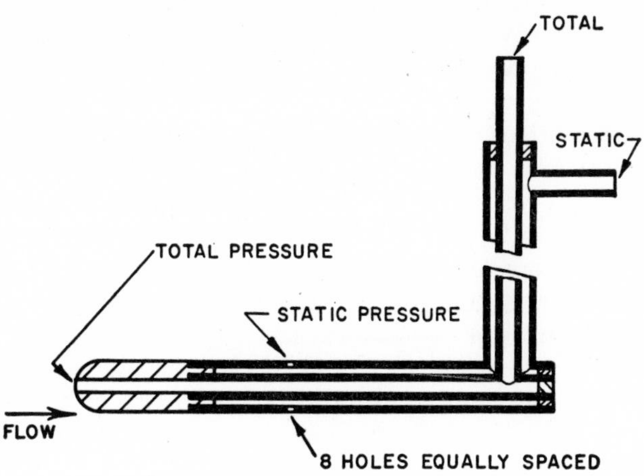

Fig. 16-3. Pitot-static tube.

In the case of supersonic flow the stream cannot be brought to rest reversibly, so the total pressure of a supersonic stream cannot be measured directly by an impact device.

Stagnation temperature (*impact* or *total* temperature) is measured by a stationary thermometer in communication solely with fluid which has been brought to rest reversibly and adiabatically from the stream velocity; this condition is approximated by a thermometer at the nose of a rod pointed directly against the flow. (With a perfect gas the deceleration need not be reversible.) Static temperature can be measured only by a thermometer moving along with the fluid stream. A thermometer in the side wall of a duct (analogous to a static pressure tap) gives a temperature nearer to the impact temperature than to the static temperature.

Stagnation properties should always be identified as such; an unqualified reference to a property always signifies the static property.

It is possible to solve (3) for the velocity at any section along the length of a nozzle; thus

$$\overline{V} = \sqrt{2g_0(h_1 - h) + \overline{V}_1^2} = \sqrt{2g_0(h_0 - h)} \qquad (4)$$

where symbols lacking subscripts refer to any specified section of the nozzle. The velocity at any point is seen to be determined by the enthalpy of the fluid at that point and the stagnation enthalpy. The flow rate is given from (1) by

$$w = \frac{a\overline{V}}{v} = \frac{a_1\overline{V}_1}{v_1} \qquad (5)$$

Equations (4) and (5) apply to any nozzle process, reversible or irreversible, subject to the assumptions made above. The given, or imposed, conditions for a nozzle process are usually the inlet state, the inlet velocity, and the exit pressure; these are not sufficient to obtain the exit velocity and the flow rate because they do not fix definite states within the nozzle. If, however, the analysis is made for a *reversible* nozzle process the states are fixed by their entropy, and solutions of (4) and (5) can be made. In Fig. 16-2 the paths of a reversible nozzle process and an irreversible nozzle process are shown on the enthalpy-entropy plot. State 2_s is a definite state fixed by p_2 and s_1, whereas state 2 depends upon the friction or irreversibility in the nozzle; for the irreversible adiabatic nozzle 2 must lie to the right of 2_s. The figure also shows the inlet stagnation state, 0, fixed by h_0 and s_1. The outlet stagnation state for the reversible nozzle is also 0, but for the irreversible nozzle it is a different state fixed by the intersection of the lines for h_0 and s_2. In Fig. 16-2, p_1 and p_2 are static pressures, while p_0 is a stagnation pressure.

16-2 The reversible nozzle. If the process in an adiabatic nozzle is frictionless or reversible, the path of the process is at constant entropy. This fact, with Eqs. (4) and (5), provides relations for enthalpy, velocity and cross-sectional area of the stream as functions of the pressure. From the tables of properties (or the equations for gases) the enthalpy may be found at any pressure, and the corresponding velocity is then given by (4); similarly, when the volume is obtained from the tables the area can be found by (5). For example, Fig. 16-4 shows the relations for a particular case of reversible flow of air. The velocity, specific volume and cross-sectional area for unit flow rate are plotted against the dimensionless ratio p/p_0. The general shape of the curves shown is the same for all gases and

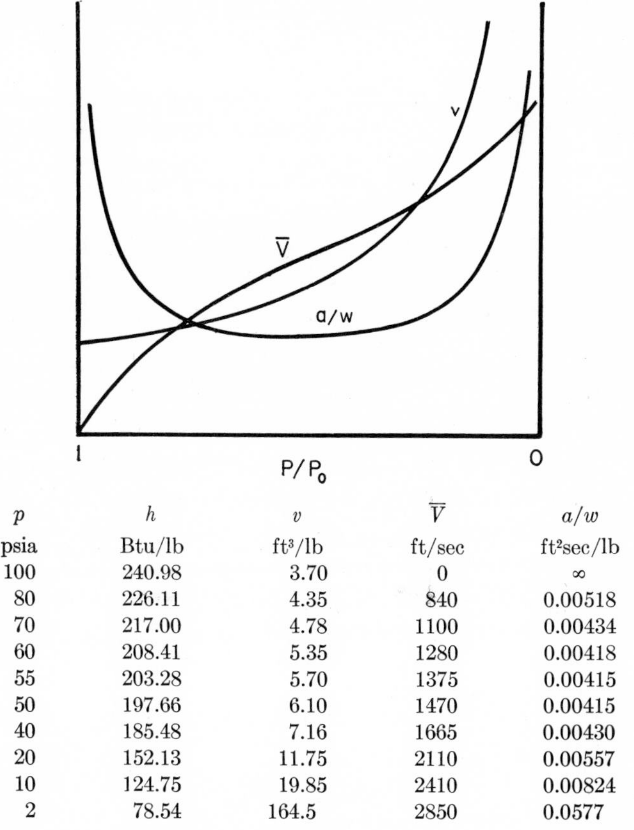

p psia	h Btu/lb	v ft³/lb	$\bar{V}$ ft/sec	a/w ft²sec/lb
100	240.98	3.70	0	∞
80	226.11	4.35	840	0.00518
70	217.00	4.78	1100	0.00434
60	208.41	5.35	1280	0.00418
55	203.28	5.70	1375	0.00415
50	197.66	6.10	1470	0.00415
40	185.48	7.16	1665	0.00430
20	152.13	11.75	2110	0.00557
10	124.75	19.85	2410	0.00824
2	78.54	164.5	2850	0.0577

Fig. 16-4. Reversible adiabatic expansion of air in a nozzle; stagnation state 100 psia, 1000° F_{abs}.

vapors when plotted in this way. It is seen that the velocity, starting from zero at the stagnation pressure, increases continuously as the pressure falls. The specific volume, however, is initially finite; it increases, first slowly, then more rapidly, as the pressure falls. The result is that the ratio of volume to velocity $v/\overline{V}$ decreases rapidly from infinity, passes a minimum, and increases again as the pressure drops from stagnation pressure toward zero. The ratio a/w, area per unit flow rate, is equal to $v/\overline{V}$ at all points, from Eq. (5). The minimum value of $v/\overline{V}$ (and of a/w) for gases and vapors always occurs in the neighborhood of $p/p_0 = 0.5$. Hence if a reversible flow is desired when the nozzle exit pressure is greater than about $0.5p_0$ the nozzle should be of converging form; if the nozzle exit pressure is less than about $0.5p_0$ the nozzle should converge to a minimum section

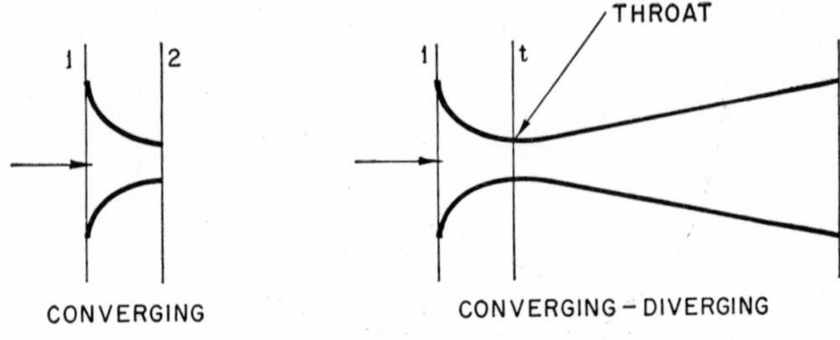

Fig. 16-5. Types of nozzles.

and then diverge to the required exit area. This is illustrated in Fig. 16-5.

The situation is different for incompressible fluids (liquids). In such cases the specific volume is constant and by (5) the cross-sectional area of the stream is inversely proportional to the velocity. Therefore a fire hose nozzle, for example, must always be a converging nozzle to obtain maximum velocity at the exit.

A more detailed analytical approach to the characteristics of nozzle flow can be made as follows. For any pure substance

$$T\,ds = dh - v\,dp \tag{9-6}$$

Then for a reversible adiabatic process

$$dh = v\,dp$$

Using this relation in (4), the velocity is given by

$$\overline{V} = \sqrt{2g_0(h_0 - h)} = \sqrt{-2g\cdot\int_{p_0}^{p} v\,dp} \tag{6}$$

To make the integration, the relation between v and p must be known for the reversible adiabatic process. For a perfect gas the relation was found

$$pv^k = \text{constant} \tag{11-17}$$

For real gases and vapors the same form of equation is a satisfactory approximation if the value of k is properly chosen. In any event the general trend is correctly shown by (11-17). Substituting from (11-17) in (6)

$$\overline{V} = \sqrt{-2g_0 \int_{p_0}^{p} p_0^{1/k} v_0 \frac{dp}{p^{1/k}}} = \sqrt{\frac{2g_0 k}{k-1} p_0 v_0 \left[1 - \left(\frac{p}{p_0}\right)^{(k-1)/k}\right]} \tag{7}$$

From (5) and (11-17)

$$\frac{a}{w} = \frac{v}{\overline{V}} = \frac{(p_0/p)^{1/k}\, v_0}{\overline{V}}$$

and substituting from (7),

$$\frac{a}{w} = \frac{1}{\sqrt{\dfrac{2g_0 k}{k-1} \dfrac{p_0}{v_0} \left[\left(\dfrac{p}{p_0}\right)^{2/k} - \left(\dfrac{p}{p_0}\right)^{(k+1)/k}\right]}} \tag{8}$$

The minimum value of a/w, corresponding to the *throat* of the nozzle, can be located by taking the derivative of the right side of (8) with respect to p/p_0 and setting it equal to zero. This gives the following result for the condition of minimum area:

$$\frac{p}{p_0} = \left(\frac{2}{k+1}\right)^{k/(k-1)} \tag{9}$$

The value of the pressure ratio for minimum area is called the *critical pressure ratio* for a nozzle. The values of k applying to gases and vapors lie between 1.1 and 1.67; the following table gives the values of critical pressure ratio corresponding to some values of k in this range.

k	p/p_0 (critical)
1.10	0.585
1.20	0.565
1.30	0.546
1.40	0.528
1.67	0.487

It is a matter of interest that the velocity corresponding to the critical pressure ratio in reversible nozzle flow is the local velocity of sound.* The velocity at higher pressures is less than the local sound velocity and the velocity at lower pressures is greater than the local sound velocity, hence the terms *sub-sonic* and *super-sonic* are applied to the flows in the respective regions.

The manner of using the nozzle equations is best shown by an example.

Example 1. (a) Find the exit area of a reversible nozzle to pass 1 lb/sec of steam if the inlet state is 100 psia, 500°F, 100 fps velocity, and the exit pressure is 10 psia. (b) Will the nozzle be converging or converging-diverging and, if the latter, what will be the throat area?

Solution: (a) The properties at the inlet state are $h_1 = 1279.1$ Btu/lb, $s_1 = 1.7085$ Btu/lb °R. At the exit state $s_2 = s_1$, $p_2 = 10$ psia; then from the tables $h_2 = 1091.7$ Btu/lb and $v_2 = 36.41$ cu ft/lb.

$$a_2 = w\frac{v_2}{\overline{V}_2} = w\frac{v_2}{\sqrt{2g_0(h_1 - h_2) + \overline{V}_1^2}}$$

Substituting values in consistent units

$$a_2 = 1\,\frac{36.4}{3{,}060} = 0.01185 \text{ sq ft} = 1.705 \text{ sq in.}$$

Note the order of magnitude of the nozzle exit velocity and the relative insignificance of the inlet velocity in this particular case. Inlet velocity of 224 fps is equivalent to 1 Btu/lb enthalpy.

(b) The exit pressure for this nozzle is well below $0.5p_0$ so the nozzle should be converging-diverging. To find the area at the throat, the critical pressure ratio must be known. Take from Keenan and Keyes, Fig. 8, at 100 psia, 500°F, the value of $k = 1.30$. From Eq. (9) the critical pressure ratio is 0.55. Then the properties at the nozzle throat are as follows:

$$p_t = 0.55(100) = 55 \text{ psia}$$

(The difference between p_1 and p_0, about 0.2 psi, is negligible.)

$$s_t = s_1$$

From the tables
$$h_t = 1221.5 \text{ Btu/lb}$$
$$v_t = 8.841 \text{ cu ft/lb}$$

* The sound velocity is given by $\sqrt{kg_0pv}$. Therefore, it has different values at different states of the fluid.

Solving as in (a) above,

$$a_t = 1\,\frac{8.841}{1700} = 0.0052 \text{ sq ft} = 0.75 \text{ sq in.}$$

It is not necessary to know the critical pressure ratio very precisely to obtain a precise value of throat area, because the area of the nozzle changes very slowly with pressure near the throat; as an example see Fig. 16-4. Thus if the critical pressure ratio had been assumed to be anything between 0.53 and 0.58 the computed area would have been the same within slide-rule accuracy. Although both the volume and the velocity would change appreciably their ratio would be almost constant.

16-3 Real nozzle coefficients. A real nozzle process is never reversible; therefore the measured characteristics of real nozzle processes are referred to the characteristics of reversible processes by means of conventional coefficients defined below.

The *velocity coefficient* $C_{\overline{V}}$ of a nozzle is defined as the ratio of the velocity of the stream at the nozzle exit section to the exit velocity for a reversible nozzle, both nozzles having the same inlet conditions and discharging into a space at the same pressure.

$$C_{\overline{V}} = \frac{\overline{V}_2}{\overline{V}_{2s}} = \frac{\overline{V}_2}{\sqrt{2g_0(h_0 - h_{2s})}} \tag{10}$$

Note that the velocity coefficient has absolutely no connection with the velocity anywhere but at the *exit* section of a nozzle.

The *efficiency* of a nozzle η_n is defined as the ratio of the kinetic energy per unit mass of the stream at the exit section of the nozzle to the exit kinetic energy per unit mass for the reversible nozzle, both nozzles having the same inlet conditions and discharging into a space at the same pressure.

$$\eta_n = \frac{\overline{V}_2^2/2g_0}{\overline{V}_{2s}^2/2g_0} = \frac{\overline{V}_2^2/2g_0}{h_0 - h_{2s}} \tag{11}$$

From (10) and (11)

$$\eta_n = (C_{\overline{V}})^2 \tag{12}$$

As in the case of the velocity coefficient, the nozzle efficiency has no connection with the kinetic energy anywhere but at the *exit* section of the nozzle.

The *coefficient of discharge* C_w of a nozzle is defined as the ratio of the actual flow rate to the flow rate of a reversible nozzle of the same *minimum* cross-section, both nozzles having the same inlet conditions and both discharging into a space at the same pressure.

$$C_w = \frac{w}{w_s} = \frac{w}{a_m \overline{V}_{sm}/v_{sm}} \tag{13}$$

where a_m is the minimum area; $\overline{V}_{sm}$ and v_{sm} are taken in the reversible nozzle at its minimum area section.

Note that the minimum cross-section may be either an exit section or a throat section, depending upon whether the nozzle is a converging nozzle or a converging-diverging nozzle. Furthermore the reversible reference nozzle is not necessarily the same as the actual nozzle in this respect. For example, if an actual converging nozzle is discharging to a region at a pressure ratio lower than the critical pressure ratio, the reversible reference nozzle must be a converging-diverging nozzle and its minimum area, which is its *throat* area, must be the same as the minimum area of the actual nozzle, which is its *exit* area.

When a well made converging nozzle is used under the proper conditions, with a pressure ratio not less than the critical, both the velocity coefficient and the discharge coefficient may be very close to unity (perhaps 0.99).*

For pressure ratios smaller than critical the discharge coefficient remains close to unity, but the velocity coefficient decreases with the pressure ratio. Under these conditions the fluid expands to the critical pressure ratio at the minimum nozzle area (the exit area), passes out of the nozzle at sonic velocity, and then expands irreversibly to the exhaust pressure in the exhaust region. The resultant dissipation of energy in pressure waves accounts for the smaller velocity coefficient.

The flow rate through a rounded entrance nozzle operating with less than critical pressure ratio depends only upon the initial state of the fluid, not upon the pressure ratio. The names *critical flow* and *choking flow* are used to describe this condition. A sharp-edged orifice, having converging flow at the exit, does not reach a limit of flow at the ordinary critical pressure ratio.

A well made converging-diverging nozzle may have a maximum velocity coefficient of the order of 0.95, but this will be reached only at the particular pressure ratio corresponding to the nozzle expansion

* The discharge coefficient may sometimes slightly exceed unity, but the velocity coefficient cannot exceed unity.

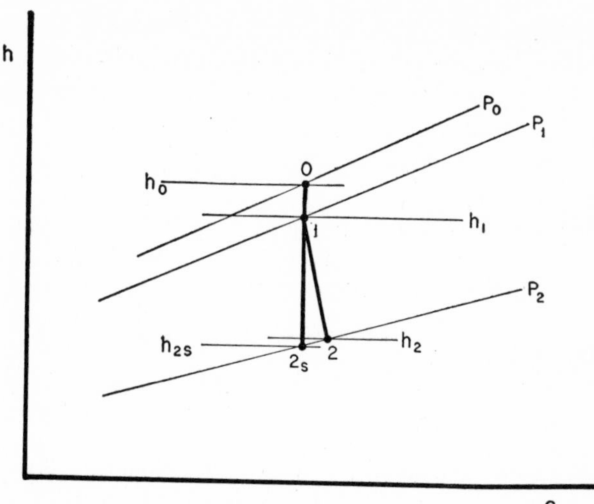

Fig. 16-6. Flow state paths, reversible and irreversible.

ratio (ratio of exit area to throat area). At greater or smaller pressure ratios the velocity coefficient falls off. The maximum flow rate through a converging-diverging nozzle occurs when the *throat* pressure ratio equals the critical; because of diffuser action (Sec. 16-5) in the diverging section, the throat pressure ratio may equal the critical ratio when the exhaust pressure ratio is greater than critical, and the discharge coefficient may then be much larger than unity. For exhaust pressure ratios smaller than critical the throat pressure ratio is critical, and the discharge coefficient is slightly less than unity.*

16-4 Real nozzle computations. A properly designed real nozzle has cross-sections proportioned to agree with the area of the stream as in the case of the reversible nozzle, but the areas are modified by the effects of friction. In an adiabatic nozzle any loss of kinetic energy through friction must be balanced by an equal increase in the enthalpy of the fluid; therefore in a real nozzle the enthalpy difference between inlet and exhaust states will be less than in a reversible nozzle by the amount of the friction loss. This is expressed by the following equations:

$$h_1 + \frac{\overline{V}_1^2}{2g_0} = h_0 = h_2 + \frac{\overline{V}_2^2}{2g_0}$$

* For experimental data see J. H. Keenan, p. 773, and H. Kraft, p. 781, *Trans. ASME*, Vol. 71, October 1949.

$$\overline{V}_2^2 = \eta_n \overline{V}_{2s}^2 = \eta_n\, 2g_0(h_0 - h_{2s})$$

$$h_0 - h_2 = \eta_n(h_0 - h_{2s}) \tag{14}$$

The states involved are plotted in Fig. 16-6.

Some examples of the manner of using nozzle coefficients in computations will now be given.

Example 2. Find the throat and exit areas for a nozzle to pass 10,000 lb/hr of steam from an initial state of 250 psia, 500°F, negligible velocity, to a final pressure of 1 psia, if the velocity coefficient is 0.949 and the discharge coefficient is unity. Find the exit jet velocity.

Solution: $h_0 = 1263.4$ Btu/lb; $s_0 = 1.5949$ Btu/lb °R. The throat pressure p_t is obtained from the critical pressure ratio as in Example 1 of this chapter.

$$p_t = 0.55p_0 = 137.5 \text{ psia}$$

At p_t and s_0, from the tables, $h_{ts} = 1208.2$ Btu/lb and $v_{ts} = 3.415$ cu ft/lb. At p_2 and s_0 find $h_{2s} = 891.0$ Btu/lb. The throat area is fixed by the desired flow and the coefficient of discharge.

$$a_t = \frac{w}{C_w} \frac{v_{ts}}{\overline{V}_{ts}}$$

$$\overline{V}_{ts} = \sqrt{2g_0\, 778(1263.4 - 1208.2)} = 223.8\sqrt{55} = 1660 \text{ fps}$$

$$w = \frac{10{,}000}{3{,}600} = 2.778 \text{ lb/sec}, \qquad a_t = 0.00572 \text{ ft}^2$$

The exit area is fixed by the desired flow and the actual volume and velocity (*not* by the discharge coefficient, which applies solely to the *minimum* section of the nozzle). The actual exit velocity is given by

$$\overline{V}_2 = C_{\overline{V}} \sqrt{2g_0(h_0 - h_{2s})} = 4{,}094 \text{ fps}$$

The actual exit volume can be found only through the actual enthalpy which is given by Eq. (14);

$$h_2 = h_0 - \eta_n(h_0 - h_{2s}), \qquad \eta_n = (C_{\overline{V}})^2 = 0.90, \qquad h_2 = 928 \text{ Btu/lb}$$

From the tables, at p_2 and h_2,

$$v_2 = 276 \text{ cu ft/lb}, \qquad a_2 = w\frac{v_2}{\overline{V}_2} = 0.187 \text{ ft}^2$$

For the frictionless nozzle, for comparison,

$$a_{2s} = 0.170 \text{ ft}^2$$

If the discharge coefficient had been 0.98 what would the throat and exit areas have been?

$$a_t = \frac{a_{ts}}{C_w} = 0.00584 \text{ ft}^2$$

The exit area is unchanged because the actual flow rate and the nozzle efficiency, which fix the exit area, are specified conditions independent of the discharge coefficient.

Example 3. Find the exit velocity and the throat and exit areas for a nozzle to pass 1 lb/sec of air from an inlet state of 150 psia, 340°F, negligible velocity, to an exhaust pressure of 15 psia if the nozzle efficiency is 88 percent and the discharge coefficient is unity.

Solution: By the perfect gas rules:

For air, $k = 1.40$; $p_t/p_0 = 0.53$

$$\frac{T_{ts}}{T_0} = \left(\frac{p_t}{p_0}\right)^{(k-1)/k} = (0.53)^{(1.4-1)/1.4} = 0.833$$

$$T_{ts} = (800)(0.833) = 666°\text{R}$$

$$\overline{V}_{ts} = \sqrt{2g_0(h_0 - h_{ts})} = \sqrt{2g_0c_p(T_0 - T_{ts})}$$

$$\overline{V}_{ts} = \sqrt{2g_0\,778(0.24)(134)} = 1270 \text{ fps}$$

$$v_{ts} = \frac{RT_{ts}}{p_t} = 3.12 \text{ cu ft/lb}$$

$$a_t = \frac{w}{C_w}\frac{v_{ts}}{\overline{V}_{ts}} = 0.00246 \text{ ft}^2$$

$$\frac{T_{2s}}{T_0} = \left(\frac{15}{150}\right)^{(1.4-1)/1.4} = 0.518$$

$$T_{2s} = (0.518)(800) = 414°\text{R}$$

From (14), since $\Delta h = c_p\Delta T$,

$$c_p(T_0 - T_2) = \eta_n c_p(T_0 - T_{2s})$$

$$T_2 = T_0 - \eta_n(T_0 - T_{2s}) = 460°\text{R}$$

$$\overline{V}_2 = \sqrt{2g_0c_p(T_0 - T_2)} = 2030 \text{ fps}$$

$$v_2 = \frac{RT_2}{p_2} = 11.4 \text{ cu ft/lb}$$

$$a_2 = \frac{wv_2}{\overline{V}_2} = 0.00560 \text{ ft}^2$$

For converging nozzles the computations are made in a similar way but the minimum area is also the exit area. Hence in this case the exit area is not computed from exit velocity and volume but comes from the discharge coefficient, which applies to the minimum area, wherever situated.

16-5 Diffusers. If a nozzle process is reversed, the result will be the compression of a fluid at the expense of its kinetic energy. A duct in which this effect is accomplished is called a *diffuser*. The same equations apply to diffusers as to nozzles, and in the reversible case the only difference in operation between the nozzle and the diffuser is the reversal in direction of all velocities.

In the real case, for subsonic flow the diffuser shape must differ from the nozzle shape because of the difficulty of causing the fluid to fill a diverging passage. The pressure in a diffuser is rising in the direction of flow; consequently the establishment of reverse flow and separation from the wall in the slower-moving fluid near the wall (boundary layer) is aided by the pressure gradient. Separation results in the formation of large eddies in the fluid, with consequent friction losses. For this reason the angle of divergence of an efficient diffuser must be small; tests indicate minimum losses when the angle between the walls is of the order of 7 degrees.

Figure 16-7 indicates the general difference in shape between a

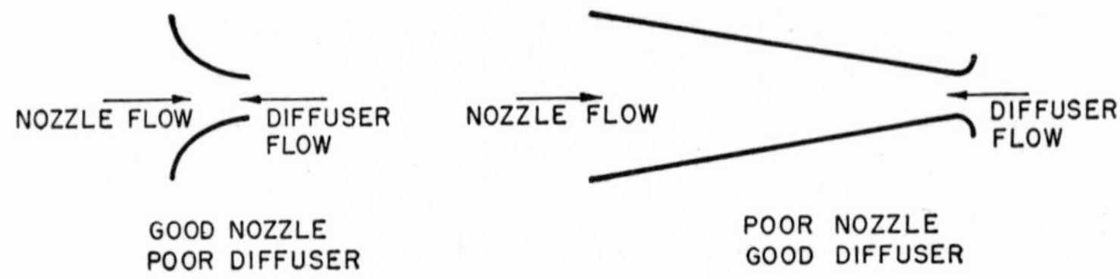

Fig. 16-7. Subsonic nozzle and diffuser.

good nozzle and a good diffuser for a pressure ratio greater than critical. For a pressure ratio smaller than critical the diffuser must converge and then diverge. The relation of supersonic diffusers to supersonic nozzles is not as simple as that between subsonic diffusers and nozzles, because of shock effects; this matter is beyond the scope of this book.

The efficiency of a diffuser is defined as the ratio of the enthalpy increase in a reversible adiabatic diffuser to the enthalpy increase in the real diffuser, when both diffusers receive fluid at the same state, and discharge to a region at the same pressure. Diffuser efficiency is related to nozzle efficiency as pump efficiency is related to engine efficiency.

Diffusers are used as adjuncts to, or integral parts of, pumps, fans, jet pumps (ejectors and injectors), venturi meters, and other apparatus where kinetic energy is available and increased pressure is desired. Except for the development of the flow equations, which is

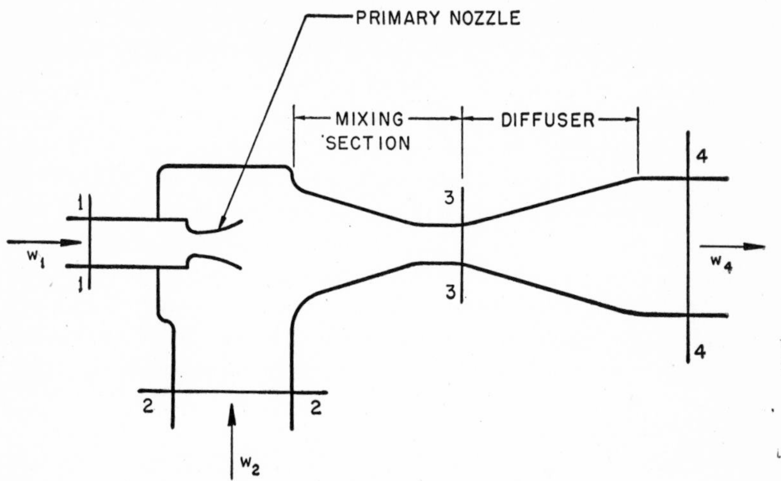

Fig. 16-8. Ejector or jet pump.

done as for the nozzle, the analysis of such devices belongs in the field of fluid mechanics.

16-6 Supersaturation in nozzle flow. When a nozzle process starts with superheated vapor and ends with a wet mixture, the states of the fluid in the process do not always correspond to the states shown on the ordinary diagrams of properties. As the fluid passes the saturation line at p_a in Fig. 16-9 it might be expected that con-

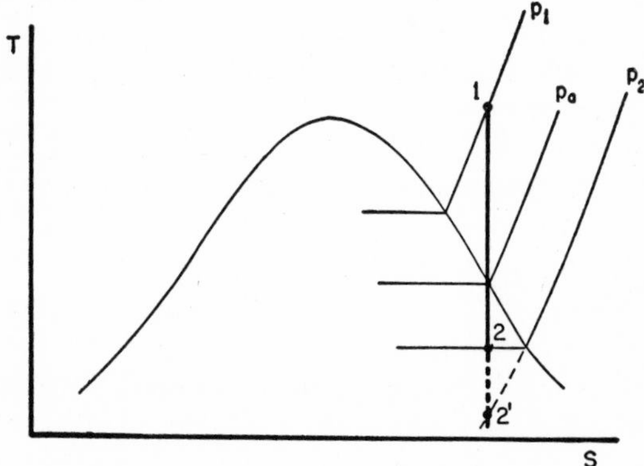

Fig. 16-9. Supersaturated vapor states.

densation would begin; however, the velocity of the fluid may be so great that before any drops of liquid have time to form the fluid has passed well below the saturation state. During this interval the fluid is not in a state of stable equilibrium, for it is a homogeneous vapor at a temperature below the saturation temperature for the existing pressure; it is said to be in a *metastable* state. The name *supersaturated* vapor is used to describe this condition but a more rational term would be *subcooled* vapor.

The properties of the supersaturated vapor may be approximated by extrapolating the curves from the superheated region into the normally wet region on a plot of properties. In Fig. 16-9 point 2 is the equilibrium point and point 2′ is the supersaturated point corresponding to pressure p_2 on an isentropic path through state 1.

If, in a nozzle process with a given pressure ratio, the fluid is supersaturated while passing through the section of minimum area, the flow rate will be different from that of a fluid following a path of stable equilibrium states. Nozzle computations for cases of this kind may be handled in two ways: (1) the computations may be based upon equilibrium properties, and the deviations in performance may be accounted for by special values of the experimental coefficients; (2) the computations may be based upon supersaturated properties, using normal values of nozzle coefficients as for superheated vapor. The latter method usually involves computing nozzle flow by Eq. (8), based on pv^k = constant, assuming that k has the same value in the supersaturated state as in the adjacent superheated states. The flow through a nozzle in which supersaturation exists at the minimum section may be several percent higher than the flow predicted on the basis of equilibrium states; the efficiency, however, is reduced because of the irreversibility of the eventual condensation from a supersaturated state.

It is beyond the scope of this book to go deeper into the subject of supersaturation.*

16-7 Nozzles and orifices as flow meters. From the relations between pressure drop and mass flow rate for a nozzle it is apparent that a nozzle could serve as a flow meter. In Fig. 16-10 are shown a *flow nozzle*, shaped like a converging nozzle, and a *venturi meter* shaped like a converging-diverging nozzle. Essentially, the venturi meter

* For a thorough explanation see J. H. Keenan, *Thermodynamics*. Wiley, 1941, chap. 25.

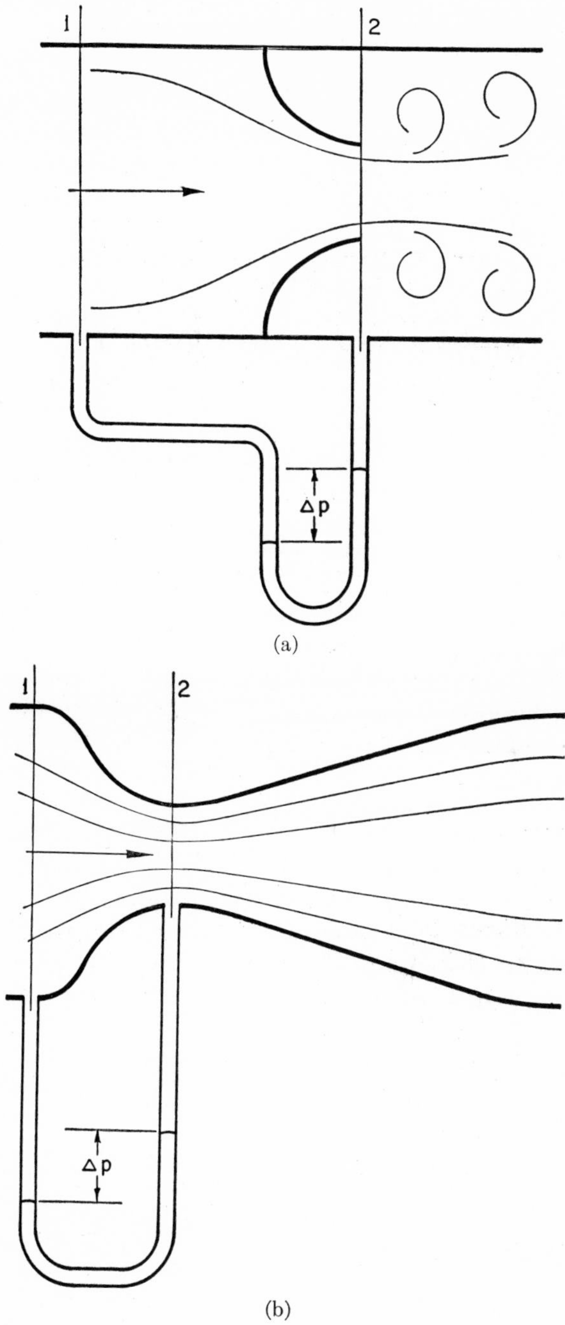

Fig. 16-10. (a) Flow nozzle. (b) Venturi flow meter.

is a flow nozzle followed by a diffuser to improve the recovery of pressure after the reduced section. The pressure loss may be reduced 50 to 75 percent by this means.

In both the flow nozzle and the venturi meter the pressure drop for metering purposes is measured between the inlet section and the section of minimum area. Then, knowing the dimensions of the device and the inlet state of the fluid, the flow rate can be calculated as a function of the pressure drop. Solving Eqs. (1) and (2) simultaneously gives, for a reversible flow nozzle or venturi meter

$$w_s = \frac{a_2}{v_1\sqrt{1 - (a_2/a_1)^2}}\sqrt{2g_0(h_1 - h_{2s})} \tag{15}$$

Inserting an experimental coefficient C, the flow through a real meter is given by

$$w = \frac{Ca_2}{v_1\sqrt{1 - (a_2/a_1)^2}}\sqrt{2g_0(h_1 - h_{2s})} \tag{16}$$

The enthalpies could be determined from the inlet state and the pressure drop; however in flow-metering practice it is customary to use, instead, an equation based upon the relation pv^k = constant. From this, by a development similar to that for Eq. (8), there is obtained

$$w = \frac{Ca_2}{\sqrt{1 - \left(\frac{a_2}{a_1}\right)^2\left(\frac{p_2}{p_1}\right)^{2/k}}}\sqrt{\frac{2g_0k}{k-1}\frac{p_1}{v_1}\left[\left(\frac{p_2}{p_1}\right)^{2/k} - \left(\frac{p_2}{p_1}\right)^{(k+1)/k}\right]} \tag{17}$$

This equation is too complicated for general use; for the special case of an incompressible fluid, however, it reduces to the *hydraulic equation*

$$w = \frac{Ca_2}{\sqrt{1 - (a_2/a_1)^2}}\sqrt{2g_0\frac{p_1 - p_2}{v_1}} \tag{18}$$

Now for compressible fluids it is possible to write

$$w = \frac{YCa_2}{\sqrt{1 - (a_2/a_1)^2}}\sqrt{2g_0\frac{p_1 - p_2}{v_1}} \tag{19}$$

where the coefficient Y is defined by setting (19) equal to (17). Y is a function of the dimensionless ratios k, a_2/a_1, and p_2/p_1; it has been evaluated and tabulated for certain frequently encountered cases and values may be found in the references at the end of this chapter.

For the special case of critical flow (exhaust pressure ratio less than critical pressure ratio) Eq. (17) is greatly simplified by substitution from (9), which gives a fixed value of the pressure ratio at the minimum section, for any given value of k. For example, for critical flow with k equal to 1.4 Eq. (17) reduces to

$$w = \frac{3.88Ca_2}{\sqrt{1 - (0.404)(a_2/a_1)^2}}\sqrt{\frac{p_1}{v_1}}$$

The relative simplicity of this type of equation is apparent; the fact that under critical flow conditions the coefficient of a well-made rounded entrance nozzle can be relied upon to be close to unity (approximately 0.99) recommends the critical flow meter for applications in which the large pressure drop is acceptable.

The thin plate orifice is a form of flow meter which uses a sharp edged hole in a plate, as shown in Fig. 16-11, for a metering passage. The flow through the orifice is similar to that through a flow nozzle but the fluid passing the orifice has not been guided into a parallel

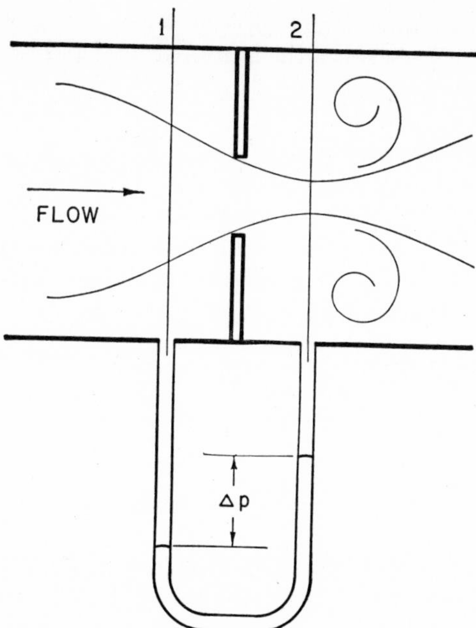

Fig. 16-11. Thin-plate orifice flow meter.

stream; it is converging on the hole from all sides, and in the absence of infinite forces cannot turn immediately around the sharp corner into a parallel stream. Hence the fluid completes its convergence at some distance downstream from the edge of the orifice plate, where the area of the stream is less than the area of the orifice. Since the orifice area is a physical dimension of the apparatus, it is used for a_2 in the flow equations, and the difference between the fluid stream area at section 2 and the orifice area is accounted for in the experimental coefficient C. The value of C will then depend upon the location of the pressure taps (measuring holes) as well as upon the ratio of orifice area to pipe area, and other factors.

The equations used for the thin plate orifice are identical with those used for the flow nozzle but the values of the coefficient C are quite different, orifice coefficients being of the order of 0.6 while nozzle coefficients are of the order of unity. The coefficient Y for compressible flow is also different for orifices from that for flow nozzles and venturi meters. Values of coefficients for standardized designs of flow nozzles and orifices may be found in the ASME references.

The pitot-static tube, Fig. 16-3, is used as a flow metering device. From the difference between the measured total and static pressures, the stream velocity can be calculated; since this is a local velocity, several measurements at selected points in the stream cross-section are averaged to give a mean velocity.

16-8 Turbines. A turbine is a machine in which the acceleration of a stream of fluid produces a turning moment on a rotating shaft. The acceleration may be a change in either magnitude or direction of the fluid velocity. Thus in the classical turbine of Hero, Fig. 16-12, the fluid entering through the axis of the sphere changes direction to escape through the tangential exhaust nozzles. If the nozzles are smaller in cross-section than the axial inlet pipes, there may also be an increase in magnitude of the velocity. In any case, by Newton's laws of motion the force required to accelerate the fluid must have an equal and opposite reaction; the reaction constitutes a tangential force on the sphere. If the sphere rotates, this force does work and the device becomes a turbine.

In modern turbines the same principle applies as in Hero's turbine; a tangential force on a rotating body is produced by a change in the tangential component of velocity of a fluid stream. In Fig. 16-13 is shown a simple type of turbine named for its inventor,

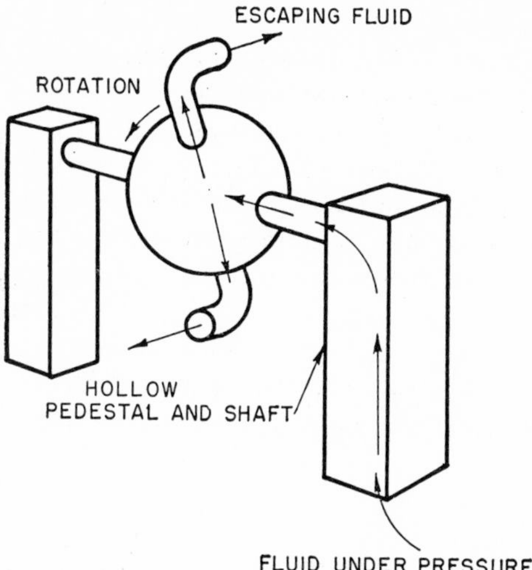

Fig 16-12. Hero's turbine.

De Laval. In this turbine a jet of fluid from the fixed nozzle strikes the buckets on the wheel in nearly the tangential direction. The curvature of the buckets is such that the stream is turned, relative to the buckets, almost 180 degrees, reversing the tangential compo-

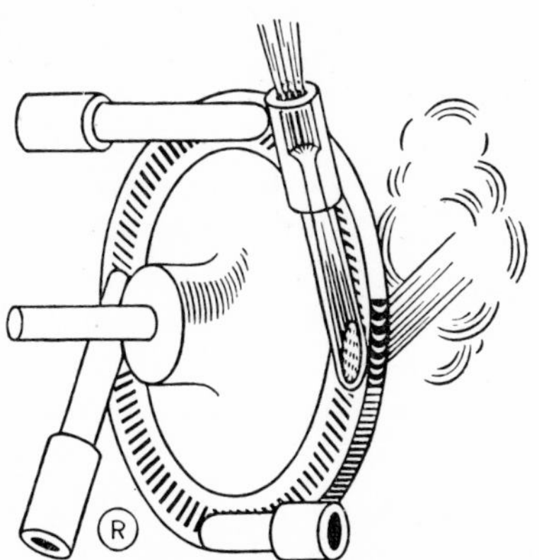

Fig. 16-13. Simple impulse turbine.
(Courtesy De Laval Steam Turbine Co.)

nent of its velocity relative to the buckets. The force required to accelerate the fluid in this manner comes from the buckets; consequently the reaction force is a tangential force upon the buckets.

16-9 Force relations for a fluid stream in steady flow. The analysis of the turbine process for compressible fluids requires the use of both mechanics and thermodynamics. A simple presentation of the basic mechanics follows. In this development a one-dimensional flow is assumed; this means that velocity does not vary over any cross-section taken normal to the stream. Also, in this chapter gravity forces are neglected.

By Newton's laws of motion, the force acting on a mass to change its velocity is proportional to the rate of change of momentum of the mass.

$$F = C\frac{d(m\overline{V})}{dt} \tag{20}$$

Consider a stream flowing steadily at rate w lb/sec through a curved passage in the xy plane, Fig. 16-14; take a control volume bounded by

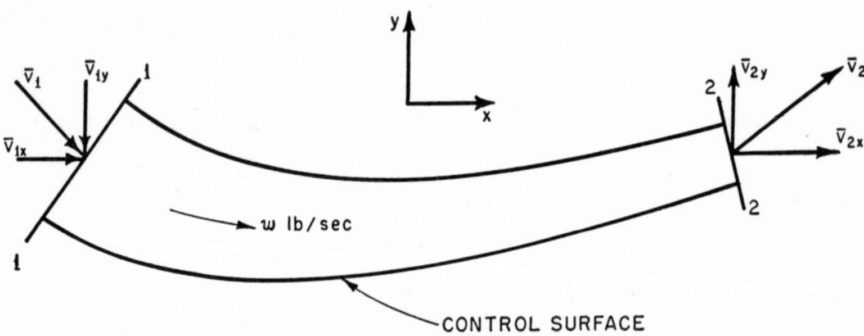

Fig. 16-14. Flow through a curved passage.

the walls of the passage and the sections 1 and 2, normal to the flow axis. It may be shown from fluid mechanics that the application of Eq. (20) to the fluid within the control volume gives the following result:

$$F = \frac{w}{g_0}(\overline{V}_2 - \overline{V}_1) \tag{21}$$

Equation (21) says that the accelerating force upon the fluid within the control volume is equal to the difference between the momentum of the fluid leaving the control volume per unit time and the momentum of the fluid entering the control volume per unit time.

Equation (21) is a vector equation. For many purposes it is most conveniently written in terms of components, thus:

$$F_x = \frac{w}{g_0}(\overline{V}_{2x} - \overline{V}_{1x}), \qquad F_y = \frac{w}{g_0}(\overline{V}_{2y} - \overline{V}_{1y}) \tag{21a}$$

The forces acting upon the fluid within the control volume are represented in Fig. 16-15. p_1a_1 and p_2a_2 are pressure forces from the adjacent fluid, upon the sections 1 and 2; s_1a_1 and s_2a_2 are shear forces on these sections; R is the resultant of pressure and shear forces from the passage walls upon the fluid; F is the overall resultant of all forces

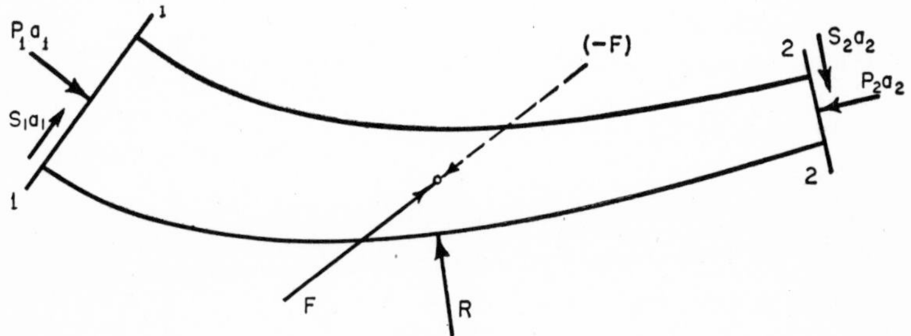

Fig. 16-15. Forces on the fluid in a curved passage.

upon the fluid, and $-F$ is the resultant reaction of the fluid. The relations of these forces may be expressed by

$$F = p_1a_1 + p_2a_2 + s_1a_1 + s_2a_2 + R \tag{22}$$

or

$$0 = (-F) + p_1a_1 + p_2a_2 + s_1a_1 + s_2a_2 + R$$

In these equations each term is a vector and the addition is by the laws of vectors. Equation (22) may also be written in terms of components, for example,

$$F_x = (p_1a_1)_x + (p_2a_2)_x + (s_1a_1)_x + (s_2a_2)_x + R_x \tag{22a}$$

In this equation, forces are positive when they act in the positive x direction; if x is positive in the direction of flow, $(p_1a_1)_x$ is a positive force, while $(p_2a_2)_x$ is negative.

16-10 Force and work on a turbine bucket. In this section it is assumed that the row of turbine buckets on the rim of a wheel is

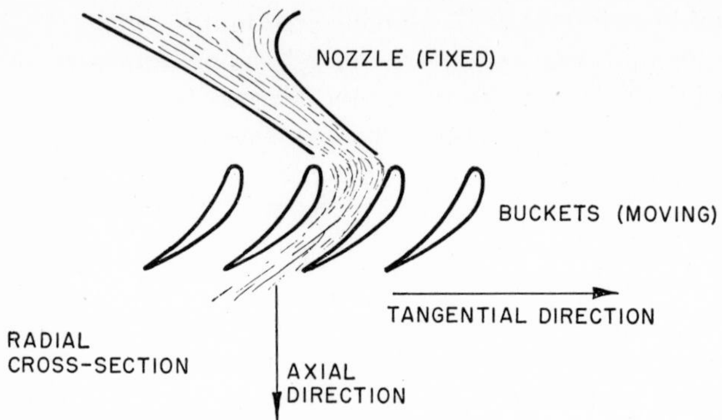

Fig. 16-16. Flow path nomenclature.

equivalent to a straight line of buckets moving in the tangential direction. Looking radially in toward the axis of rotation, the cross-section of the flow path is as shown in Fig. 16-16.

The velocities of the fluid at the entrance and exit from the buckets are shown by a velocity diagram, Fig. 16-17. The velocity of the nozzle jet $\overline{V}_1$ is the *absolute* velocity entering the bucket; the angle α between $\overline{V}_1$ and the tangential direction is the *nozzle angle*. Since the bucket is moving with velocity $\overline{V}_b$ the *relative* velocity entering the bucket $\overline{V}_{1R}$ is obtained by subtracting $\overline{V}_b$ from $\overline{V}_1$; this is a vector subtraction as shown in the figure.

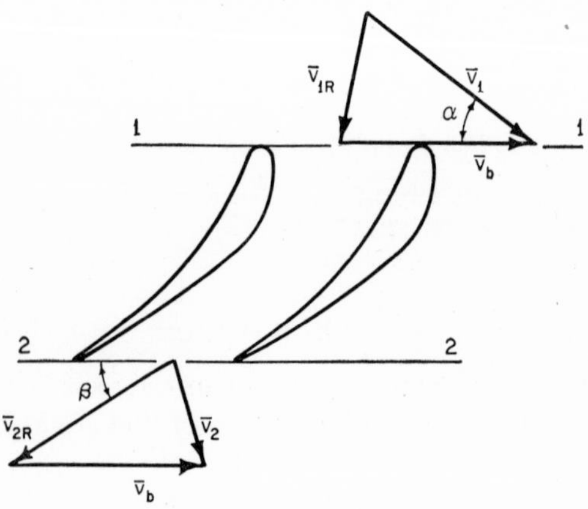

Fig. 16-17. Velocity diagram.

The *relative* velocity of the stream leaving the bucket $\overline{V}_{2R}$ is fixed in direction by the *bucket exit angle* β. The magnitude of $\overline{V}_{2R}$ is determined by considering the bucket passage as a nozzle. Applying Eq. (10) with a velocity coefficient C_b,

$$\overline{V}_{2R} = C_b\sqrt{2g_0(h_1 - h_{2s}) + \overline{V}_{1R}^2} \tag{23}$$

The *absolute* velocity leaving the bucket $\overline{V}_2$ is obtained by adding $\overline{V}_b$ to $\overline{V}_{2R}$ vectorially as shown in Fig. 16-17.

The tangential and axial components of the several velocities are designated respectively by y and z with appropriate subscripts. The two velocity triangles of Fig. 16-17 are shown complete with components in Fig. 16-18.

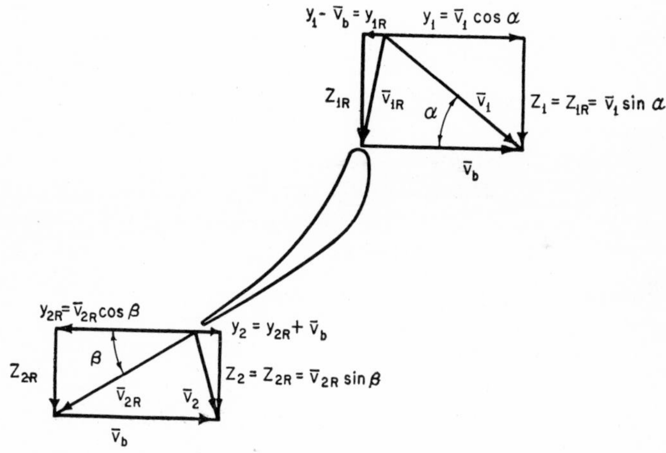

Fig. 16-18. Velocity components.

From Eq. (22a), neglecting shear forces on sections 1 and 2, the y component of force upon the fluid from the bucket passage walls is

$$R_y = F_y - (p_1a_1)_y - (p_2a_2)_y \tag{24}$$

Now the tangential components of pressure force on the surfaces 1 and 2 are zero because of the direction of the surfaces. Therefore if the tangential component of force upon the bucket from the fluid is F_t,

$$F_t = -R_y = -F_y \tag{25}$$

where y is in the tangential direction. Then from (21a) and (25)

$$F_t = \frac{w}{g_0}(y_1 - y_2) \tag{26}$$

The axial direction is designated the z direction; then the axial force on the bucket F_A is the negative of R_z, the z component of bucket force on the fluid. The surfaces 1 and 2 are normal to the z direction and the pressure on a_1 is in the positive sense, while the pressure on a_2 is in the negative sense. Then from (22a), since the shear forces have no axial component,

$$F_z = p_1a_1 - p_2a_2 + R_z$$

and

$$F_A = -R_z = -F_z + p_1a_1 - p_2a_2$$

or

$$F_A = \frac{w}{g_0}(z_1 - z_2) + p_1a_1 - p_2a_2 \tag{26a}$$

In actual turbines the effective areas against which the pressures p_1 and p_2 act are not limited to the bucket passage areas but include all the areas of the buckets and their supporting disc or drum which are subjected to those pressures.

The work done per second on a turbine bucket is given by the product of tangential force on the bucket and tangential velocity of the bucket, or

$$wW_x = F_t\overline{V}_b = \frac{w}{g_0}(y_1 - y_2)\overline{V}_b \tag{27}$$

where W_x is the work done per pound of fluid. Then

$$W_x = \frac{(y_1 - y_2)\overline{V}_b}{g_0}\,\frac{\text{ft lbf}}{\text{lbm}} \tag{28}$$

By substituting for y_1 its equal $(y_{1R} + \overline{V}_b)$, and for y_2 its equal $(y_{2R} + \overline{V}_b)$, (26) becomes

$$F_t = \frac{w}{g_0}(y_{1R} - y_{2R}) \tag{29}$$

and (28) becomes

$$W_x = \frac{(y_{1R} - y_{2R})\overline{V}_b}{g_0} \tag{30}$$

Equations (28) and (30) may be written in different forms by geometric reasoning. Referring to Fig. 16-18, by the properties of right triangles,

$$y_1^2 = \overline{V}_1^2 + z_1^2, \qquad y_{1R}^2 = \overline{V}_{1R}^2 + z_{1R}^2$$
$$y_2^2 = \overline{V}_2^2 + z_2^2, \qquad y_{2R}^2 = \overline{V}_{2R}^2 + z_{2R}^2 \tag{a}$$

Also, by the properties of relative and absolute velocities,

$$\overline{V}_b = y_1 - y_{1R} = y_2 - y_{2R} \tag{b}$$

Substituting from (b) in (28) and (30) and adding (28) to (30) gives

$$W_x = \frac{y_1^2 - y_{1R}^2 + y_{2R}^2 - y_2^2}{2g_0} \tag{31}$$

Substituting from (a) in (31) gives

$$W_x = \frac{\overline{V}_1^2 - \overline{V}_{1R}^2 + \overline{V}_{2R}^2 - \overline{V}_2^2}{2g_0} \tag{32}$$

Equation (32) may also be derived by thermodynamic reasoning, as follows. Consider the turbine bucket passage in Fig. 16-17. Applying the steady flow energy equation to a stationary control volume extending through the bucket region from section 1 to section 2 gives, for an adiabatic turbine,

$$h_1 + \frac{\overline{V}_1^2}{2g_0} = h_2 + \frac{\overline{V}_2^2}{2g_0} + W_x \tag{c}$$

The steady flow energy equation can also be applied to a control volume moving with the buckets at constant speed; in this case relative velocities must be used. Since no mechanism moves through the control surface there is no shaft work (on the basis of a control volume moving with the buckets the flow through the buckets is simply duct flow). The enthalpies at 1 and 2, however, are independent of the basis of velocities. Therefore on the relative basis

$$h_1 + \frac{\overline{V}_{1R}^2}{2g_0} = h_2 + \frac{\overline{V}_{2R}^2}{2g_0} \tag{d}$$

Eliminating h_1 and h_2 between (c) and (d) gives (32).

The relations derived in this section are based on the assumption of uniform velocities across the fluid stream at all sections and uniform bucket velocity. In real turbines the fluid velocity will vary across each section, the bucket velocity will vary in the radial direction, and the mean radius of the bucket (and its mean velocity) often increase in the direction of flow. Some turbines are arranged for radial flow instead of the axial flow considered here. These variations are accounted for by more detailed methods of analysis, which do not fall within the scope of this book.

16-11 Nozzle-bucket efficiency. The effectiveness of a nozzle and bucket combination in obtaining shaft work from a fluid stream is stated in terms of the *nozzle-bucket efficiency*, η_{nb}, defined by

$$\eta_{nb} = \frac{W_x}{h_a - h_{2s}} \tag{33}$$

where W_x is the work done by the fluid on the buckets, h_a is the enthalpy of the fluid approaching the nozzle, and h_{2s} is the enthalpy at a state having the exit pressure and the entropy of the fluid approaching the nozzle. The nozzle-bucket efficiency compares the

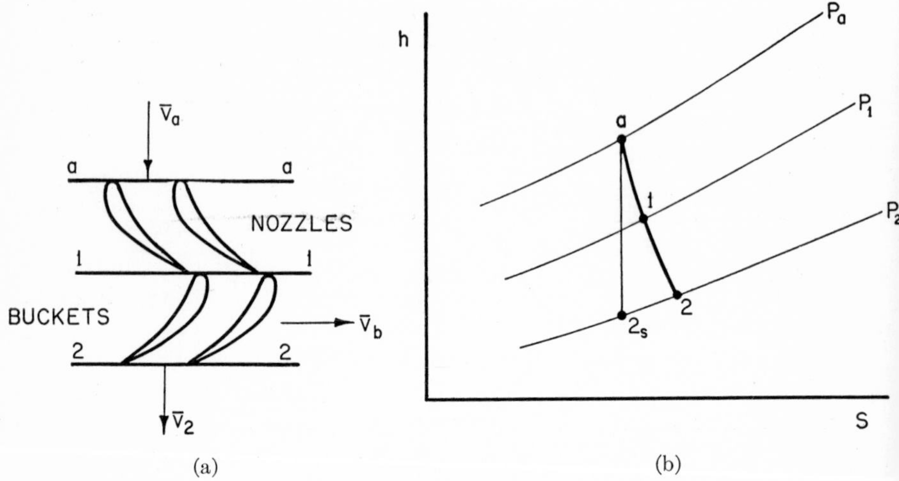

(a) (b)

Fig. 16-19. (a) Turbine stage flow diagram. (b) State path plot.

work of the actual combination to the shaft work of a reversible adiabatic expansion in steady flow, beginning at the nozzle inlet state and ending at the bucket exit pressure. In Fig. 16-19 are plotted the state paths for a nozzle-bucket process and the corresponding reversible adiabatic process.

It is sometimes convenient to substitute for the *isentropic enthalpy drop* $(h_a - h_{2s})$ an equivalent kinetic energy in terms of a *characteristic velocity* $\overline{V}_c$ defined by

$$\overline{V}_c = \sqrt{2g_0(h_a - h_{2s})} \tag{34}$$

Then

$$\eta_{nb} = \frac{W_x}{\overline{V}_c^2/2g_0} \tag{35}$$

The nozzle-bucket efficiency depends upon the geometry of the passages, and upon irreversibility due to friction and shock effects. The nozzle-bucket efficiency for two hypothetical reversible combinations, or *stages*, will be determined in the following sections.

16-12 The simple impulse stage. The classical distinction among turbine stages is that between impulse and reaction stages. As a matter of convenience this distinction will be followed here, but it should be understood that in practice the two types differ more in degree than in kind. In the simple impulse stage the entire pressure drop occurs in the nozzle; this can be made to happen simply by providing free passages (other than the bucket passages) from one side of the buckets to the other so that the pressure is equalized. The changes of pressure and absolute velocity during flow through such a stage are indicated in Fig. 16-20.

A velocity diagram for an impulse stage is shown in Fig. 16-21.

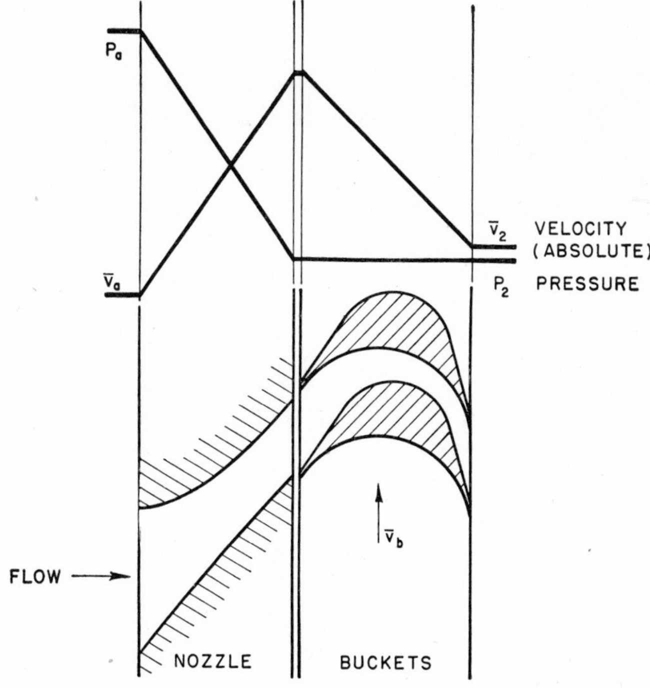

Fig. 16-20. Impulse stage.

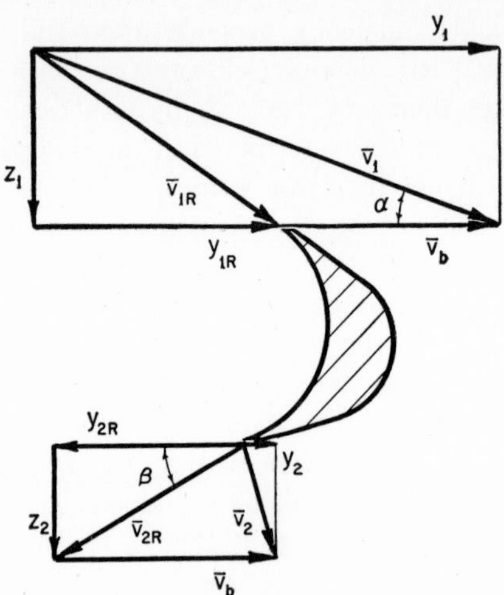

Fig. 16-21.

The forces and the work on the bucket are determined by the relations derived in Sec. 16-10. The various velocities are determined as follows. The nozzle jet velocity $\overline{V}_1$ is fixed by the fluid state before the nozzle, the exit pressure, and the nozzle velocity coefficient, as shown in Sec. 16-4. The bucket velocity $\overline{V}_b$ is fixed by the speed and size of the wheel. Then the other velocities can be calculated if the angles α and β and the bucket velocity coefficient are known. The relations used are summarized below.

$$y_1 = \overline{V}_1 \cos \alpha, \qquad y_{1R} = y_1 - \overline{V}_b$$
$$\overline{V}_{1R} = \sqrt{z_1^2 + y_{1R}^2}, \qquad \overline{V}_{2R} = C_b \overline{V}_{1R}$$
$$y_{2R} = \overline{V}_{2R} \cos \beta \qquad y_2 = y_{2R} + \overline{V}_b$$
$$z_2 = \overline{V}_{2R} \sin \beta \qquad \overline{V}_2 = \sqrt{z_2^2 + y_2^2}$$

It is to be noted that all additions are algebraic and that y_{2R} is always in the negative direction; hence y_2 may be either positive or negative depending upon the relative magnitudes of y_{2R} and $\overline{V}_b$. It is also to be noted that there is no nozzle action in the bucket; the difference in magnitude of $\overline{V}_{2R}$ from $\overline{V}_{1R}$ is due to friction. (In real impulse stages there may be some pressure drop in the buckets.)

The bucket entrance angle does not enter the above computations at all, but from physical considerations it is apparent that the bucket entrance should be aligned with the relative velocity $\overline{V}_{1R}$. There is, however, nothing to prevent the operation of a turbine at such conditions that the angle of $\overline{V}_{1R}$ is not equal to the bucket entrance angle, in which case there will be some *shock loss* due to the incorrect angle. In the consideration of hypothetical stages under different design conditions it is assumed in each case that the correct entrance angle is used. In the consideration of different operating conditions for a given design, however, the reduction of efficiency due to shock at the bucket entrance must be taken into account for conditions other than the design condition. The distinction between design characteristics and operating characteristics is often overlooked by students, but it is of fundamental importance in the study of machinery of any type.

It is possible to show the general nature of the design characteristics of an impulse stage by means of a hypothetical reversible stage having variable bucket angles such that the bucket entrance angle is always kept equal to the angle of the entering relative velocity $\overline{V}_{1R}$ and the bucket exit angle β is kept equal to the bucket entrance angle. For such a case $\overline{V}_{2R} = \overline{V}_{1R}$ (no friction) and $z_2 = z_1$. $y_{2R} = y_{1R}$ in magnitude (similar triangles), but algebraically $y_{2R} = -y_{1R}$. Then

$$y_2 = \overline{V}_b + y_{2R} = 2\overline{V}_b - y_1 \tag{a}$$

The tangential force is

$$F_t = \frac{w}{g_0}(y_1 - y_2) = \frac{2w}{g_0}(y_1 - \overline{V}_b) \tag{b}$$

The work per pound of fluid is

$$W_x = \frac{F_t\overline{V}_b}{w} = \frac{2}{g_0}(y_1\overline{V}_b - \overline{V}_b^2) \tag{c}$$

From (34)
$$y_1 = \overline{V}_1 \cos\alpha = \overline{V}_c \cos\alpha \tag{d}$$

From (35), (c) and (d) the nozzle-bucket efficiency for the hypothetical stage is

$$\eta_{nb} = \frac{4}{\overline{V}_c^2}(\overline{V}_b\overline{V}_c\cos\alpha - \overline{V}_b^2) = 4\left[\frac{\overline{V}_b}{\overline{V}_c}\cos\alpha - \left(\frac{\overline{V}_b}{\overline{V}_c}\right)^2\right] \tag{e}$$

Plotting efficiency as a function of $\overline{V}_b/\overline{V}_c$ for nozzle angles of zero and 20 degrees gives the parabolas of Fig. 16-22. By differentiation it can be shown that the maximum efficiency for the hypothetical stage occurs at $\overline{V}_b/\overline{V}_c = 0.5\cos\alpha$ and is equal to $\cos^2\alpha$. Thus, to obtain maximum efficiency, the nozzle angle should be as small as

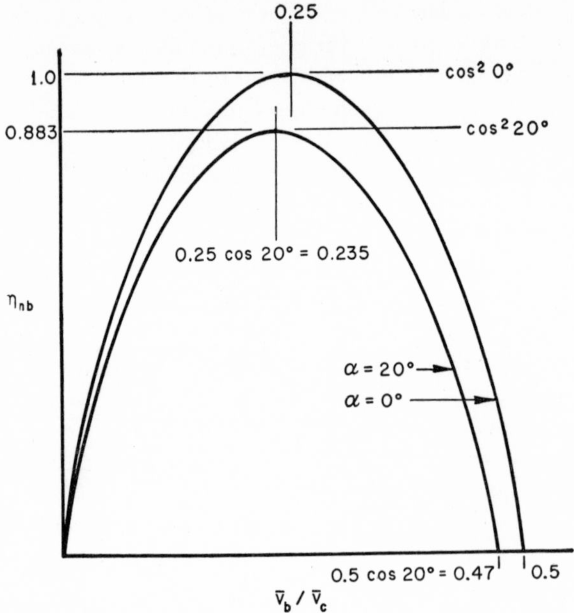

Fig. 16-22. Impulse stage efficiency; for reversible flow with symmetrical buckets.

possible. Since zero angle would give zero area for flow, the nozzle angle must have some finite value; it may be of the order of 15 or 20 degrees in actual turbines.

The relations for the hypothetical reversible impulse stage with equal entrance and exit angles show qualitatively what happens in real impulse stages as the design conditions are altered. For maximum nozzle-bucket efficiency the bucket velocity must be in the vicinity of half the nozzle jet velocity, but the exact point of best efficiency depends upon the angles and the friction and shock losses

16-13. The reaction stage. It is often desirable to operate turbine stages in such a manner that there is a pressure drop in the bucket passages as well as in the nozzles; in this case the bucket passages become moving nozzles. A turbine stage operating in this way is called a reaction stage. The overall pressure drop through a reaction stage may be divided in various proportions between nozzle pressure drop and bucket pressure drop. Hero's turbine, Fig. 16-12, is one

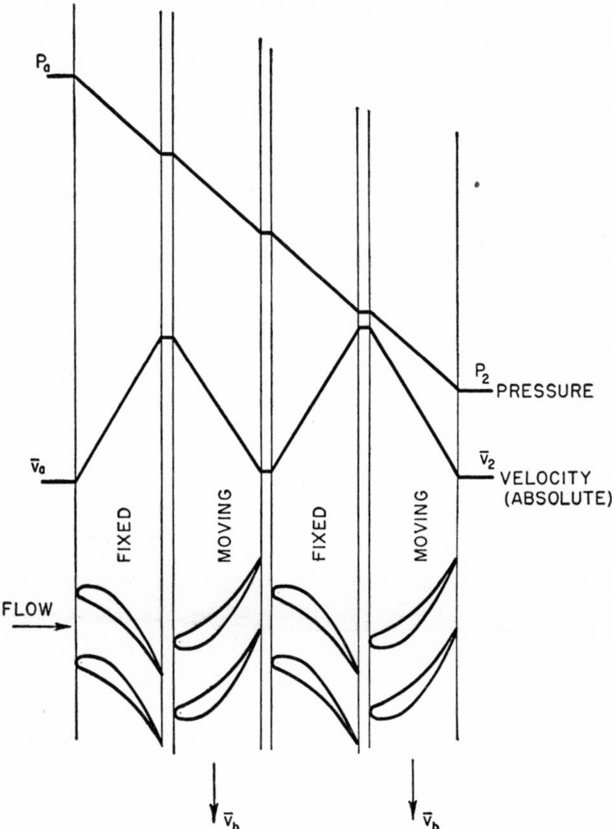

Fig. 16-23. Reaction stages.

limiting case in which the total pressure drop occurs in the moving nozzles; the simple impulse turbine may be considered as the other limiting case in which all the pressure drop occurs in the fixed nozzles. In many modern reaction turbines approximately half the pressure drop occurs in the moving nozzles or buckets.

The changes of pressure and absolute velocity of the fluid during flow through a series of reaction stages are indicated in Fig. 16-23.

A velocity diagram for a reaction stage is shown in Fig. 16-24. In the reaction stage the nozzle jet velocity is determined by the fluid state before the stage, the exit pressure, the division of pressure

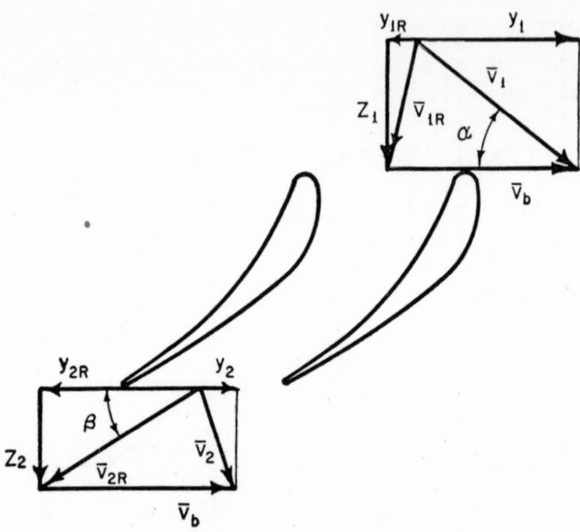

Fig. 16-24. Reaction stage velocity diagram.

drop between fixed and moving elements, and the velocity coefficient of the nozzle. The bucket velocity is fixed by the speed and size of the wheel. Then the other velocities can be calculated if the angles α and β and the bucket velocity coefficient are known.

For example, in a so-called pure reaction stage it is specified that the isentropic enthalpy drop for the whole stage shall be divided

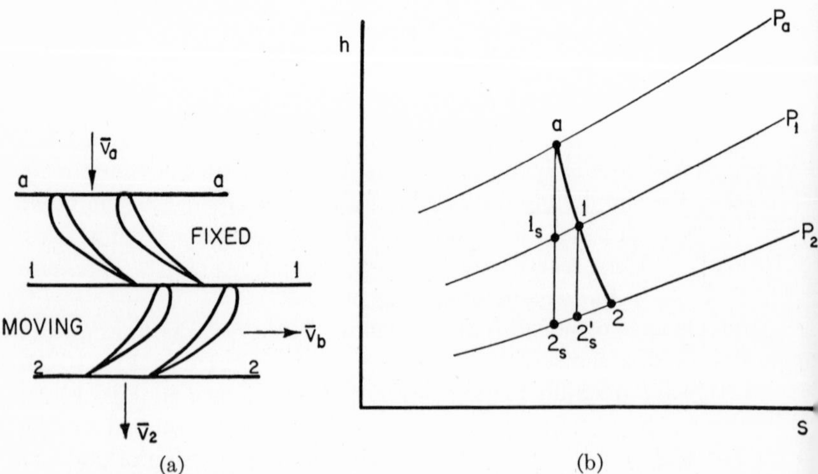

Fig. 16-25. Reaction stage: (a) flow diagram, and (b) state path plot.

equally between the fixed and moving elements. Referring to Fig. 16-25, the inlet state a and the exit pressure p_2 are given; the intermediate pressure p_1 is then fixed by the relation

$$h_a - h_{1s} = \tfrac{1}{2}(h_a - h_{2s}) \tag{a}$$

Then

$$\begin{aligned}
\overline{V}_1 &= C_{\overline{V}}\sqrt{2g_0(h_a - h_{1s}) + \overline{V}_a^2} \\
y_1 &= \overline{V}_1 \cos\alpha, \qquad y_{1R} = y_1 - \overline{V}_b \\
z_1 &= \overline{V}_1 \sin\alpha \\
\overline{V}_{1R} &= \sqrt{z_1^2 + y_{1R}^2} \\
\overline{V}_{2R} &= C_b\sqrt{2g_0(h_1 - h_{2s}) + \overline{V}_{1R}^2} \\
y_{2R} &= -\overline{V}_{2R}\cos\beta, \qquad y_2 = y_{2R} + \overline{V}_b \\
z_2 &= \overline{V}_{2R}\sin\beta \\
\overline{V}_2 &= \sqrt{z_2^2 + y_2^2}
\end{aligned}$$

It is possible to show the general design characteristics of a pure reaction stage by means of a hypothetical reversible stage having equal angles α and β, and a negligible approach velocity $\overline{V}_a$. For such a stage

$$\overline{V}_1 = \sqrt{\frac{2g_0(h_a - h_{2s})}{2}} = \frac{\overline{V}_c}{\sqrt{2}}$$

$$y_1 = \frac{\overline{V}_c}{\sqrt{2}}\cos\alpha$$

$$y_{1R} = y_1 - \overline{V}_b = \frac{\overline{V}_c}{\sqrt{2}}\cos\alpha - \overline{V}_b$$

By the law of cosines

$$\overline{V}_{1R}^2 = \overline{V}_1^2 + \overline{V}_b^2 - 2\overline{V}_1\overline{V}_b\cos\alpha = \frac{\overline{V}_c^2}{2} + \overline{V}_b^2 - \sqrt{2}\,\overline{V}_c\overline{V}_b\cos\alpha$$

Also

$$\overline{V}_{2R} = \sqrt{\frac{\overline{V}_c^2}{2} + \overline{V}_{1R}^2} = \sqrt{\overline{V}_c^2 + \overline{V}_b^2 - \sqrt{2}\,\overline{V}_c\overline{V}_b\cos\alpha}$$

$$y_{2R} = -\overline{V}_{2R}\cos\beta = -\overline{V}_{2R}\cos\alpha$$

The work is

$$\begin{aligned}
W_x &= \frac{y_{1R} - y_{2R}}{g_0}\overline{V}_b \\
&= \frac{\overline{V}_b}{g_0}\left[\frac{\overline{V}_c}{\sqrt{2}}\cos\alpha - \overline{V}_b + (\cos\alpha)\sqrt{\overline{V}_c^2 + \overline{V}_b^2 - \sqrt{2}\,\overline{V}_c\overline{V}_b\cos\alpha}\right]
\end{aligned}$$

The nozzle-bucket efficiency is

$$\eta_{nb} = \frac{W_x}{\overline{V}_c^2/2g_0}$$

$$= 2\frac{\overline{V}_b}{\overline{V}_c}\left[\frac{\cos\alpha}{\sqrt{2}} - \frac{\overline{V}_b}{\overline{V}_c} + (\cos\alpha)\sqrt{1 + \left(\frac{\overline{V}_b}{\overline{V}_c}\right)^2 - \sqrt{2}\frac{\overline{V}_b}{\overline{V}_c}\cos\alpha}\right]$$

Plotting efficiency as a function of $\overline{V}_b/\overline{V}_c$ for nozzle angles of zero and 20 degrees gives the curves of Fig. 16-26. The curves show quali-

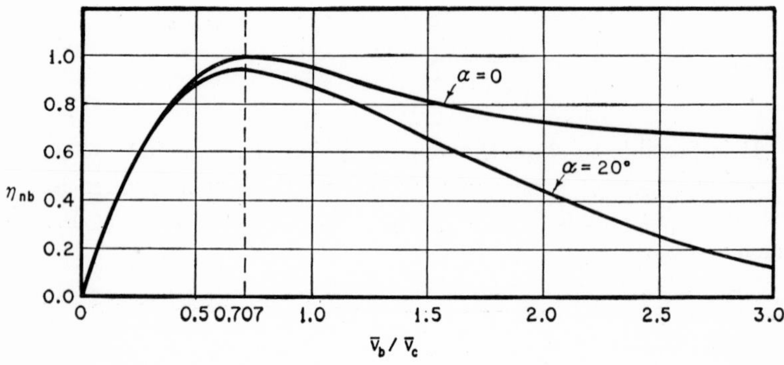

Fig. 16-26. Reaction stage efficiency; for reversible flow with equal angles and equal enthalpy drops in nozzles and buckets.

tatively the characteristics of real reaction stages of equal enthalpy drop as the design conditions are altered. The maximum efficiency occurs for $\alpha = 0$ at $\overline{V}_b/\overline{V}_c = 1/\sqrt{2}$, and for useful values of α at slightly lower speeds; its magnitude approaches unity as the nozzle angle and bucket angle approach zero. For any fraction of isentropic enthalpy drop in the bucket other than five-tenths, the curves are similar but the maximum is at a different value of $\overline{V}_b/\overline{V}_c$. For any given nozzle angle the maximum efficiency of the hypothetical reaction stage is greater than that of the hypothetical impulse stage discussed in the preceding sections. In practice, however, either type of stage may be better, depending upon the operating conditions, and each type is used where it is better. The reaction stage in general has lower fluid velocities than the impulse stage, resulting in lower friction losses. The reaction stage has the disadvantage of a higher ratio of bucket velocity to characteristic velocity than the impulse stage.

PROBLEMS

16-1. (a) In reversible adiabatic nozzle flow under the assumptions made in this chapter, what will happen to the stagnation temperature and stagnation pressure as the static pressure falls? (b) In irreversible adiabatic nozzle flow of a perfect gas what will happen to the stagnation temperature and stagnation pressure as the static pressure falls?

16-2. (a) Steam is flowing in a duct at 50 psia, 300°F, 1000 fps; find the stagnation enthalpy, stagnation pressure, and stagnation temperature of the steam. (b) Repeat (a) for a velocity of 200 fps. (c) At low velocities the static and impact properties differ negligibly. For air at total temperature of 140°F find the velocity at which the static and impact pressures differ by 1 percent. (d) Repeat (c) for a difference of 1 percent in static and impact temperatures (1 percent of absolute temperature).

16-3. Make a plot like Fig. 16-4 for the reversible adiabatic expansion of steam in a nozzle from a stagnation state at 160 psia, 600°F, to atmospheric pressure, taking property values from the steam tables. Estimate the critical pressure ratio from your plot.

16-4. Anozzle is to be designed to pass 10,000 lb/hr of liquid water from a pipe to a tank. The water in the pipe is at 350 psia, 70°F, and 10 fps velocity; the static pressure in the tank is 15 psia. Consider the water incompressible (v = constant). Sketch a suitable nozzle shape and calculate the minimum diameter of the nozzle.

16-5. A turbine nozzle is to pass 1 lb/sec of hydrogen from 100 psia, 70°F, to a region at 10 psia. If the process is frictionless and adiabatic find the nozzle throat and exit areas, the gas exit velocity, and the gas temperature at exit. State any assumptions made.

16-6. A steam turbine nozzle is to receive steam at 800 psia, 800°F, 450 fps, and to discharge to a region at 700 psia. The coefficient of discharge is 1.01, and the nozzle efficiency is 0.98. Find the minimum area, and the exit area, of the nozzle if the flow rate is 1 lb/sec.

16-7. Gaseous combustion products having properties, $c_p = 0.26$ Btu/lb °F, $k = 1.36$, and $R = 53.2$ ft lb/lb °R, are to flow at the rate of 100 lb/sec through a nozzle having inlet conditions of 15 psia, 1500°F, 600 fps, and discharge pressure of 5 psia. The discharge coefficient is 0.96, and the nozzle efficiency is 0.88. Find the proper throat and exit areas and the exit temperature.

16-8. Solve Problem 16-7 if the gas is air having the properties given in the Gas Tables of Keenan and Kaye.

16-9. A nozzle receives 10,000 lb/hr of steam at 250 psia, 500°F, 200 fps, and discharges to a region at 100 psia. The average exit velocity is measured as 1850 fps. The nozzle exit area is 1.00 sq in. Find: (a) the nozzle efficiency; (b) the throat area, assuming reversible flow in the nozzle as far as the throat; (c) whether the stream expands more or less than is proper for the nozzle.

16-10. Steam enters a diffuser at 1000 fps, 20 psia, 300°F, and is diffused to 50 fps. (a) If the diffuser is reversible find the exit pressure and the ratio of exit area to inlet area. (b) If the diffuser efficiency is 80 percent find the exit pressure and the ratio of exit area to inlet area.

16-11. An adiabatic diffuser is to be used to recover kinetic energy from the discharge of an exhaust fan. The gas flowing is a mixture of combustion products having properties $c_p = 0.25$ Btu/lb °F, $k = 1.38$, and $R = 54$ ft lb/lb °R. The gas leaves the fan at 400 fps, 540°F, and flows through the diffuser which reduces the velocity to 100 fps at the discharge to atmosphere. If the diffuser efficiency is 90 percent how much below the atmospheric pressure of 15 psia will the diffuser inlet pressure be? What will be the ratio of diffuser exit area to inlet area?

16-12. An ejector, Fig. 16-8, has a primary flow, w_1, of 1.0 lb/sec of air at 100 psia, 340°F; the secondary air flow, w_2, enters at 5 psia, 340°F. The discharge at section 4 is to the atmosphere at 15 psia. The efficiency of the ejector as a pump is defined by

$$\eta = \frac{w_2(h_4 - h_2)_s}{w_1(h_1 - h_4)_s}$$

where the enthalpy differences are taken on isentropic paths from the respective initial states to the pressure at 4. (a) If the efficiency of the ejector is 20 percent find the flow rate, w_2. (b) If the entrance and exit velocities are all 80 fps, and the efficiency is 20 percent, find the diameter at section 4.

16-13. Steam from a large chamber at 180 psia, 400°F, expands to atmospheric pressure through an adiabatic nozzle of 0.00270 sq ft throat area; the flow rate is 1.00 lb/sec. Find the coefficient of discharge based upon each of two ideal state paths: (a) the fluid follows an equilibrium isentropic path (steam table data); (b) the fluid follows a metastable isentropic path in the mixture region. (Supersaturated properties obtained by using pv^k = constant; from Fig. 8 of Keenan and Keyes, k is about 1.30 for the given conditions.)

16-14. When a hot liquid passes below the saturation pressure during a nozzle process some of the liquid may "flash" into vapor. Such "flashing flow" requires much more area than liquid flow. Assuming reversible adiabatic flow, find from the steam table data the nozzle exit area required for the conditions of Problem 16-4, if the initial temperature is 300°F. (Actually, the vaporization will be delayed, and the flow rate per unit area may be much greater than calculated. This is analogous to the supersaturation phenomenon with vapor.)

16-15. A flow nozzle of 6.00 in. diameter in an 8.00 in. diameter pipe is used to meter steam which arrives at the upstream pressure tap at 40 psia, 300°F. If the differential pressure ($p_1 - p_2$) is 80 in. of water, the nozzle coefficient, C, is 0.985, and the expansion factor, Y, is 0.94, find the flow rate in lb/hr.

16-16. A sharp-edged orifice, 0.75 in. diameter in a 3.00 in. pipe, is calibrated with air at 14.8 psia and 75°F; when the air flow rate (measured by a direct volumetric method) is 7.75 cu ft/min, the differential pressure is 1.00 in. of water. This is so small (about $\frac{1}{4}$ percent of the absolute pressure) that the flow may be considered incompressible. What is the coefficient, C, for the orifice?

16-17. A well-known empirical formula for critical flow of saturated steam through a rounded-entrance nozzle with negligible approach velocity is Napier's formula,

$$w = \frac{ap_1}{70}$$

where w is flow rate, lb/sec; a is nozzle throat area, sq in.; and p_1 is upstream pressure (stagnation), psia.

Compute the rate of flow through a nozzle of 1.00 sq in. area, with steam supplied at 100 psia, saturated, for critical flow, using: (a) Napier's formula; (b) the thermodynamic nozzle equations, taking the flow to be isentropic. (Note: Although good agreement is found between Napier's formula and the thermodynamic equations for a considerable range of pressures, using saturated steam, the formula does not apply for other conditions. Empirical formulas are useful, but care should be taken not to mis-use them.)

16-18. Fliegner's empirical formula for critical flow of air through a rounded-entrance nozzle with negligible approach velocity is

$$w = \frac{0.53ap_1}{\sqrt{T_1}}$$

where w is flow rate, lb/sec; a is nozzle throat area, sq in.; p_1 is upstream pressure (stagnation), psia; and T_1 is upstream temperature (stagnation), F_{abs}.

Taking air as a perfect gas of specific heat ratio 1.4, and gas constant 53.34 ft lb/lb °R, derive from the thermodynamic flow equation an equation in the form of Fliegner's formula and compare the values of the numerical coefficients in the derived equation and in the empirical formula.

16-19. In a rocket the driving fluid starts from rest, relative to the rocket, and is accelerated out through the tail pipe or nozzle. In a certain rocket 2 lb/sec of combustion products are discharged at a relative velocity of 4000 fps. Find the thrust on the rocket.

16-20. A flexible fire hose 2.5 inches in diameter runs through a 90-degree bend to a nozzle having an exit diameter of 1.00 in. Water flows through the hose at the rate of 0.341 cfs. How much force must the fireman exert, parallel to the axis of the nozzle, to hold the nozzle? In which direction must he exert the force?

16-21. Referring to Problem 16-20, for each case listed below, what is the magnitude of the total stress on the connection between the nozzle and the hose, and is it tension or compression? Case (a), the fireman holds the nozzle itself. Case (b), the fireman holds the hose adjacent to the nozzle. Neglect gravity forces.

16-22. Steam enters a turbine bucket passage at 1400 fps absolute, with a nozzle angle of 14 degrees. The bucket velocity is 600 fps. The bucket velocity coefficient is unity, and there is no pressure change through the bucket passage; the bucket exit angle is 32 deg. Find, per pound of fluid flow per second: (a) the power supplied to the buckets; (b) the tangential force on the buckets; (c) the axial force on the buckets.

16-23. Air enters a turbine bucket passage at 20.0 psia, 1000°F, 850 fps absolute velocity, with a nozzle angle of 20 deg. The pressure drop through the bucket passage is 2.00 psi, the bucket velocity is 650 fps, and the bucket exit angle is 20 deg. (a) If the bucket velocity coefficient is unity find the power supplied to the buckets per pound of air flow per second.

16-24. In an impulse turbine stage the nozzle exit velocity is 1200 fps and the nozzle angle is 14 deg. The bucket exit angle is 25 deg and the bucket velocity coefficient is 0.90. If the bucket velocity is 600 fps find: (a) the work done per pound of fluid flow; (b) the kinetic energy per pound of fluid in the stream leaving the buckets; (c) the angle between the absolute velocity leaving the buckets and the tangential direction. Sketch the velocity diagram.

16-25. For the conditions of Problem 16-24, assuming the bucket velocity coefficient stays constant, find the bucket velocity for maximum work (assume bucket velocities, find the work, and make a plot). At this bucket velocity find the angle of the relative fluid velocity entering the bucket.

16-26. In an impulse air-turbine stage the air supply is at 105 psia, 80°F, negligible velocity, and the exhaust pressure is 15 psia. The nozzle angle is 15 deg and the bucket exit angle is 30 deg. The pitch diameter of the wheel is 8 in. The nozzle velocity coefficient is 0.96, and the bucket velocity coefficient is 0.88; the bucket velocity is half the tangential component of the nozzle exit velocity. Find: (a) the turbine speed (rpm); (b) the nozzle-bucket efficiency; (c) the nozzle exit area required to do work on the buckets at the rate of 100 hp; (d) the temperature of the air leaving the buckets.

16-27. In a hypothetical reversible reaction turbine stage with zero angles the enthalpy drop is $\frac{2}{3}$ in the nozzle and $\frac{1}{3}$ in the bucket passage. Find the ratio of bucket velocity to characteristic velocity for maximum efficiency.

16-28. A reversible reaction stage has equal enthalpy drops in the moving and stationary rows. Air enters the stage at 1500°F, 30 psia, 500 fps, and leaves at 20 psia. The nozzle and bucket passages have the same exit angle,

25 deg. The speed is such that the relative velocity entering the buckets is at 90 deg to the bucket velocity.

Find: (a) the pressure at the nozzle exit; (b) the bucket velocity; (c) the absolute velocity leaving the buckets; (d) the nozzle-bucket efficiency (can you explain why this is higher than for the hypothetical stage of Sec. 16-13?); (e) the ratio of bucket velocity to characteristic velocity for the stage; (f) the power supplied to the buckets per pound per second of air flow; (g) the ratio of bucket exit area to nozzle exit area (this indicates how much the radial length of the bucket exit must exceed that of the nozzle exit).

REFERENCES

ASME Research Report, "Fluid Meters, Theory, and Application."

ASME Power Test Codes, "Flow Measurement by Orifices and Flow Nozzles."

Binder, R. C., *Fluid Mechanics*. New York: Prentice-Hall, 1949.

Hunsaker, J. C. and B. G. Rightmire, *Engineering Applications of Fluid Mechanics*. New York: McGraw-Hill, 1948.

Keenan, J. H., *Thermodynamics*. New York: Wiley, 1941, chaps. 11, 18.

Shapiro, A. H., *The Dynamics and Thermodynamics of Compressible Fluid Flow*. New York: Ronald, 1953.

Stodola, A. (tr., L. Loewenstein), *Steam and Gas Turbines*. New York: Peter Smith, 1945, Vols. I and II.

Salisbury, J. K., *The Steam Turbine and Its Cycle*. New York: Wiley.

Gaffert, G. A., *Steam Power Stations*. New York: McGraw-Hill, 1952, chap. 3.

Moss, S. A., C. W. Smith, and W. R. Foote, "Energy Transfer Between a Fluid and a Rotor for Pump and Turbine Machinery." *Trans. ASME*, August 1942, p. 567.

Reese, H. R. and J. R. Carlson, "Thermal Performance of Modern Turbines," *Mech. Eng.*, March 1952, p. 205.

Warren, G. B. and P. H. Knowlton, "Engine Efficiencies from Modern Steam-Turbine Generators." *Trans. ASME*, Vol. 63, February 1941, p. 125.

Elston, C. W. and P. H. Knowlton, "Comparative Efficiencies of Central-Station Reheat and Non-reheat Steam-Turbine-Generator Units," *Trans. ASME*, Vol. 74, November 1952, p. 1389.

Gas turbine practice is developing so rapidly that specific references soon become obsolete. The *ASME Transactions, Mechanical Engineering* and *Power* should be consulted for current developments.

Chapter 17

GAS COMPRESSION

The compression of gases is an important process in many power plants, refrigeration plants, and industrial plants of which power plants and refrigeration plants are discussed elsewhere in this book. Industrial uses of gas compression occur in connection with compressed air motors for tools, air brakes for vehicles, servo-mechanisms, metallurgical and chemical processes, conveying of materials through ducts, transportation of natural gas, and production of bottled gases. The term *gas compression* is usually applied only to processes involving appreciable change of gas density; this excludes ordinary ventilation and furnace draft processes.

The machinery used in gas compression may be turbine type, such as centrifugal and axial flow machines; or positive displacement type, such as reciprocating machines, meshing rotor or gear machines, and vane-sealed machines. Insofar as it operates under steady flow conditions, any of these types of machine may have its energy analysis written in the form of the steady flow energy equation; in this chapter some general deductions will be made upon this basis.

17-1 Compression processes. A gas compression process may be designed either to be adiabatic or to involve heat transfer, depending upon the purpose for which the gas is compressed. If the compressed gas is to be used promptly in an engine or in a combustion process, adiabatic compression may be desirable in order to obtain the maximum available energy in the gas at the end of compression. In many applications, however, the gas is not used promptly but is stored in a tank for use as needed. The gas in the tank transfers heat to the surroundings, so that when finally used it is at room temperature. In these cases the overall effect of the compression and storage process is simply to increase the pressure of the gas without change of temperature. Now it will be shown below that if the gas is cooled *during* compression, instead of after that process, the work required will be less than for adiabatic compression. A further ad-

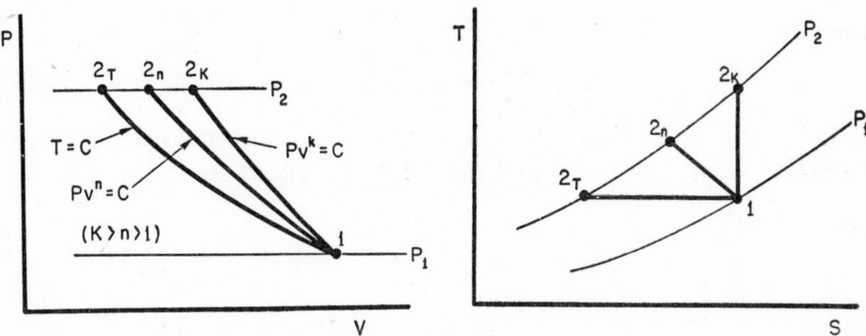

Fig. 17-1. Compression processes.

vantage of cooling is the reduction of volume and the consequent reduction of pipe line friction losses. For this reason, since cooling during the compression process is not very effective, *after-coolers* are often used to cool the gas leaving a compressor.

Because of the situation described above it is of interest to consider the effect of cooling upon a compression process. It is customary to investigate two particular idealized cases, namely reversible adiabatic and reversible isothermal, as well as a general case of a reversible polytropic process (pv^n = constant). The paths of such processes are plotted in Fig. 17-1 for a perfect gas compression process from state 1 to pressure p_2.

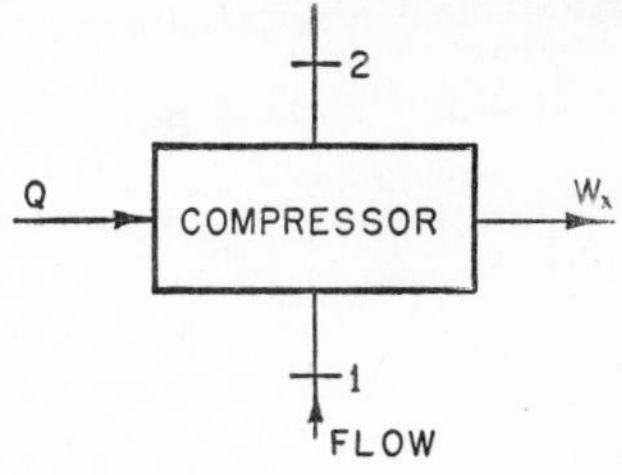

Fig. 17-2.

17-2 Work of compression in steady flow. The steady flow energy equation for a compression process may be written, referring to Fig. 17-2, as

$$h_1 + Q = h_2 + W_x \tag{1}$$

assuming that changes of potential and kinetic energy are negligible. It is often convenient to write the equation in terms of pressure and volume rather than enthalpy; this is done by using the property relation

$$T\,ds = dh - v\,dp \tag{9-6}$$

For a reversible process this gives

$$Q = \Delta h - \int v\,dp \tag{2}$$

then for any of the idealized cases of Fig. 17-1, from (1) and (2)

$$W_x = -\int v\,dp \tag{3}$$

It is assumed that for any gas a compression process may be represented with sufficient accuracy by an equation of the form pv^n = constant. Then

$$v = \frac{p_1^{1/n} v_1}{p^{1/n}}$$

and

$$\begin{aligned} W_x &= -p_1^{1/n} v_1 \int_1^2 \frac{dp}{p^{1/n}} \\ &= -p_1^{1/n} v_1 \frac{1}{1 - 1/n} \left[p_2^{1-1/n} - p_1^{1-1/n}\right] \\ &= -\frac{n}{n-1} p_1 v_1 \left[\left(\frac{p_2}{p_1}\right)^{(n-1)/n} - 1\right] \end{aligned}$$

Now the *work of compression*, or the steady flow work input to the gas, is the negative of the shaft work W_x. Designating the work of reversible polytropic compression by W_n, and the work of reversible adiabatic compression by W_k,

$$W_n = \frac{n}{n-1} p_1 v_1 \left[\left(\frac{p_2}{p_1}\right)^{(n-1)/n} - 1\right] \tag{4}$$

and

$$W_k = \frac{k}{k-1} p_1 v_1 \left[\left(\frac{p_2}{p_1}\right)^{(k-1)/k} - 1\right] \tag{5}$$

For isothermal compression of a perfect gas, pv = constant; then the work of reversible isothermal compression for a perfect gas is

$$W_t = -W_x = p_1 v_1 \int_1^2 \frac{dp}{p} = p_1 v_1 \ln \frac{p_2}{p_1} \tag{6}$$

In the p–v plot of Fig. 17-1 the work of compression for each type of process is represented by the area between the path of that process and the axis of pressures. It is evident that the work of reversible isothermal compression is less than the work of reversible adiabatic compression; the work of reversible polytropic compression is intermediate between the others if n lies between k and unity. This will be the case if the polytropic process involves some cooling but not enough to obtain isothermal compression; such conditions obtain in actual reciprocating compressors. In a real compressor the work will be greater than the work of the reversible compression process because of friction. In such cases the path of compression may be represented by pv^n = constant, but the work of compression is not given by $\int v\,dp$; the shaft work cannot be determined solely from the properties of the fluid. The friction effects in a reciprocating compressor are often small so that the work may be computed by the integral of $v\,dp$ without great error.

Example 1. Tests on reciprocating air compressors with water-cooled cylinders show that it is practical to cool the air sufficiently during compression to correspond to a polytropic exponent n in the vicinity of 1.3. Compare the work per pound of air compressed from 15 psia, 80°F to 90 psia, according to three processes: reversible adiabatic, reversible isothermal, and reversible $pv^{1.3}$ = constant. Find the heat transferred from the air in each case.

Solution: For air, considered as a perfect gas,

$$p_1v_1 = RT_1 = (53.34)(540)$$

Then
$$W_k = \frac{1.4}{0.4}(53.34)(540)\left[\left(\frac{90}{15}\right)^{(1.4-1)/1.4} - 1\right]$$

$$= (101{,}000)(1.67 - 1) = 67{,}700 \text{ ft lb/lb}$$

$$W_n = \frac{1.3}{0.3}(53.34)(540)\left[\left(\frac{90}{15}\right)^{(1.3-1)/1.3} - 1\right]$$

$$= (125{,}000)(1.52 - 1) = 65{,}000 \text{ ft lb/lb}$$

$$W_t = (53.34)(540)\ln\frac{90}{15} = 51{,}900 \text{ ft lb/lb}$$

The heat transferred in the adiabatic process is zero. In the polytropic process, using Eq. (11-24),

$$Q = c_v\frac{k-n}{1-n}(T_2 - T_1)$$

$$= c_v\frac{k-n}{1-n}T_1\left(\frac{T_2}{T_1} - 1\right)$$

$$= c_v\frac{k-n}{1-n}T_1\left[\left(\frac{p_2}{p_1}\right)^{(n-1)/n} - 1\right]$$

$$= (0.171)(778)\frac{1.4-1.3}{1-1.3}(540)(1.52-1) = -16.0 \text{ Btu/lb}$$

Then the heat transferred from the air is 16.0 Btu/lb. In the isothermal process with a perfect gas the heat transfer is equal to the work; then the heat transferred from the air is 51,900 ft lb/lb or 66.7 Btu/lb.

The usefulness of cooling for work reduction in the idealized compression process is clearly shown by Example 1. In real compression processes the desired cooling can only be approximated because it is impractical to build a compressor with sufficient heat transfer capacity without sacrificing other desired characteristics.

17-3 Efficiency of a compressor. The efficiency of a compressor working in a steady flow process may be defined as

$$\eta_C = \frac{h_{2s} - h_1}{W_C} = \frac{W_k}{W_C} \tag{7}$$

where W_C is the shaft work supplied to the actual compressor per pound of gas passing through, and W_k is the shaft work supplied to a reversible adiabatic compressor per pound of gas compressed from the same initial state to the same final pressure as in the actual compressor.

In the case of gas compression, as explained above, the desirable idealized process may sometimes be a reversible isothermal process rather than a reversible adiabatic process. Therefore the compressor efficiency is sometimes defined as

$$\eta_C = \frac{W_t}{W_C} \tag{8}$$

where W_t is the work of reversible isothermal compression from the actual initial state to the actual final pressure.

The two efficiencies of Eqs. (7) and (8) are called respectively the *adiabatic efficiency* and the *isothermal efficiency*. For general thermodynamic purposes it is well to use the reversible adiabatic basis, but the existence in practice of the other concept should be recognized and if the possibility of confusion exists, the basis of any stated value of efficiency should be specified.

Because of the effects of cooling, the adiabatic efficiency of a real compressor may be greater than unity. This should cause no logical difficulty if the definition of the machine efficiency as an arbitrary ratio is understood.

Many turbine-type compressors are essentially adiabatic machines. For an adiabatic machine the work of compression is equal to the enthalpy rise of the gas;

$$W_C = h_2 - h_1$$

Then for an adiabatic compressor the efficiency is

$$\eta_C = \frac{h_{2s} - h_1}{h_2 - h_1} \tag{9}$$

Observe the distinction between the adiabatic efficiency of any machine, Eq. (7), and the efficiency of an adiabatic machine, Eq. (9). For an adiabatic compressor working with a perfect gas the efficiency may be obtained from (9) if only the pressures and temperatures at inlet and outlet are known. The actual temperature rise gives the actual enthalpy change per pound, and the pressure ratio, with the initial temperature, gives the isentropic enthalpy change per pound.

Compressor efficiency of a reciprocating machine may be on an indicated work basis or a brake work basis, depending upon where the work input is measured.

Example 2. For the conditions of Example 1 find the adiabatic efficiency and the isothermal efficiency of the reversible polytropic compressor.

Solution: For the given conditions W_C is the W_n of Example 1.

Isothermal: $$\eta_C = \frac{W_t}{W_C} = \frac{51{,}900}{65{,}000} = 0.80$$

Adiabatic: $$\eta_C = \frac{W_k}{W_C} = \frac{67{,}700}{65{,}000} = 1.04$$

Example 3. For the same initial state and final pressure as in Example 1, a real compressor has an efficiency of 95 percent on the adiabatic basis. The initial and final states correspond to a polytropic compression with $n = 1.3$. Find the heat transferred per pound of air.

Solution: Since this is not a reversible compression process the heat transferred cannot be obtained from the polytropic equation. By the steady flow equation

$$h_1 + Q = h_2 + W_x, \qquad Q = h_2 - h_1 + W_x$$

From Example 1, $W_k = 67{,}700$ ft lb/lb. Then

$$W_x = -\frac{W_k}{\eta_C} = -\frac{67{,}700}{0.95} = -71{,}200 \text{ ft lb/lb}$$

$$h_2 - h_1 = c_p(T_2 - T_1) = c_pT_1\left(\frac{T_2}{T_1} - 1\right)$$

From Example 1, for the process between the same end states,

$$T_1\left(\frac{T_2}{T_1} - 1\right) = 540(1.52 - 1)$$

then $$h_2 - h_1 = (0.240)(540)(1.52 - 1) = 67.7 \text{ Btu/lb}$$

$$Q = 67.7 - \frac{71{,}200}{778} = -24.0 \text{ Btu/lb}$$

Compare the work, the heat, and the enthalpy change in this example with the corresponding quantities for the reversible polytropic process in Example 1.

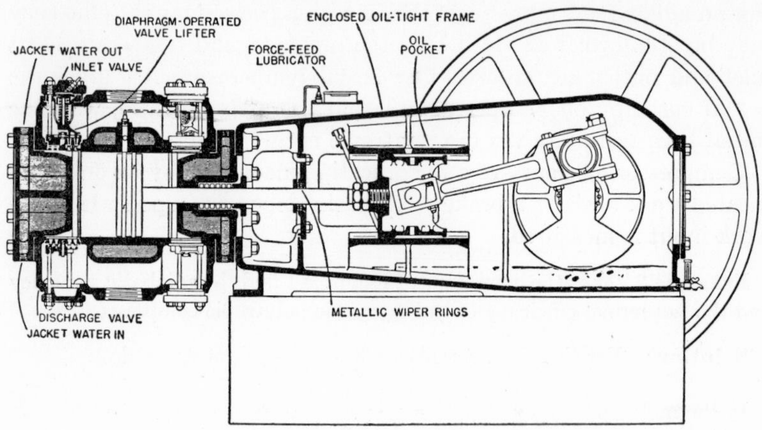

Fig. 17-3. Reciprocating compressor. Note the cooling-water jackets. The valve lifter is an unloading device which stops the pumping action by holding the inlet valves open when the air reservoir is filled to capacity. (Courtesy Socony-Vacuum Oil Co.)

17-4 Reciprocating compressors—work of compression. A reciprocating compressor is shown in section in Fig. 17-3, and a typical indicator card from such a machine is shown in Fig. 17-4. The sequence of operations in the cylinder is as follows:

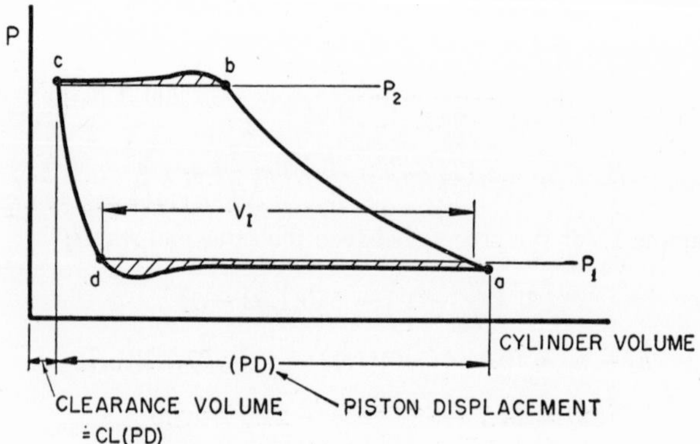

Fig. 17-4. Compressor indicator diagram.

(1) *Compression*: Starting at maximum cylinder volume, point a, slightly below the inlet pressure p_1, as the volume decreases the pressure rises until it reaches p_2 at b; the discharge valve does not open until the pressure in the cylinder exceeds p_2 by enough to overcome the valve spring force.

(2) *Discharge*: Between b and c gas flows out at a pressure higher than p_2 by the amount of the pressure loss through the valves; at c, the point of minimum volume, the discharge valve is closed by its spring.

(3) *Expansion*: From c to d, as the volume increases, the gas remaining in the clearance volume expands and its pressure falls; the suction valve does not open until the pressure falls sufficiently below p_1 to overcome the spring force.

(4) *Intake*: Between d and a gas flows into the cylinder at a pressure lower than p_1 by the amount of the pressure loss through the valve.

The mass of fluid in the clearance space at point c is called the *clearance fluid* m_c. The mass of fluid taken in during the process d–a and expelled during the process b–c is called the *flow fluid* m_f. Then during the compression process a–b the cylinder contains a mass of fluid $(m_c + m_f)$. During the intake and discharge processes the mass in the cylinder is varying.

The total area of the diagram represents the actual work of the compressor on the gas. The cross-hatched areas of the diagram above p_2 and below p_1 represent work done solely because of pressure drop through the valves and port passages; this work is called the valve loss.

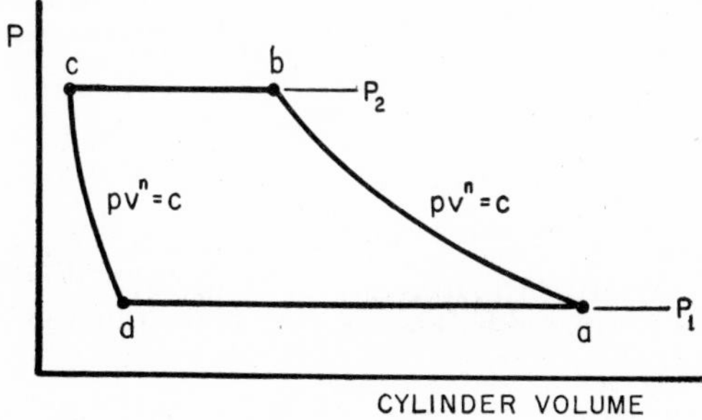

Fig. 17-5.

The idealized machine to which the actual machine is compared has an indicator diagram like Fig. 17-5, in which there are no pressure loss effects, and the processes *a–b* and *c–d* are reversible polytropic processes. Assuming no state change in the intake and discharge processes, *d–a* and *b–c*, and assuming equal values of the exponent n in the compression and expansion processes, *a–b* and *c–d*, the ideal work of compression can be found by taking the integral of $p\,dV$ around the diagram. Let the flow fluid mass be m_f (this is the mass of fluid taken in and discharged per machine cycle). Then using the known work values for the constant-pressure and polytropic processes,

$$\begin{aligned}
W &= W_{a-b} + W_{b-c} + W_{c-d} + W_{d-a} \\
&= \frac{p_bV_b - p_aV_a}{1-n} + p_2(V_c - V_b) + \frac{p_dV_d - p_cV_c}{1-n} + p_1(V_a - V_d) \\
&= \frac{p_2(V_b - V_c)}{1-n} - p_2(V_b - V_c) + \frac{p_1(V_d - V_a)}{1-n} - p_1(V_d - V_a) \\
&= \frac{n}{1-n}\left[p_2(V_b - V_c) + p_1(V_d - V_a)\right] \\
&= \frac{n}{1-n}\left[p_2m_fv_2 - p_1m_fv_1\right] = m_f\frac{n}{1-n}\left[p_2v_2 - p_1v_1\right] \\
&= m_f\frac{n}{1-n}p_1v_1\left[\frac{p_2v_2}{p_1v_1} - 1\right]
\end{aligned}$$

Since pv^n = constant

$$\frac{p_2v_2}{p_1v_1} = \left(\frac{p_2}{p_1}\right)^{(n-1)/n}$$

Thus it is seen that the work per pound of fluid flow is the same as obtained from the steady flow analysis, and given in Eq. (4). It is therefore unnecessary to make any further analysis of the work of the idealized reciprocating compressor since all desired results have already been obtained by the steady flow analysis.

One point to be noted here is that the mass of clearance fluid does not appear in the work equation; that is to say the work per pound of flow fluid *in the idealized compressor* is independent of the clearance volume of the compressor. This is only approximately true for real compressors.

17-5 Volumetric efficiency of reciprocating compressors. The flow capacity of positive displacement compressors (and internal combustion engines) is expressed in terms of a quantity called the *volumetric efficiency* η_v. This is defined as the ratio of the actual volume of fluid taken into the compressor per machine cycle, to the displacement volume of the machine. For a reciprocating compressor

$$\eta_v = \frac{m_f v_1}{\text{(PD)}} \tag{10}$$

where m_f is the flow fluid mass per machine cycle, v_1 is the specific volume of the fluid approaching the inlet valve, and (PD) is the piston displacement per machine cycle.

The true volumetric efficiency can be determined only by measuring the flow through the machine. An approximate or *apparent* volumetric efficiency may be obtained from the indicator diagram as shown in Fig. 17-4. Here the volume V_I is the volume between the point where the cylinder pressure reaches p_1 during the expansion process and the point where it reaches p_1 during the compression process. If the gas remained at constant temperature during the intake process, the volume V_I would be the actual volume taken in at state 1; then the ratio $V_I/\text{(PD)}$ would be the volumetric efficiency. In an actual compressor, because of heat transfer from the cylinder walls, the gas is at higher temperature after entering the cylinder than at state 1. Consequently the volume V_I is greater than the volume taken in from the supply pipe, and the ratio $V_I/\text{(PD)}$ is larger than the true volumetric efficiency.

17-6 Volumetric efficiency and clearance. The volumetric efficiency of an idealized compressor having an indicator diagram like Fig. 17-5 can be derived as follows:
the piston displacement is given by

$$\text{(PD)} = V_a - V_c$$

the clearance is given as a fraction of the piston displacement by

$$\text{Cl} = \frac{V_c}{\text{(PD)}}$$

the mass of flow fluid in the cylinder is given by

$$m_f = \frac{V_a - V_d}{v_1}$$

where v_1 is the specific volume during intake; but

$$V_a = (1 + \text{Cl})\text{(PD)}$$

and $$V_d = V_c \frac{v_d}{v_c} = \text{Cl(PD)} \frac{v_1}{v_2}$$

Then $$m_f = \frac{\text{(PD)}}{v_1}\left[1 - \text{Cl}\left(\frac{v_1}{v_2} - 1\right)\right]$$

and $$\eta_v = \frac{m_f v_1}{\text{(PD)}} = 1 - \text{Cl}\left(\frac{v_1}{v_2} - 1\right)$$

Since it is usually convenient to deal with pressure ratio rather than volume ratio this equation may be rewritten as follows for a compressor working on an ideal polytropic process;

$$\eta_v = 1 - \text{Cl}\left[\left(\frac{p_2}{p_1}\right)^{1/n} - 1\right] \tag{11}$$

17-7 Volumetric efficiency and pressure ratio—multistage compression. It is evident from Eq. (11) that as the pressure ratio is increased the volumetric efficiency of a compressor of fixed clearance decreases, eventually becoming zero. This can also be seen in an indicator diagram, Fig. 17-6. As the discharge pressure is increased, the volume V_I, taken in at p_1, decreases. At some pressure p_{2c} the compression line intersects the line of clearance volume and there is

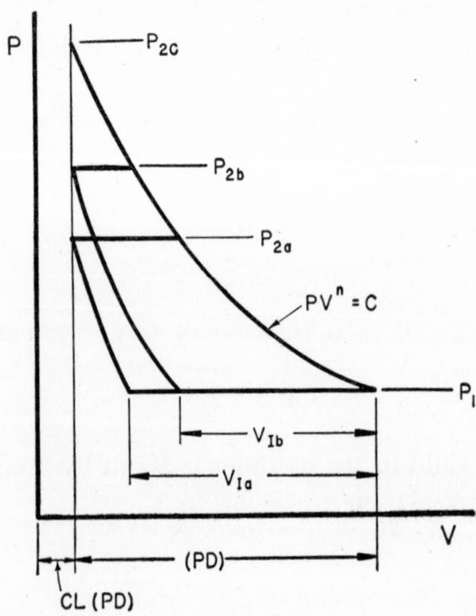

Fig. 17-6. Effect of pressure-ratio on capacity.

no discharge of gas. An attempt to pump to p_{2c} (or any higher pressure) would result in compression and re-expansion of the same gas repeatedly, with no flow in or out.

The maximum pressure ratio attainable with a reciprocating compressor cylinder is then seen to be limited by the clearance. There are practical and economic limits to the reduction of clearance; when these limits interfere with the attainment of the desired discharge pressure, it is necessary to use multistage compression. In a multistage compressor the gas is passed in series through two or more compressors, or *stages*, each of which operates on a small pressure ratio. Disregarding pressure losses between stages, the overall pressure ratio is the product of the pressure ratios of the stages. A two-stage compressor is illustrated in Fig. 17-7.

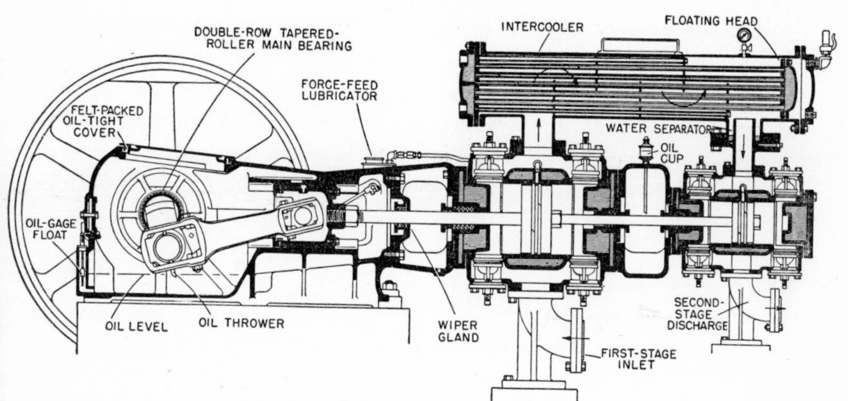

Fig. 17-7. Two-stage reciprocating compressor with intercooler. (Courtesy Socony-Vacuum Oil Co.)

Example 4. A gas is to be compressed from 5 psia to 83.5 psia. It is known that cooling corresponding to a polytropic exponent of 1.25 is practical and the clearance of the available compressors is 3 percent. Compare the volumetric efficiencies to be anticipated for (a) single-stage compression, and (b) two-stage compression with equal pressure ratios in the stages. (A reasonable *comparison* can be made on the idealized basis even though the actual volumetric efficiencies may be lower than the ideal.)

Solution: For the single-stage machine

$$\eta_v = 1 - 0.03[(16.7)^{1/1.25} - 1] = 0.744$$

For the two-stage machine the pressure ratio in each stage is $\sqrt{16.7}$ and the volumetric efficiency is that of the first stage.

$$\eta_v = 1 - 0.03[(4.09)^{1/1.25} - 1] = 0.934$$

17-8 Intercooling. The advantage of multistage compression in itself is primarily that of increased flow capacity or volumetric efficiency for a given pressure ratio. In the idealized case no reduction is obtained in the work of compression per pound of gas passed through the machine; in actual cases some work may be saved, but considering the added mechanical complication of the two-stage machine such a saving is likely to be small or even negative. In both ideal and actual cases, however, an appreciable saving of work may be obtained by utilizing the opportunity for effectively cooling the gas between the stages of compression. The cooling is usually done by a water-cooled tubular heat exchanger which also serves as a receiver or reservoir between the stages.

The work saved by intercooling in the idealized two-stage reciprocating compressor is illustrated on the indicator diagram of Fig. 17-8. Cooling by the cylinder water jackets is never very effective; the compression curve is always closer to adiabatic than to isothermal.

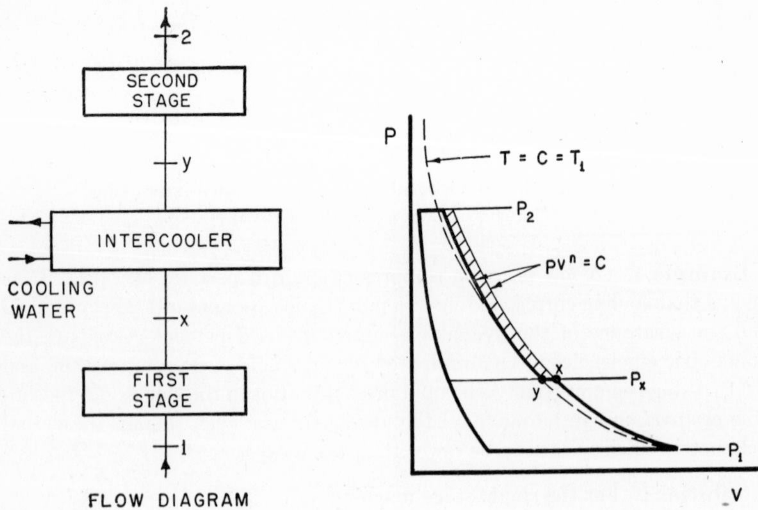

Fig. 17-8. Two-stage compression with intercooling.

Therefore the gas discharged from the first stage at state x is at a higher temperature than the inlet temperature T_1; if the gas is then cooled to state y at temperature T_1, the volume entering the second stage will be less than the volume leaving the first stage. The compression in the second stage will then proceed along a new polytropic curve at smaller volume. The cross-hatched area between the two polytropic curves in Fig. 17-8 represents work saved by interstage cooling to the initial temperature. Actual cooling might be to some other temperature, but it is conventional to discuss cooling to T_1.

The saving of work by two-stage compression with intercooling will depend upon the interstage pressure p_x chosen. Obviously, as p_x approaches either p_1 or p_2 the process approaches single-stage compression. Any saving of work must increase from zero to a maximum and return to zero as p_x varies from p_1 to p_2.

17-9 Minimum work in two-stage compression with intercooling. The conditions affecting the work of compression may be studied by use of the steady flow system and T–s diagram of Fig. 17-9. As shown, a perfect gas is compressed from the initial state p_1, T_1 to p_x; it is then cooled at constant pressure to T_y, and then compressed from p_x, T_y to p_2. Given p_1, T_1, T_y, and p_2, it is desired to find the value of p_x which gives minimum work. Let the adiabatic

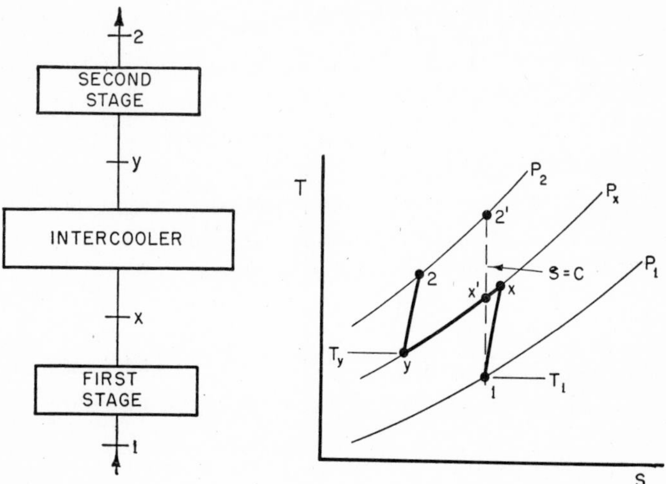

Fig. 17-9. T–s plot of two-stage compression process.

compression efficiencies of the two stages be respectively η_{C1} and η_{C2}; then the work of compression is

$$W_C = W_1 + W_2$$

where
$$W_1 = \frac{1}{\eta_C} \frac{k}{k-1} RT_1 \left[\left(\frac{p_x}{p_1} \right)^{(k-1)/k} - 1 \right]$$

and
$$W_2 = \frac{1}{\eta_{C2}} \frac{k}{k-1} RT_y \left[\left(\frac{p_2}{p_y} \right)^{(k-1)/k} - 1 \right]$$

But
$$\left(\frac{p_x}{p_1} \right)^{(k-1)/k} = \frac{T'_x}{T_1}$$

and
$$\left(\frac{p_2}{p_y} \right)^{(k-1)/k} = \left(\frac{p_2}{p_x} \right)^{(k-1)/k} = \frac{T'_2}{T'_x}$$

Then
$$W_C = \frac{kR}{k-1} \left[\frac{T_1}{\eta_{C1}} \left(\frac{T'_x}{T_1} - 1 \right) + \frac{T_y}{\eta_{C2}} \left(\frac{T'_2}{T'_x} - 1 \right) \right]$$

Taking the derivative with respect to T'_x and setting it equal to zero (noting that T_1, T'_2, and T_y are constant),

$$\frac{dW_C}{dT'_x} = 0 = \frac{kR}{k-1} \left[\frac{1}{\eta_{C1}} + \frac{T_y T'_2}{\eta_{C2}} \left(\frac{-1}{(T'_x)^2} \right) \right]$$

Then
$$(T'_x)^2 = \frac{\eta_{C1}}{\eta_{C2}} T_y T'_2$$

and
$$\frac{T'_x}{T_1} = \sqrt{\frac{\eta_{C1}}{\eta_{C2}} \frac{T_y}{T_1} \frac{T'_2}{T_1}}$$

for minimum work. Now

$$\frac{T'_x}{T_1} = \left(\frac{p_x}{p_1} \right)^{(k-1)/k} \quad \text{and} \quad \frac{T'_2}{T_1} = \left(\frac{p_2}{p_1} \right)^{(k-1)/k}$$

therefore for minimum work in two-stage compression of a perfect gas with intercooling to a fixed temperature T_y,

$$\frac{p_x}{p_1} = \sqrt{\left(\frac{\eta_{C1}}{\eta_{C2}} \frac{T_y}{T_1} \right)^{k/(k-1)} \left(\frac{p_2}{p_1} \right)} \tag{12}$$

For the special case of $T_y = T_1$ and $\eta_{C1} = \eta_{C2}$, which is often taken as a standard of comparison, the requirement for minimum work is

$$\frac{p_x}{p_1} = \sqrt{\frac{p_2}{p_1}} \tag{13}$$

Also for this special case the condition of minimum work is the condition of equal work in the two stages.

When three stages of equal efficiency are used, with intercooling to the initial temperature at two points as shown in Fig. 17-10, the condition of minimum work, and of equal division of work among stages, is

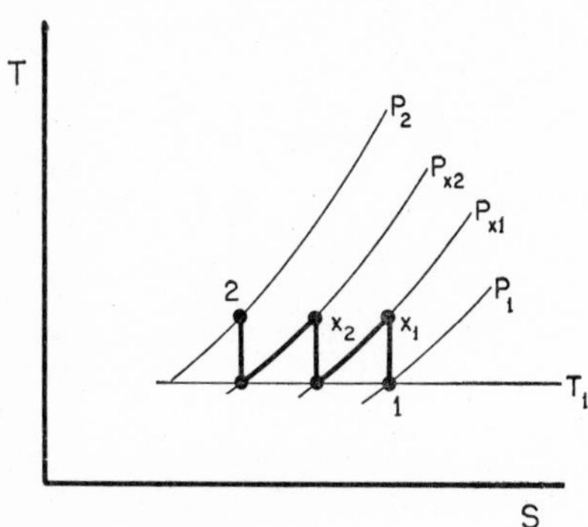

Fig. 17-10. Three-stage compression with intercooling to initial temperature.

$$\frac{p_{x1}}{p_1} = \frac{p_{x2}}{p_{x1}} = \frac{p_2}{p_{x2}} = \sqrt[3]{\frac{p_2}{p_1}} \tag{14}$$

PROBLEMS

17-1. Compare the work of steady flow compression of helium from 15 psia, 80°F, to 1000 psia by two processes: (a) reversible adiabatic; (b) isothermal.

17-2. Solve Problem 17-1 for carbon dioxide.

17-3. For the processes of Problem 17-1, find for (a) the temperature rise, and for (b) the heat transfer per pound.

17-4. Solve Problem 17-3 for the processes of Problem 17-2.

17-5. A refrigeration compressor requires an input of 4.5 hp when compressing 90 lb/hr of ammonia in steady flow from 35 psia, saturated vapor, to 170 psia. Find the compressor adiabatic efficiency. If the heat transferred from the compressor to cooling water and the surroundings is 5000 Btu/hr, find the final state of the ammonia.

17-6. Assume that the compression process of Problem 17-5 can be represented by the equation pv^n = constant. Find the value of n for the process in the problem; find n for a reversible adiabatic process.

17-7. A well-insulated centrifugal compressor takes in air at 14 psia, 70°F, and discharges it at 28 psia, 230°F. Find the adiabatic efficiency of the machine; find the value of the exponent n if the process is represented by the equation pv^n = constant.

17-8. An air compressor cylinder has 8 in. diameter and 8 in. stroke, and the clearance is 5 percent. The machine operates between 14.5 psia, 80°F, and 95 psia; the polytropic exponent is 1.32. Sketch the idealized indicator diagram and find: (a) the total cylinder volume at each corner of the diagram; (b) the flow mass and the clearance mass of air; (c) the capacity, cu ft/min, at 300 rpm, based upon the idealized diagram.

17-9. For the compressor of Problem 17-8 find: (a) the ideal volumetric efficiency; (b) the mean effective pressure; (c) the heat transferred, stated as a fraction of the indicated work.

17-10. Find the necessary piston displacement per minute for the compressor of Problem 17-5 if the actual volumetric efficiency is 0.9 of the volumetric efficiency based on the idealized indicator diagram for the process; the exponent n is 1.20, and the compressor clearance is 3 percent.

17-11. An air compressor has a volumetric efficiency of 70 percent when tested; the discharge state is 75 psia, 300°F, and the inlet state is 15 psia, 60°F. If the clearance is 4 percent, predict the new volumetric efficiency when the discharge pressure is increased to 100 psia. (Assume the ratio of real to ideal volumetric efficiency stays constant, and the exponent n stays constant.)

17-12. Find the percentage saving in work by compressing nitrogen in two stages, compared to single-stage compression, for the following conditions: initial state 25 psia, 0°F; final pressure 400 psia. Compression in each stage is reversible and adiabatic; in the two-stage process the pressure ratios of the stages are equal, and there is intercooling to 90°F.

17-13. Air is compressed in three stages of 2.5 compression ratio per stage. Each stage is adiabatic and has an efficiency of 75 percent. A cooler after each of the stages reduces the air temperature to 100°F; there is 2 psi pressure drop through each cooler. The air enters the first stage at 14 psia, 70°F. Find: (a) the pressure, temperature, and specific volume at the entrance and exit of each of the three coolers; (b) the heat transferred, per pound of air, in each cooler.

17-14. In Problem 17-13 the air flow rate is 50 lb/sec. Find: (a) the power required for each stage, and the total power; (b) the power required to reach the same discharge pressure in a single adiabatic stage of 75 percent efficiency.

17-15. Carbon dioxide is to be compressed in two stages from 1 atm, −60°F, to 16 atm. The compression in the first stage will be at 80 percent efficiency, and in the second stage at 85 percent. Intercooling to 80°F will be obtained. Find the interstage pressure for minimum work.

17-16. For the conditions of Problem 17-15 assume an interstage pressure of 100 psia; find the work of each stage and the heat transferred in the intercooler per pound of gas.

17-17. Derive an expression for the interstage pressure for minimum work in two-stage compression of a perfect gas when there is a pressure drop in the intercooler, the other conditions being the same as in the derivation given in Sec. 17-9. The result should be an expression like Eq. (17-12), but including a pressure-drop factor F defined by $P_y = FP_x$.

17-18. Derive an expression for the interstage pressure for minimum work in two-stage compression of a perfect gas under the following conditions: process as in Fig. 17-9; adiabatic stages of fixed efficiencies, η_{C1} and η_{C2}; no pressure drop in the intercooler; a fixed intercooler effectiveness η_x defined by

$$\eta_x = \frac{T_x - T_y}{T_x - T_1}$$

The resulting expression should be in terms of p_1, p_2, k, η_{C1}, η_{C2}, and η_x.

REFERENCES

Compressed Air Handbook. Compressed Air and Gas Institute, Cleveland, Ohio.

Church, A. H., *Centrifugal Pumps and Blowers*. New York: Wiley, 1944.

Chapter 18

REFRIGERATION

Refrigeration is the cooling of a system below the temperature of its surroundings. This may be accomplished by non-cyclic processes such as the melting of ice or the sublimation (vaporization) of solid carbon dioxide. Of greater interest, however, are methods in which the cooling substance is not consumed and discarded but used in a thermodynamic cycle. Such methods, called *mechanical refrigeration processes*, will be discussed in this chapter.

18-1 Reversed heat engine cycles. In Sec. 7-8 it was pointed out that a reversible heat engine cycle might be visualized as operating in reverse, receiving heat from a low-temperature region, discharging heat to a high-temperature region, and receiving a net inflow of work. Under such conditions the cycle is called a *heat pump cycle*, or a refrigeration cycle.* Figure 18-1 indicates, schematically, a reversed heat engine A operating as a refrigerator or heat pump. By the First Law, if the heat engine works in a cycle, Q_1 must equal the sum of Q_2 and W. By the Second Law and the definition of the temperature scale, if A is a reversible heat engine such as a Carnot engine, working in reverse between the fixed temperatures T_2 and T_1, the relationship between the work and heat quantities is

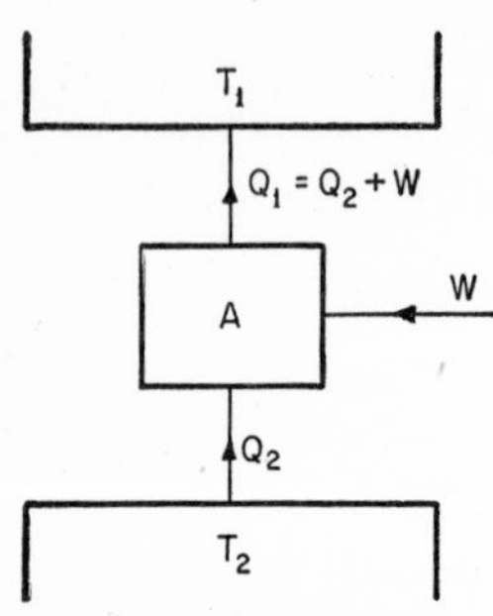

Fig. 18-1. Refrigerator or heat pump.

$$W = \frac{T_1 - T_2}{T_1} Q_1 = \frac{T_1 - T_2}{T_2} Q_2 \qquad (1)$$

* The conventional distinction between a heat pump and a refrigerator is entirely arbitrary. If the primary purpose is to discharge heat *to* a certain high-temperature region, the system is called a *heat pump*. If the purpose is to absorb heat *from* a certain low-temperature region, the system is called a *refrigerator*.

The work of a reversible heat pump as given by Eq. (1) is the minimum work for any heat pump working in a cycle between the fixed temperatures T_2 and T_1. This may be shown by the method used in Chap. 8, that is, assume the statement untrue and show that a violation of the Second Law follows.

As an index of the perfection of a heat pump or refrigerator a quantity called the *coefficient of performance* (CP) is defined as follows:

For a heat pump, $$(\text{CP}) = \frac{Q_1}{W_{\text{net}}} \tag{2}$$

For a refrigerator, $$(\text{CP}) = \frac{Q_2}{W_{\text{net}}} \tag{3}$$

These coefficients correspond to the efficiency of a heat engine; in each case the numerator is the measure of the desired effect, or output, of the apparatus, while the denominator is the input. In the case of a heat pump the desired effect is the heat transferred *from* the system, while in the case of the refrigerator it is the heat transferred *to* the system. In both cases the input is the work.

The maximum coefficient of performance with fixed limits of temperature T_1 and T_2 is that of a Carnot cycle.

For a heat pump, $$(\text{CP})_{\text{max}} = \frac{T_1}{T_1 - T_2} \tag{4}$$

For a refrigerator, $$(\text{CP})_{\text{max}} = \frac{T_2}{T_1 - T_2} \tag{5}$$

It will be observed that these maximum coefficients may be greater than unity, and that they become greater as the temperature difference decreases.

18-2 Vapor compression refrigeration cycles. Practical refrigeration cycles, like practical power cycles, are designed to represent with reasonable fidelity the operations carried out in an actual plant. The basic operations involved in a vapor compression refrigeration plant are illustrated in the flow diagram, Fig. 18-2, and property diagrams, Fig. 18-3. The operations represented are as follows for an idealized plant:

Compression. A reversible adiabatic process 1–2, starting with saturated vapor, and ending with superheated vapor.

Cooling and Condensing. A frictionless constant-pressure process, 2–3, ending with saturated liquid. Heat is transferred out.

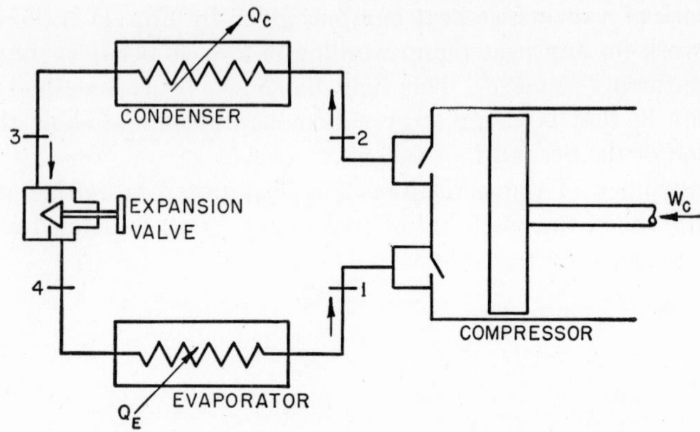

Fig. 18-2. Vapor-compression refrigerator flow diagram.

Expansion. A throttling process 3–4, for which the enthalpy is unchanged. (The throttling process is not a *constant* enthalpy process as shown in the diagram, but it has the same final effect and is commonly represented as a constant-enthalpy process.) There is no heat transfer.

Evaporation. A constant-pressure process 4–1, which completes the cycle. This is the process in which the refrigerating effect occurs, as heat is transferred to the evaporating fluid.

If a real plant is to be analyzed, the irreversibilities of the various processes and the undesired transfers of heat would be accounted for

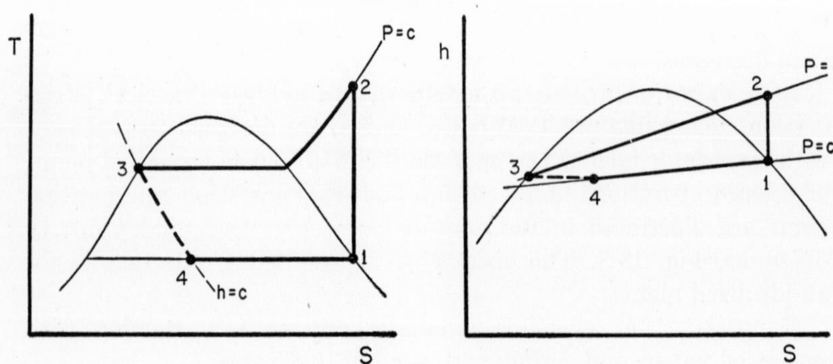

Fig. 18-3. Vapor-compression refrigerator cycle.

in the same way as in the case of power cycles; the manner of doing this has been amply demonstrated in previous chapters.

The compressor of a refrigerating plant is often a reciprocating machine as sketched, but it may be a rotary positive displacement machine or a turbine type machine if such is suitable for the conditions involved. The condenser is usually a tubular heat exchanger which may transfer heat from the system to the atmosphere or to cooling water. The expansion valve may be a manually operated valve as sketched, but is usually some form of automatic regulating valve, such as a pressure regulator, a temperature regulator, or a liquid-level regulator, according to the particular control scheme in use. The evaporator is usually a tubular heat exchanger which is in contact with the substance being cooled by the refrigerator. For example in a water cooler the tubes might be submerged in water, while in a cold storage room the evaporator tubes could be suspended in the air of the room, receiving heat directly from the air.

The choice of a working fluid, or *refrigerant*, for a given cycle depends upon considerations similar to those which determine the desirability of a fluid for a power cycle, as explained in Sec. 13-11. For given condenser and evaporator temperatures the fluid should not require an extremely large pressure range; it is desirable that the pressure should always be above one atmosphere to avoid leakage of air into the apparatus. The latent heat should be large to minimize the quantity of fluid circulated. The fluid should, if possible, be low in cost, non-toxic, stable, and inert with respect to materials of construction. It must not freeze at the lowest temperature of the cycle. Several other characteristics may be of importance to the refrigeration engineer, but those mentioned should suffice to show why no single fluid is suitable for all installations. A few of the refrigerants commonly used are *ammonia*, NH_3; *Freon-12*, CCl_2F_2; *methyl chloride*, CH_3Cl; *sulfur dioxide*, SO_2; and *water*, H_2O. Brief tables of properties of some of these substances will be found in the Appendix; charts for some of them are also provided in the back cover envelope.

In some cases the compressor of a refrigeration plant may be replaced by a jet-pump, as in steam-jet plants, or by a process of absorption and fractional distillation, as in an absorption plant. For descriptions of such plants see the references at the end of the chapter.

Example 1. A refrigeration plant is to operate with an evaporator saturation temperature of 0°F while removing 10,000 Btu/hr from a cold room. The condenser is to be cooled by water so that the saturation temperature can be kept at 76°F. The refrigerant is ammonia.

(a) Assuming the plant works on a cycle like that of Fig. 18-3, find its coefficient of performance and compare this with the coefficient of performance for a Carnot refrigerating machine to do the same job.

(b) If the volumetric efficiency of the compressor is 70 percent how much piston displacement per minute will be needed?

Solution: (a) Referring to Fig. 18-3, on a basis of one pound of fluid the heat transferred in the evaporator is

$$Q_E = h_1 - h_4 \text{ Btu/lb}$$

The net work to the cycle is the compressor work, given by

$$W_C = h_2 - h_1 \text{ Btu/lb}$$

The enthalpy values are obtained from the ammonia tables:

$$h_1 = h_g \text{ at } 0°\text{F} = 611.8 \text{ Btu/lb}$$

$$h_2 = h \text{ at } s_1 \text{ and } p_2$$

$$s_1 = s_g \text{ at } 0°\text{F} = 1.3352 \text{ Btu/lb °R}$$

$$p_2 = \text{saturation pressure at } 76°\text{F} = 143.0 \text{ psia}$$

$$h_2 = 704.4 \text{ Btu/lb}$$

$$h_3 = h_f \text{ at } 76°\text{F} = 127.4 \text{ Btu/lb}$$

$$h_4 = h_3 = 127.4 \text{ Btu/lb}$$

$$(\text{CP}) = \frac{Q_E}{W_C} = \frac{h_1 - h_4}{h_2 - h_1} = \frac{484.4}{92.6} = 5.23$$

For a Carnot refrigerator

$$(\text{CP}) = \frac{T_2}{T_1 - T_2} = \frac{460}{76} = 6.05$$

(b) The piston displacement required is given by

$$(\text{PD}) = \frac{wv_1}{\eta_v}$$

where w is the refrigerant flow rate, lb/min. The flow rate is found from the time rate of heat flow to the refrigerant, and the value of Q_E;

$$w = \frac{10{,}000/60}{484.4} = 0.344 \text{ lb/min}$$

$$v_1 = v_g \text{ at } 0°\text{F} = 9.116 \text{ cu ft/lb}$$

$$(\text{PD}) = \frac{(0.344)(9.116)}{0.70} = 4.48 \text{ cu ft/min}$$

If the cycle involved a real (irreversible) compressor the calculation would be handled as in Chap. 17; also, as explained in that chapter, for cycles requiring a wide pressure range multistage compression might be desirable. Another refinement sometimes used for refrigeration through a wide range of temperature is a *cascade* arrangement in which the condenser of one plant is the evaporator of another, so the heat-pumping process takes place in two stages. This is the reverse of the binary vapor power cycle process.

18-3 Capacity of a vapor compression plant. The capacity of a refrigerating plant, or the rate at which it can absorb heat from some region, is expressed in *tons*. One *refrigeration ton* is defined as the transfer of heat to the plant at the rate of 200 Btu per minute. This is approximately the rate of cooling obtained by melting ice at the rate of one ton (2000 lb) per day.

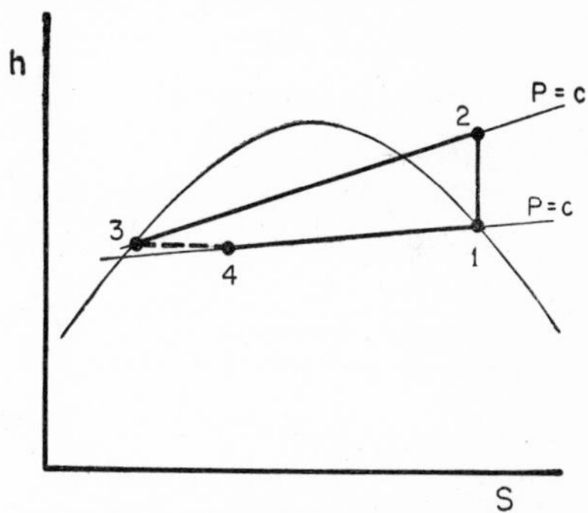

Fig. 18-4.

Consider the cycle of Fig. 18-4; the heat received from the cold region is

$$Q_E = h_1 - h_4 \text{ Btu/lb of refrigerant circulated}$$

Converting this into terms of tonnage

$$(\text{Ref}) = w\frac{h_1 - h_4}{200} \tag{6}$$

where (Ref) is refrigeration rate in tons, and w is refrigerant flow rate in lb/min. Then for a plant using a reciprocating compressor

$$(\text{Ref}) = \frac{V_1(h_1 - h_4)}{v_1\, 200} = \frac{N(\text{PD})\eta_v}{v_1}\frac{(h_1 - h_4)}{200} \tag{7}$$

where N is compressor speed in rpm, (PD) is compressor piston displacement in cu ft, η_v is compressor volumetric efficiency, V_1 is refrigerant volume flow rate entering the compressor in cu ft/min, v_1 is refrigerant specific volume entering the compressor in cu ft/lb.

From Eq. (7) it appears that the capacity of a plant using a reciprocating compressor is dependent upon the following factors:

cycle temperatures; these affect h_1, h_4, v_1, η_v.
refrigerant used; this affects h_1, h_4, v_1, η_v.
compressor size; this affects (PD).
compressor speed; this affects N, η_v.
compressor clearance; this affects η_v.
heat transfer and fluid friction in the compressor; these affect η_v.

In the above reasoning no account is taken of the heat transfer capacities of the evaporator and condenser; these capacities will affect the saturation temperatures actually attainable in any given case, as explained in Sec. 18-5.

The capacity of a plant depends upon so many factors determined by cycle conditions that it is meaningless to specify a plant rating of a certain number of tons unless the cycle conditions at which this rating can be obtained are also specified. Since, for many years, the most important uses of mechanical refrigeration were in ice manufacturing and food storage, a set of standard cycle conditions suited to these applications was established as follows: evaporator saturation temperature 5°F; condenser saturation temperature 86°F; compressor inlet state, vapor superheated 9°F at evaporator pressure; expansion valve inlet state, liquid subcooled 9°F at condenser pressure (for practical purposes this is equivalent to saturated liquid at 77°F). For each refrigerant these conditions will give certain definite values of h_1, h_4, and v_1.

Many modern applications of refrigeration have requirements quite different from those of ice making or food storage plants. Air conditioning often requires much higher evaporator temperatures, while many industrial processes require cooling to much lower temperatures. When a plant is intended to operate at conditions which differ greatly from those specified in the standard rating cycle, the plant should be rated for its designed operating conditions, and these conditions must then be specified in each individual case.

18-4 Power consumption of a vapor compression plant. The power required to operate a vapor compression plant of unit capacity is generally expressed in one or the other of two forms, as a coefficient of performance, or as *horsepower per refrigeration ton*.

The coefficient of performance was defined above for a cycle; for a refrigeration plant the same general definition holds,

$$(\text{CP}) = \frac{Q_E}{W}$$

The work may be measured as compressor indicated work, compressor brake work, or electrical input to the motor, according to the purpose for which the coefficient of performance is to be used. Similarly the heat absorbed by the plant may be measured by the enthalpy change of the refrigerant between expansion valve inlet and compressor inlet, or it may be measured as the heat transferred in the evaporator alone. The use of different bases for computing performance coefficient is analogous to the use of different bases for the efficiency of power plants.

Horsepower per refrigeration ton is a ratio of a work flow rate to a heat flow rate; its relation to coefficient of performance may be seen from the following equations.

$$\frac{\text{hp input}}{\text{refrig. tons}} = \frac{(W \text{ Btu/lb} \times w \text{ lb/min})/(42.42 \text{ Btu/hp min})}{(Q_E \text{ Btu/lb} \times w \text{ lb/min})/(200 \text{ Btu/ton min})}$$

$$\text{hp/ton} = \frac{W}{Q_E}\frac{200}{42.42} = \frac{4.175}{(\text{CP})} \tag{8}$$

For ordinary food storage purposes actual refrigeration plants require approximately 1 hp/ton.

18-5 Effect of irreversible heat transfer on plant performance. Up to this point it has been assumed that the saturation temperatures in the evaporator and condenser are fixed, for a given application, by the surrounding conditions. Actually, however, it is the desired temperature in the region being cooled, and the available temperature of condenser cooling water (or air) that are fixed. The evaporator saturation temperature must be lower than the desired cold-region temperature and the condenser saturation temperature must be higher than the available cooling water temperature by suffi-

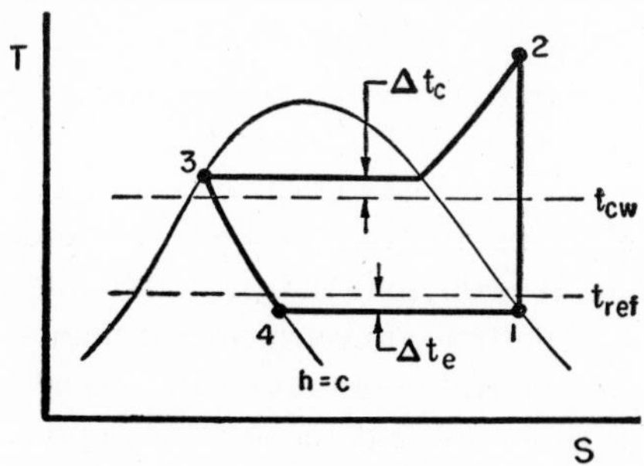

Fig. 18-5. Cycle with temperature differences in condenser and evaporator.

cient amounts to obtain the necessary rates of heat transfer. This situation is illustrated in Fig. 18-5, in which t_{cw} is cooling water temperature, t_{ref} is the cold-region temperature, Δt_c is the temperature difference in the condenser, and Δt_e is the temperature difference in the evaporator.

The temperature differences Δt_c and Δt_e are fixed by economic considerations, not by thermodynamics; for if there were no limits to the size of the heat transfer apparatus and the quantity of cooling water used, the temperature differences might be made to approach zero. Considering only the performance coefficient and the capacity of the plant it would be desirable to make both temperature differences as small as possible for this would reduce the temperature range of the cycle and increase the performance coefficient and capacity. This would mean reduced expense both for initial cost of the compressor and for power supply. The reduction of temperature difference, however, can only be obtained at an increased expense for the heat transfer apparatus and for the cooling water supply; these expenses are small when Δt is large, but go toward infinity as Δt approaches zero. At some value for each Δt, depending upon the relative magnitudes of the various elements of cost, the total cost for compressor power, heat exchanger installation, and cooling water supply will be a minimum; these particular values of Δt are what the designer tries to use.

Practice seems to indicate that temperature differences of the order of 5 to 25°F are economical under various conditions. It is clear that such differences may have an appreciable effect on performance. Consider a case of a cold region at 32°F, with cooling water available at 72°F; if a Carnot refrigerator worked between these limits it would have a performance coefficient of 12.3. Now if the temperature difference in both the evaporator and the condenser is taken to be 10°F, since the external conditions remain the same, the cycle must work between 22°F and 82°F, and have a performance coefficient of 8.03. Although the effect on a real cycle would not be identical to the effect on the Carnot cycle, the order of magnitude would be the same. It is obvious that irreversibility in heat transfer can be of primary importance in refrigeration plants.

18-6 Expansion engines—gas cooling. The common vapor compression refrigeration cycle described in the preceding sections is not, even in the ideal case, a reversible cycle because the process in the expansion valve is irreversible. It is possible, however, to substitute for the expansion valve an expansion engine, which in the ideal case can effect a reversible adiabatic expansion of the fluid. The resulting flow diagram and state plot are shown in Fig. 18-6.

In this cycle it is clear that the net work input is less than the work of the compressor by the work of the expansion engine; moreover, the end state 4 of the expansion in the engine is a state of lower enthalpy than the end state 4′ of the expansion in a throttle valve. Thus by use of the engine it is possible to increase the refrigerating effect $h_1 - h_4$ and at the same time decrease the net work required.

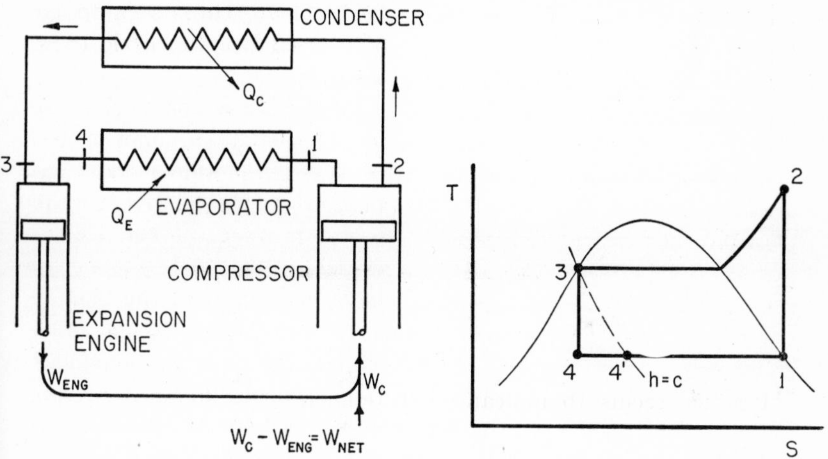

Fig. 18-6. Vapor-compression plant with expansion engine.

The magnitude of the improvement in capacity and performance obtained by use of the expansion engine depends upon the properties of the particular refrigerant involved. The enthalpy change associated with the isentropic expansion of a saturated liquid is small, so the advantage of the expansion engine with vapor cycles is not great, even in the ideal case. In actual cases, after allowing for the irreversibility of the real engine process, the gain by use of the expansion engine is usually negligible and such machines are not used in modern *vapor* refrigerating plants. When *gas* is used as a refrigerant little if any cooling can be obtained by throttling since, for a perfect gas, the temperature remains constant in a constant-enthalpy process, and for real gases the temperature change is small. The expansion engine is often used, therefore, in plants involving gaseous refrigerants and in gas liquefaction plants.

An ideal gas refrigeration cycle using an expansion engine is plotted in Fig. 18-7. The gas is compressed reversibly and adiabatically from 1 to 2, cooled at constant pressure to the cooling water temperature t_{cw} at 3, expanded reversibly and adiabatically to the initial pressure at 4, and heated at constant pressure to the temperature of the cold region t_{ref} at 1. This cycle is seen to be a reversed Brayton cycle (Sec. 15-6).

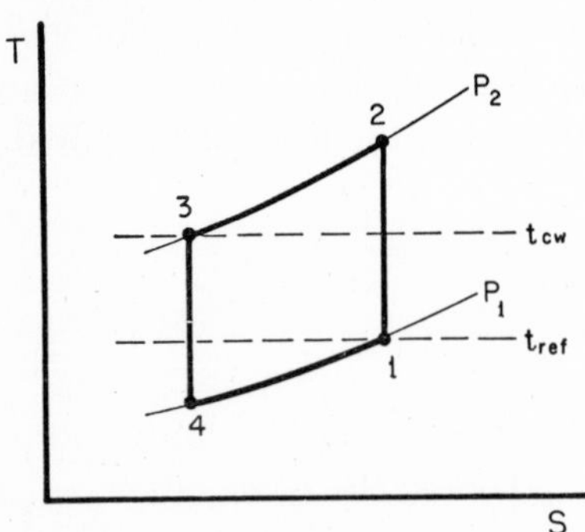

Fig. 18-7. Gas compression cycle; reversed Brayton cycle.

Gas refrigeration cycles are less efficient than vapor cycles, because of the wide overall temperature range relative to the useful range $t_{cw} - t_{ref}$; this is evident in Fig. 18-7. Gas cycles have been used in the past in place of vapor cycles for applications where a harmless gas such as air was considered safer in case of leakage than an obnoxious and toxic substance like ammonia. The principal application was on board ship. With the development of less noxious vapor refrigerants such as the various Freon substances, this type of application has become obsolete. Gas refrigeration is still used, however, for two types of application, air cooling, and gas liquefaction.

Air cooling by expansion in a turbine is particularly useful in the cooling of aircraft cabins. Here the weight and bulk of a vapor plant with its heat exchangers are extremely costly because they deprive the aircraft of equivalent load-carrying capacity and its corresponding revenue. Plants in which the air supply itself serves as the refrigerant can save as much as half the weight of vapor compression plants for the same cooling load. A flow diagram and state diagram for an air expansion cooling plant are shown in Fig. 18-8 for an idealized case. Air is supplied at 1 from the engine-driven supercharger; this air would be necessary for ventilation in any case, but it is supplied here at p_1, a somewhat higher pressure than the cabin

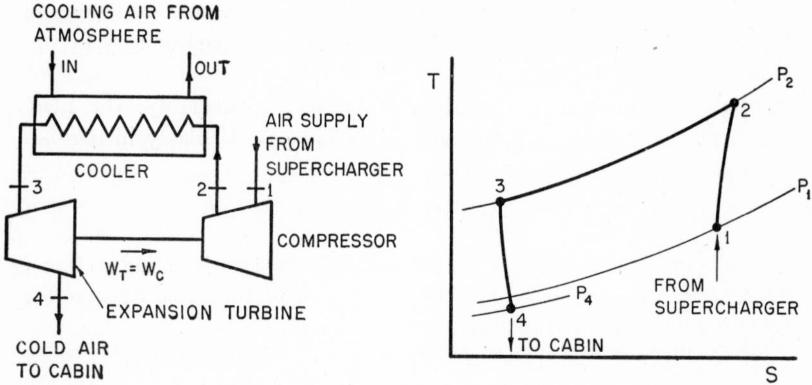

Fig. 18-8. Air expansion cooling plant.

pressure p_4 and at a relatively high temperature. The air is compressed from 1 to 2, in an adiabatic process; the temperature at 2 is then higher than the atmospheric temperature so the air can be cooled in a heat exchanger by means of cooling air taken in from the atmosphere. After cooling to state 3 the air is expanded in the turbine, the work from which is used to drive the compressor. The cool air leaving the turbine goes into the cabin, providing cooling and ventilation simultaneously. In order to supply the energy to overcome friction the pressure drop in the turbine is greater than the pressure rise in the compressor, that is, p_4 is less than p_1. This, in effect, is equivalent to taking a net work input from the supercharger to drive the cooling plant.

18-7 Heat pumps. Every cyclic refrigeration plant is a heat pump, but this name has come to be applied specifically to plants in which the heat flow *from* the plant is the desired effect. Much effort has been expended to popularize heat pumps for building heating. In such installations the evaporator takes heat from the atmosphere, a natural body of water, or the earth, and the condenser discharges heat at a higher temperature for the heating application. The operation is analyzed in the same way as is the refrigeration plant; in fact many installations use the same plant to cool a building in summer and heat it in winter. Technically, the heat pump is a perfectly practical device, but up to the present its economic justification has depended upon special circumstances, for example, a large cooling system that was to be installed in any event and would be available for use in winter as a heat pump. In other cases particularly favorable power costs might exist. If the relative cost of heating fuel compared to electric power should increase appreciably, the heat pump might become economically attractive, particularly in regions of relatively mild winter climate.

PROBLEMS

18-1. Plot the coefficient of performance of Carnot refrigerating cycles vs. the refrigerator temperature, for a hot-region temperature of 90°F; cover the range of refrigerator temperatures from 70°F to −100°F.

18-2. Plot the coefficient of performance of Carnot heat pump cycles vs. the hot-region temperature, for a heat source temperature of 30°F; cover the range of hot-region temperatures from 60°F to 200°F.

18-3. An ideal refrigerating plant, Fig. 18-2, operates between a cold region at 10°F and a hot region at 80°F; saturated vapor enters the compressor, and saturated liquid enters the expansion valve. The plant operates at the rate of 10 tons (2000 Btu/min). The refrigerant is ammonia. Find: (a) the coefficient of performance; (b) the refrigerant flow rate, lb/min; (c) the volume flow rate entering the compressor, cu ft/min; (d) the maximum and minimum pressures of the cycle.

18-4. If the compressor for the plant of Problem 18-3 has a constant volumetric efficiency, what refrigeration capacity (tons) can the compressor serve when the hot- and cold-regions are respectively at: (a) 80°F and 40°F; (b) 80°F and −20°F?

18-5. If the refrigerant in the plant of Problem 18-3 is changed to Freon-12, but the piston displacement and volumetric efficiency remain unchanged, what will be the new capacity in tons?

18-6. Solve Problem 18-3 for a plant using Freon-12 as a refrigerant.

18-7. Solve Problem 18-3 for a plant using carbon dioxide as a refrigerant.

18-8. Solve Problem 18-3 if there is a 10°F temperature difference in both the evaporator and the condenser. Compare with results of Problem 18-3.

18-9. Solve Problem 18-3 if the compressor efficiency is 80 percent.

18-10. Solve Problem 18-8 if the compressor efficiency is 80 percent.

18-11. Solve Problem 18-3 if a reversible expansion engine is used in place of the expansion valve.

18-12. Solve Problem 18-7 if a reversible expansion engine is used in place of the expansion valve.

18-13. A gas refrigeration system using air as a refrigerant is to work at the rate of 10 tons between 10°F and 80°F using an ideal reversed Brayton cycle of pressure ratio 5 and minimum pressure 1 atm. Find: (a) the coefficient of performance; (b) the air flow rate, lb/min; (c) the volume flow rate entering the compressor, cu ft/min; (d) the maximum and minimum temperatures of the cycle.

18-14. An air-expansion cooling system, Fig. 18-8, receives air at 17 psia, 130°F (state 1). The air is compressed adiabatically to 22 psia (state 2) and cooled to 120°F at 21 psia (state 3). The air then expands through the adiabatic turbine to 12.5 psia (state 4). The efficiency of the compressor is 75 percent. Find: (a) the temperature at state 4; (b) the heat transferred in the intercooler per pound of air cooled; (c) the efficiency of the turbine; (d) the work transferred from turbine to compressor per pound of air cooled.

18-15. A heat pump is to use a Freon-12 cycle to operate between outdoor air at 30°F and air in a domestic heating system at 105°F; the temperature difference in the evaporator and in the condenser is 15°F. The compressor efficiency is 80 percent, and the compression begins with saturated vapor; the expansion begins with saturated liquid. The combined efficiency of the motor and belt drive is 75 percent. If the required heat supply to the warm air is 150,000 Btu/hr what will be the electrical load in kw?

18-16. In a New England town in 1952 electric power for domestic use cost 3 cents per kwhr, and domestic fuel oil, heating value 140,000 Btu/gal, cost 13.4 cents per gallon. On this basis compare the cost of electric power for the heat pump of Problem 18-15 with the cost of oil fuel for a heater of 65 percent efficiency to supply the same heat.

REFERENCES

Macintire, H. J., *Refrigeration Engineering.* New York: Wiley, 1940.

Jordan, R. C., and G. B. Priester, *Refrigeration and Air Conditioning.* New York: Prentice-Hall, 1948.

Refrigerating Data Book. New York: American Society of Refrigerating Engineers, published periodically.

Chapter 19

HEAT TRANSMISSION

In the preceding chapters we have discussed heat transfer processes without paying much attention to the time element, except to note that a reversible process would take infinite time. From a practical engineering standpoint it is necessary to know the relations between heat transfer rates, temperature differences, and apparatus sizes; these relations are not part of thermodynamics but belong to the field of *transport phenomena*, which includes heat transfer, mass transfer (diffusion) and momentum transfer (fluid friction). The rates at which transport processes occur depend upon the details of the processes; therefore any analysis must take account of the physical mechanism involved in each case.

Heat transfer processes are so complicated that even a book-length treatment must be selective and abbreviated. Therefore this chapter can include only a simple outline of the methods used in heat transfer calculations, together with a limited amount of data to indicate orders of magnitude and provide practice material. The references at the end of the chapter should be consulted for more complete information. Anyone seriously interested in heat transfer should become familiar with *Heat Transmission* by W. H. McAdams. This book, outstanding in scope and authority, will be referred to in this chapter as "McAdams."

Heat transfer calculations are subject to considerable uncertainty, particularly under actual operating conditions; the results obtained should be treated as good estimates, not precise facts.

19-1 Mechanisms of heat transfer. Heat is transmitted by three different mechanisms: conduction, convection, and radiation.

Conduction is the transfer of energy by forces between adjacent molecules of a body, without transfer of the molecules from their locations. The process may be pictured as a series of impacts between vibrating molecules, with net energy flow from the faster to the slower molecules. Conduction may occur between parts of a single body or between bodies in contact, and may occur in solids, liquids, or gases.

Convection is the transfer of energy in a fluid body by transfer of fluid mass between the regions of high and low temperature. Evaporation and condensation processes, from the heat transfer viewpoint, may be considered as special types of convection processes.

Radiation is the transfer of heat by electromagnetic waves in a manner similar to the transmission of light.

Almost every engineering heat transfer process involves more than one mechanism; the different mechanisms may operate in parallel or in series: for example, convection and radiation in parallel from a gas flame to the bottom of a pan, followed in series by conduction through the bottom of the pan, and convection through the body of water inside. Because of the many factors involved, measurements of overall temperature differences and heat flow rates for engineering processes are of limited use in predicting the performance of new processes. Instead it is generally necessary to divide the new process into its elements of conduction, convection, and radiation, predict each of these elements on the basis of data from laboratory tests, and then combine the predicted elements to obtain an overall prediction.

19-2 Heat transfer coefficients and resistances. It has long been known that the rate of heat transfer by conduction or convection is, to a first approximation, proportional to the temperature difference responsible for the heat flow. Thus, for heat flow through a solid wall, Fig. 19-1, we may write

$$q = \alpha A(t_1 - t_2) \qquad (1)$$

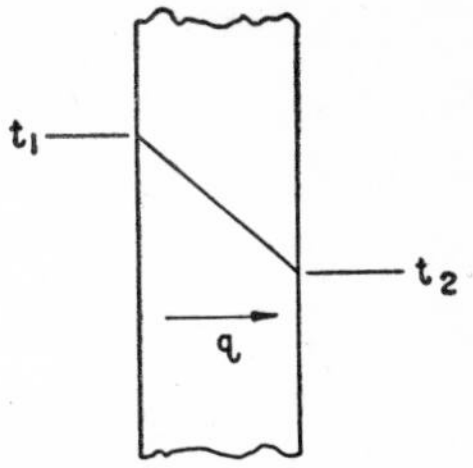

Fig. 19-1. Temperature gradient through a wall.

where q is the rate of heat flow, Btu/hr;
A is the area of the wall normal to the direction of heat flow, sq ft;
t_1 and t_2 are the respective temperatures of the hot and cold surfaces of the wall, °F;
and α is the coefficient of heat transfer for the wall, Btu/hr sq ft °F.

The coefficient α is a characteristic of the particular wall, depending upon its materials, shape, and dimensions, and upon the temperatures. Methods of predicting α for simple cases will be given in the next section. Values of α for building walls and other structural components may be found in handbooks such as the *Heating Ventilating and Air-Conditioning Guide*, published by the American Society of Heating and Ventilating Engineers.

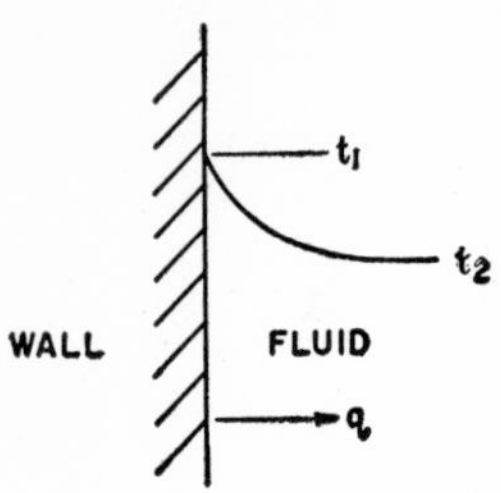

Fig. 19-2. Temperature gradient in a fluid near a wall.

In the case of convection from a solid surface to a fluid body, Fig. 19-2, the temperature in the fluid near the wall will be a function of distance from the wall, but at a great distance the fluid will approach a uniform temperature t_2. According to Newton's law of cooling

$$q = hA(t_1 - t_2) \tag{2}$$

where t_1 is the temperature of the surface of the solid wall, °F;
t_2 is the fluid temperature far from the wall, °F;
h is the cofficient of heat transfer for convection (sometimes called *film coefficient*), Btu/hr sq ft °F;
and q and A are as for Eq. (1).

Methods of predicting h for various types of processes will be given later in this chapter.

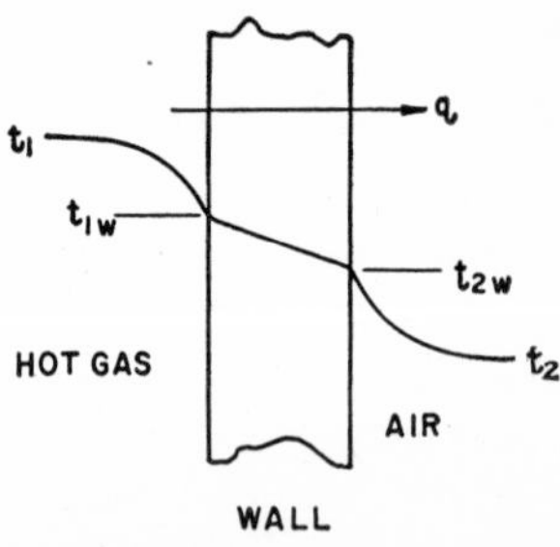

Fig. 19-3. Temperature gradient for convection and conduction in series.

When heat flows by convection and conduction processes in series, the concept of the coefficient of heat transfer may be applied to the overall process, as well as to the individual processes. Consider a wall of area A through which heat is transferred from a hot gas to the atmosphere, Fig. 19-3. Here we may write for the individual processes:

hot gas to wall, $$q = h_1A(t_1 - t_{1w}) \tag{a}$$

wall to cold gas, $$q = h_2A(t_{2w} - t_2) \tag{b}$$

hot side to cold side of wall,

$$q = \alpha A(t_{1w} - t_{2w}) \tag{c}$$

Considering the overall process of Fig. 19-3, an equation may be written to define an *overall coefficient of heat transfer U* as follows:

$$q = UA(t_1 - t_2) = UA\,\Delta t \tag{3}$$

In all of these equations, q is the same because of the series flow.

The overall coefficient may be obtained from the individual coefficients as follows. From Fig. 19-3,

$$t_1 - t_2 = (t_1 - t_{1w}) + (t_{1w} - t_{2w}) + (t_{2w} - t_2) \tag{d}$$

then from (a), (b), (c), and (3)

$$\frac{q}{UA} = \frac{q}{h_1A} + \frac{q}{\alpha A} + \frac{q}{h_2A}$$

or

$$\frac{1}{U} = \frac{1}{h_1} + \frac{1}{\alpha} + \frac{1}{h_2} \tag{4}$$

It is often convenient in dealing with series processes to use resistances to heat transfer, rather than coefficients of heat transfer. The heat transfer resistance is defined by

$$R = \frac{\Delta t}{q} \tag{5}$$

where R is in (hr °F)/Btu. This is analogous to Ohm's law in electricity, with Δt corresponding to the potential difference and q to the current. Then, from (a), (b), (c), and (3),

for the hot-side convection process:

$$R_1 = \frac{t_1 - t_{1w}}{q} = \frac{1}{h_1A}$$

for the wall conduction process:

$$R_w = \frac{t_{1w} - t_{2w}}{q} = \frac{1}{\alpha A}$$

for the cold-side convection process:

$$R_2 = \frac{t_{2w} - t_2}{q} = \frac{1}{h_2A}$$

and for the overall process:

$$R_0 = \frac{t_1 - t_2}{q} = \frac{1}{UA}$$

From Eqs. (d) and (5) the overall resistance is the sum of the resistances in series:

$$R_0 = R_1 + R_w + R_2 \tag{6}$$

The resistance concept is most useful in problems involving many resistances, or in cases of variable area in the direction of heat flow. Some of these problems will be discussed later.

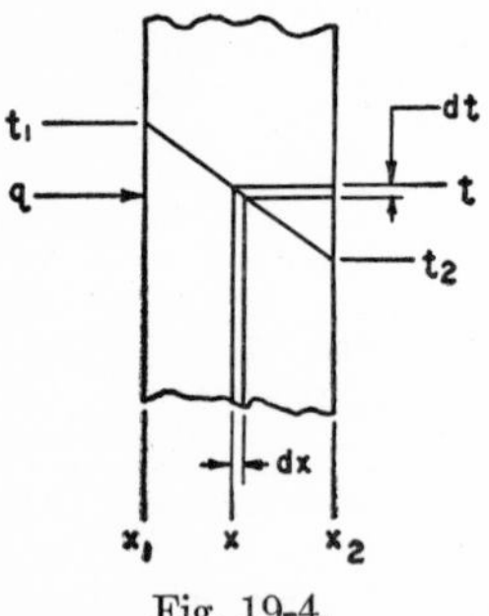

Fig. 19-4.

19-3 Conduction in solids. The general problem of conduction in solid bodies involves heat flow in any direction, with variable geometry and variable time rates of heat flow. In this book we shall consider only steady state problems which can be handled by one-dimensional equations.

Consider a slab of homogeneous material, Fig. 19-4, through which heat is flowing in the x direction. At the location x, the rate of heat transfer is given by Fourier's law of heat conduction:

$$q = -kA\frac{dt}{dx} \tag{7}$$

where q is heat flow rate, Btu/hr;
A is cross-sectional area normal to the direction of heat flow, sq ft;
dt/dx is the temperature gradient in the direction of heat flow, °F/ft;
k is the *thermal conductivity* of the material at x, Btu/hr ft °F.

The negative sign in Eq. (7) takes account of the fact that heat flows in the direction of decreasing temperature.

THERMAL CONDUCTIVITY

For a pure substance the thermal conductivity is a property and, like other properties, is a function of any two independent properties. For a given substance the thermal conductivity varies primarily with temperature, but there may be great differences between substances, even when they appear superficially alike. Inpurities in metals may affect the conductivity out of all proportion to the amount of impurity. For example, data on copper alloys show thermal conductivities ranging from 0.1 to 0.6 of the conductivity of pure copper, for alloys with more than 95 percent copper. The thermal conductivities of metals appear to be roughly proportional to their electrical conductivities.

Many non-metallic building materials, including most heat insulation materials, are more or less porous, and the conductivity of any particular sample depends upon its porosity and other structural peculiarities. Wood and laminated plastics have directional properties.

In tables of properties the values of thermal conductivity are generally given for particular temperatures; in addition to the temperature, attention should be paid to the detailed composition of metals, and to the density and structural characteristics of nonmetallic materials.

The thermal conductivities of liquids and gases are primarily functions of temperature. Although simple conduction is rarely encountered in fluids, thermal conductivities enter into the correlations of experimental data used in predicting convection heat transfer rates.

Some conductivity data are given in the Appendix. The references at the end of the chapter may be consulted for more complete data.

Equation (7) applies to a single location along the path of heat flow; the overall flow equation for the slab is obtained by integrating (7):

$$q = -\int_1^2 kA\,(dt/dx) \tag{8}$$

In general, both k and A may vary along the path of heat flow, and the nature of the variation must be known before the integration can be performed. A is purely geometrical, and is a function of x, while k is a physical property of the material and is a function of the temperature. Therefore the variables may be separated as follows:

$$q \int_{x_1}^{x_2} (dx/A) = -\int_{t_1}^{t_2} k\,dt \tag{9}$$

By using Eqs. (1) and (9), it is possible to compute the heat transfer coefficient for a particular body in terms of its geometry and properties.

Example 1. Find the heat transfer coefficient and the resistance for a rectangular slab of area A and thickness x, if the material has a constant thermal conductivity k.

Solution: From (9),

$$q \int_{x_1}^{x_2} (dx/A) = -\int_{t_1}^{t_2} k\, dt$$

Since A and k are constant,

$$q(x_2 - x_1)/A = -(t_2 - t_1)k$$

From (1) $$q = \alpha A(t_1 - t_2)$$

Then $$\alpha = k/x$$

and $$R = \Delta t/q = x/kA$$

Example 2. A large flat wall consists of two well-bonded layers of material,* 8 in. thick and 1 in. thick, respectively. The 8 in. thick layer is concrete having a thermal conductivity of 0.50 Btu/hr ft °F, and the 1 in. layer is Fiberglas having a thermal conductivity of 0.02 Btu/hr ft °F. The surface temperature of the Fiberglas is −20 °F, and the surface temperature of the concrete is 60° F. Find the rate of heat flow per unit area, Btu/hr sq ft, the temperature at the interface between the two layers, and the resistances per unit area of the two layers.

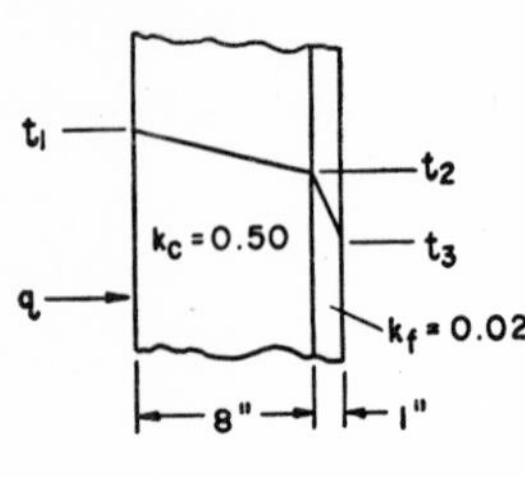

Example 2.

Solution: Sketch the process. It is most convenient to find the resistances first; for the two layers,

$$q = \alpha A\, \Delta t = (k/x)A\, \Delta t$$

Then the resistance is $$R = \Delta t/q = x/kA$$

in which A is to be 1 sq ft.

For the concrete: $$R_c = \frac{8/12}{0.50(1)} = 1.33 \text{ hr °F/Btu}$$

For the Fiberglas: $$R_f = \frac{1/12}{0.02(1)} = 4.17 \text{ hr °F/Btu}$$

The overall resistance is

$$R_0 = R_c + R_f = 1.33 + 4.17 = 5.50 \text{ hr °F/Btu}$$

* If the layers of a composite wall are not well bonded, the thin layer of stagnant gas between them may have an appreciable resistance.

The heat transfer rate for 1 sq ft of area is

$$q = \frac{t_1 - t_3}{R_0} = \frac{60 - (-20)}{5.50} = 14.5 \text{ Btu/hr}$$

The interface temperature is obtained from the heat transfer rate and the resistance of either layer.

$$q = \frac{t_1 - t_2}{R_c} = \frac{t_2 - t_3}{R_f}$$

$$t_1 - t_2 = 14.5(1.33) = 19.4$$

$$t_2 = 40.6°\text{F}$$

or

$$t_2 - t_3 = 14.5(4.17) = 60.6$$

$$t_2 = 40.6°\text{F}$$

In these two examples the area was constant; a variable area process will be discussed in the next section. The use of a constant value of k is common practice, because the accuracy of available data is rarely great enough to justify the complication of treating k as a variable. However, the particular constant value of k used will depend upon the average temperature of the substance in the particular problem.

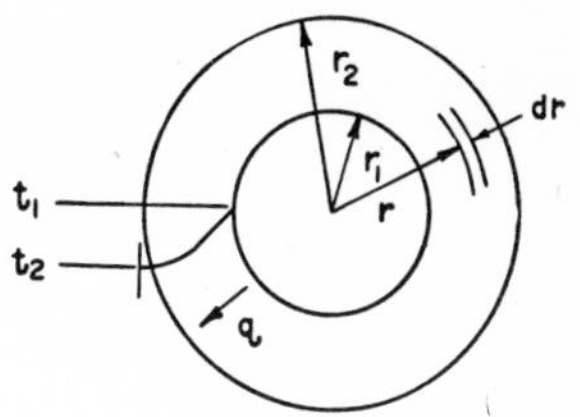

Fig. 19-5. Conduction through a cylinder.

19-4 Conduction through cylindrical walls. Two common conduction problems concern the insulation of pipes against heat transfer, and the use of a tube wall as a heat conductor between two fluids, one inside and one outside the tube. Consider the pipe shown in section in Fig. 19-5; if it has an inner wall temperature t_1 and an outer wall temperature t_2, the heat transfer rate through the pipe wall at any radius r is given by Fourier's law:

$$q = -kA\ (dt/dr)$$

In this case, if the pipe is of length L feet,

$$A = 2\pi rL$$

and

$$q = -k2\pi rL\ (dt/dr)$$

Then

$$\frac{q}{2\pi L}\frac{dr}{r} = -k\ dt$$

Taking k constant, and integrating between 1 and 2,

$$\frac{q}{2\pi L}(\ln r_2 - \ln r_1) = k(t_1 - t_2)$$

or

$$q = 2\pi Lk \frac{t_1 - t_2}{\ln (r_2/r_1)} \tag{10}$$

The heat transfer in the radial direction through a hollow cylinder of given length thus depends on the ratio of the outer and inner radii, but not upon the radius of the cylinder.

If it is desired to define a heat transfer coefficient for the cylinder an ambiguity arises; by Eq. (1),

$$q = \alpha A(t_1 - t_2) \tag{1}$$

but, since A is variable in this case, α may have various values according to the value of A. Thus we may write

$$q = \alpha_1 A_1(t_1 - t_2)$$

or

$$q = \alpha_2 A_2(t_1 - t_2)$$

or

$$q = \alpha_m A_m(t_1 - t_2)$$

where α_1 is the coefficient referred to the inside area;
α_2 is the coefficient referred to the outside area;
α_m is the coefficient referred to some mean value of area A_m.

It is clearly necessary in every case of variable area to state the reference area for a coefficient when stating the coefficient.

For most purposes coefficients for hollow cylinders are based upon either the inside or the outside area. However, it is sometimes convenient to use as a mean area A_m the area of a flat slab which would have the same thickness and the same resistance to heat flow as the hollow cylinder. This mean area is obtained as follows: for the flat slab of thickness $(r_2 - r_1)$ and area A_m, integration of Eq. (9) gives

$$q\frac{r_2 - r_1}{A_m} = k(t_1 - t_2)$$

or

$$R = \frac{\Delta t}{q} = \frac{r_2 - r_1}{kA_m} \tag{11}$$

For the cylinder, from (10),

$$R = \frac{\ln (r_2/r_1)}{2\pi Lk} \tag{11a}$$

Setting the two resistances equal, and solving for A_m,

$$A_m = 2\pi L \frac{r_2 - r_1}{\ln (r_2/r_1)} \tag{12}$$

The area defined by (12) is called the *logarithmic mean area.* A corresponding logarithmic mean radius may also be derived:

$$r_m = \frac{A_m}{2\pi L} = \frac{r_2 - r_1}{\ln (r_2/r_1)} \tag{13}$$

The logarithmic mean area is used mainly with thick cylinders such as pipe insulation. For thin-walled cylinders the arithmetic mean area is used; it differs from the logarithmic mean by less than one percent when r_1 is greater than $0.7r_2$. The logarithmic mean is always less than the arithmetic mean.

For the case of a hollow sphere, the area of the flat slab having the same thickness and resistance as the sphere is the geometric mean; thus

$$A_m = \sqrt{A_1 A_2} \tag{14}$$

The derivation is left to the reader.

Example 3. A steel pipe of 10 in. inside diameter and 0.375 in. wall thickness carries steam at 500°F. The pipe is covered by 2 in. of insulation to reduce heat losses to the surrounding atmosphere at 80°F. It is known from tests that the convection coefficients for the inside surface of the pipe and the outside surface of the insulation are respectively 2500 Btu/sq ft hr °F (condensing steam in the pipe) and 1.6 Btu/sq ft hr °F (natural convection to the atmosphere). To protect personnel it is desired that the outside surface temperature of the insulation should not exceed 140°F. If the thermal conductivity of the steel is 26 Btu/ft hr °F, and of the insulation is 0.045 Btu/ft hr °F, will the 2 in. thickness of insulation meet the requirement?

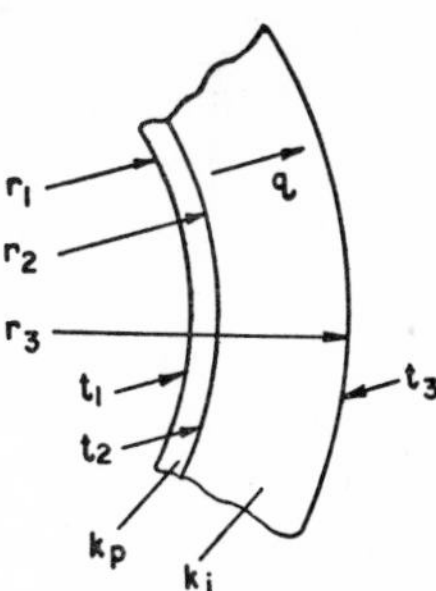

Example 3.

Solution: Sketch the process. Compute the four resistances, for the inside film R_1, pipe wall R_p, insulation R_i, and outside film R_3. The sum of these will be the overall resistance R_0 if there is no contact resistance at r_2. Using Eqs. (5) and (11a), for a unit length of pipe,

$$R_1 = \frac{1}{h_1 A_1} = \frac{1}{(2500)\pi(10/12)} = 0.000153\text{°F hr/Btu}$$

$$R_p = \frac{\ln (r_2/r_1)}{2\pi k_p} = \frac{\ln (5.375/5.000)}{2\pi(26)} = 0.000143°\text{F hr/Btu}$$

$$R_i = \frac{\ln (r_3/r_2)}{2\pi k_i} = \frac{\ln (7.375/5.375)}{2\pi(0.045)} = 1.12°\text{F hr/Btu}$$

$$R_3 = \frac{1}{h_3 A_3} = \frac{1}{(1.6)\pi(14.750/12)} = 0.162°\text{F hr/Btu}$$

R_p and R_i could also have been found from Eq. (11), using the arithmetic mean areas. Adding resistances,

$$R_0 = 1.28°\text{F hr/Btu}$$

Since the same heat flow passes through all the resistances, and $R = \Delta t/q$, each resistance is proportional to the corresponding temperature difference. Then if t_s is the steam temperature and t_a the air temperature,

$$\frac{t_s - t_a}{R_0} = \frac{t_3 - t_a}{R_3}$$

$$\frac{500 - 80}{1.28} = \frac{t_3 - 80}{0.162}$$

$$t_3 = 80 + \frac{0.162}{1.28}(500 - 80) = 133°\text{F}$$

Therefore the 2 in. thickness is sufficient.

19-5 Heat exchangers. One of the most common types of mechanical equipment is the heat exchanger, often disguised under a special name; everyday examples are automobile radiators, refrigerator evaporators and condensers, heating and cooling coils in air-conditioning equipment, and radiators for room heating. Some examples from industry are steam boilers and condensers, feed water heaters, evaporators, and chemical process heaters. A heat exchanger ordinarily comprises two independent flow paths separated by a wall through which heat flows from one fluid to the other. One of the simplest forms is the double-tube heat exchanger, Fig. 19-6. If the fluid in the central duct flows in the same direction as the fluid in the

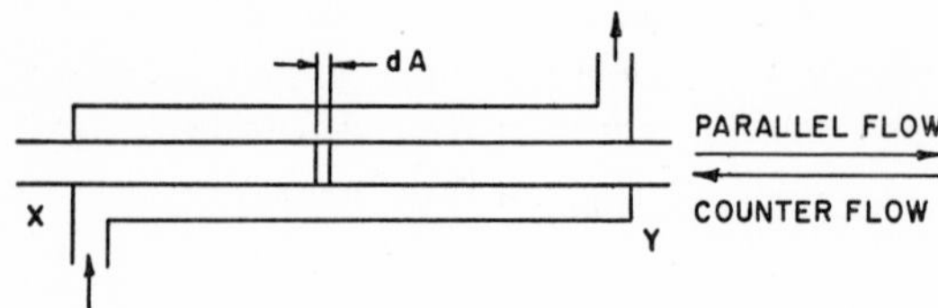

Fig. 19-6. Double-tube heat exchanger, showing parallel and counter flow.

annulus, the arrangement is called *parallel flow;* if the two fluids flow in opposite directions, it is called *counter flow.*

The rate of heat transfer between the fluids in the exchanger of Fig. 19-6 may be written in the form of Eq. (3),

$$q = UA\ \Delta t$$

but the temperature difference between the fluids may vary along the length of the exchanger, so a suitable mean, or average, Δt must be developed. The area is usually defined arbitrarily as the outside area of the inner tube, and U is assumed constant along the length. Then

$$q = UA\ \Delta t_m \tag{a}$$

where Δt_m is the mean temperature difference.

To derive the mean Δt, consider the element of heat exchanger area dA at a location where the temperature difference is Δt; the element of heat transfer rate at this location is

$$dq = U\ dA\ \Delta t \tag{b}$$

To integrate (b), q must be obtained as a function of Δt. In Fig. 19-7 the temperatures of the two fluids are plotted against heat transfer surface, which increases from zero at X to A at Y; the plot at a is for parallel flow, and at b is for counter flow. The hot fluid, entering at t_{h1} and leaving at t_{h2}, is in the annulus, Fig. 19-6, while the cold fluid, entering at t_{c1} and leaving at t_{c2}, is in the inner tube.

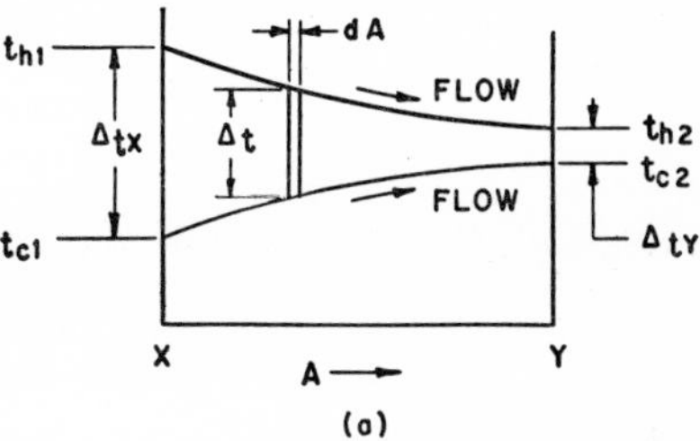

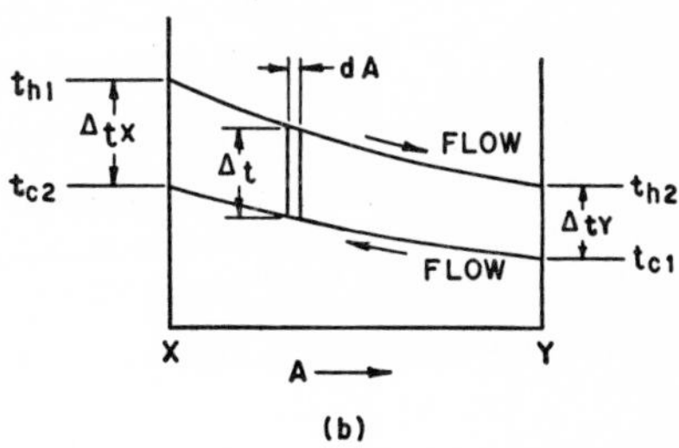

Fig. 19-7. Plots of temperature vs. heat transfer surface. (a) Parallel flow. (b) Counter flow.

Under the usual assumptions for steady flow, the heat transfer rate through the area dA may be written in an energy balance; taking dq as inherently positive,

$$dq = -w_h c_{ph}\ dt_h = \pm w_c c_{pc}\ dt_c \tag{c}$$

where w is mass flow rate, lb/hr, and the subscripts refer to hot and cold fluid. The plus sign applies for parallel flow, and the minus sign for counter flow. The change in Δt between the two ends of the area dA is

$$d(\Delta t) = dt_h - dt_c = \left(\frac{-1}{w_h c_{ph}} \pm \frac{-1}{w_c c_{pc}}\right) dq \quad \text{(d)}$$

The energy balance for the entire length from X to Y gives

$$q = w_h c_{ph}(t_{h1} - t_{h2}) = w_c c_{pc}(t_{c2} - t_{c1}) \quad \text{(e)}$$

Substituting from (e) in (d), in referring to Fig. 19-7 for relations between temperature differences in parallel and counter flow, we obtain for *both* parallel and counter flow,

$$d(\Delta t) = \frac{\Delta t_Y - \Delta t_X}{q}\, dq \quad \text{(f)}$$

Substituting from (b),

$$\frac{d(\Delta t)}{\Delta t} = \frac{\Delta t_Y - \Delta t_X}{q}\, U\, dA \quad \text{(g)}$$

Integrating between X and Y and rearranging,

$$q = \frac{\Delta t_Y - \Delta t_X}{\ln (\Delta t_Y/\Delta t_X)}\, UA \quad \text{(h)}$$

Equating (a) and (h),

$$\Delta t_m = \frac{\Delta t_Y - \Delta t_X}{\ln (\Delta t_Y/\Delta t_X)} \quad \text{(15)}$$

Equation (15) defines the *logarithmic mean temperature difference* which is commonly used in heat exchanger computations.

Consideration of the physical arrangement of a heat exchanger such as shown in Fig. 19-6 will make clear that the counter flow operation will permit the greatest possible temperature change in each fluid. The Second Law will not permit the cold fluid to become hotter than the hot fluid, nor the hot fluid colder than the cold fluid, at any location where heat is flowing from the hot fluid to the cold fluid; it follows that the leaving end of each stream must be adjacent to the entering end of the other stream if the greatest possible temperature change is to be obtained. When a parallel flow exchanger is used the the final temperatures of both fluids approach the same limit, which must lie between the respective inlet temperatures. In addition to the Second Law limitation the First Law must also be satisfied; if there is no external energy transfer, and the kinetic and potential energy changes are negligible, the enthalpy changes per unit time in the two fluids must be equal and opposite. Because of this, in general only one of the fluids may approach the entering temperature of the other in counter flow; the final temperature of the other fluid will be determined by the energy balance.

Example 4. Lubricating oil is to be cooled from 170°F to 120°F by water supplied at 100°F. In order to minimize the water requirement it is desired, if possible, to have the water rise to 140°F, the highest permissible temperature for other reasons. The oil flow rate is 40,000 lb/hr, the oil specific heat is 0.5 Btu/lb °F, and the overall heat transfer coefficient is 120 Btu/sq ft hr °F. (a) Explain why the arrangement should not be parallel flow. (b) Find the water flow rate for counter flow operation. (c) Find the heat transfer surface required for counter flow. (d) With the water flow rate found in (b), to what temperature could the oil be cooled in a parallel flow exchanger of unlimited area? (e) In the exchanger of (d), how much water flow would be required to cool the oil to 120°F?

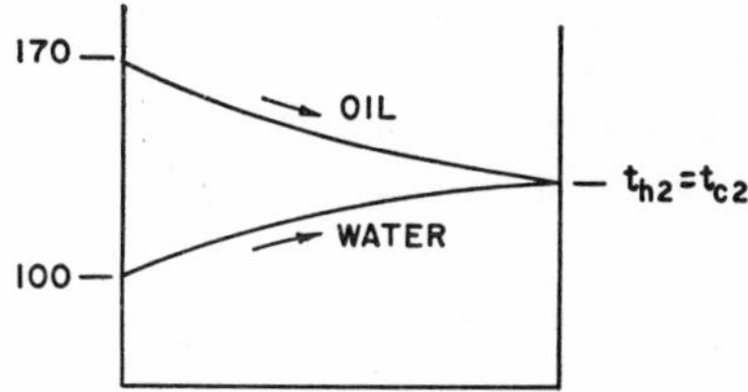

PARALLEL FLOW-INFINITE SURFACE

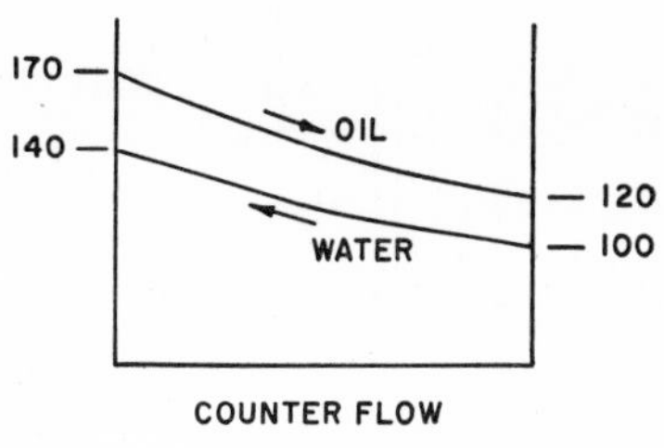

COUNTER FLOW

Example 4.

Solution: Sketch the temperature plots.

(a) If the oil is cooled to 120°F, the water cannot rise above 120°F in parallel flow, even with infinite surface area.

(b) Use the energy balance.

$$w_h c_{ph} (t_{h1} - t_{h2}) = w_c c_{pc} (t_{c2} - t_{c1})$$

$$40{,}000(0.5)(170 - 120) = w_c(1.0)(140 - 100)$$

$$w_c = 25{,}000 \text{ lb/hr}$$

(c)
$$q = UA\Delta t_m$$

$$\Delta t_m = \frac{\Delta t_Y - \Delta t_X}{\ln (\Delta t_Y/\Delta t_X)} = \frac{20 - 30}{\ln (20/30)} = \frac{-10}{-0.405} = 24.7°\text{F}$$

$$q = w_h c_{ph}(t_{h1} - t_{h2}) = 40{,}000(0.5)(170 - 120)$$
$$= 1{,}000{,}000 \text{ Btu/hr}$$

$$A = \frac{q}{U\,\Delta t_m} = \frac{1{,}000{,}000}{120(24.7)} = 337 \text{ sq ft}$$

(d) In a parallel flow exchanger of infinite area, $t_{c2} = t_{h2}$.

$$40{,}000(0.5)(170 - t_{h2}) = 25{,}000(1.0)(t_{h2} - 100)$$

$$t_{h2} = 295/2.25 = 131°\text{F}$$

(e)

$$t_{c2} = t_{h2} = 120°\text{F}$$

$$40{,}000(0.5)(170 - 120) = w_c(1.0)(120 - 100)$$

$$w_c = 1{,}000{,}000/20 = 50{,}000 \text{ lb/hr}$$

Example 4 shows that there are inherent disadvantages in parallel flow operation as compared with counter flow. However, parallel flow is sometimes desirable; for example, when very hot fluids are to be cooled, a parallel flow exchanger has lower tube wall temperature at the hot end, which may be a very important factor in design. When one of the fluids in an exchanger stays at constant temperature (condensing vapor or boiling liquid at constant pressure) there is no difference between parallel flow and counter flow.

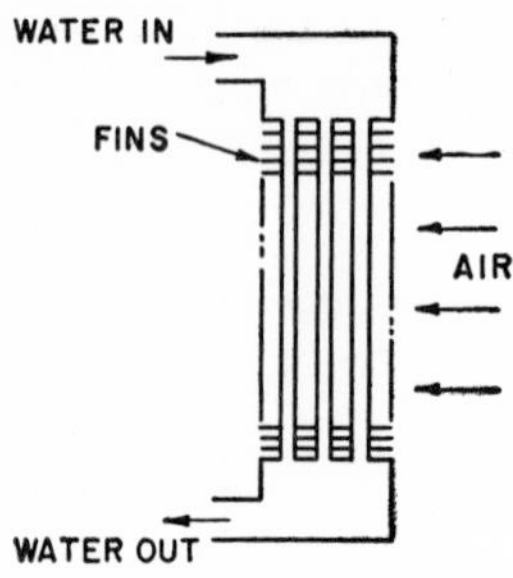

Fig. 19-8. Automobile radiator: an example of both cross flow and extended surface.

Many exchangers are neither strictly parallel flow nor counter flow but involve *cross flow* or multi-pass combinations of parallel, counter, and cross flow. A typical cross flow exchanger is the automobile radiator in which water flows vertically through the tubes while air passes horizontally over the tubes, Fig. 19-8. The air which passes through the upper part of the radiator receives heat only from the hottest water, while the air which passes through the lower part receives heat only from the coldest water. Thus the air is not all heated to the highest possible temperature, and more air must be circulated than would be the case in counter flow. This is acceptable because on the one hand the great difference in the volume of the air and water would make it difficult to design a compact counterflow exchanger with low pressure drop, and on the other hand air is readily available and cheap. In some industrial applications several cross flow exchangers are placed in series in the gas stream, with the

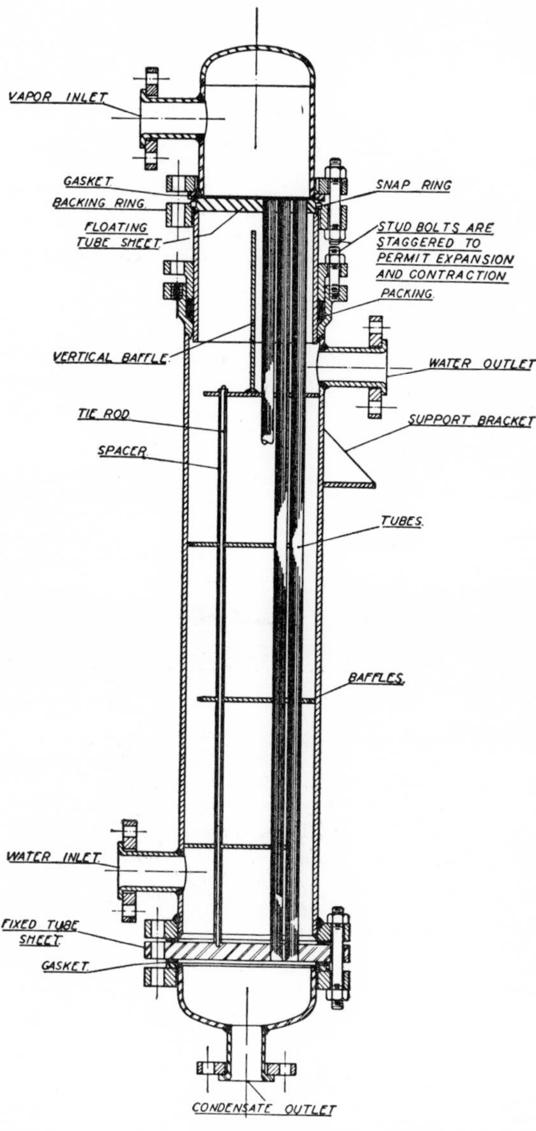

Fig. 19-9. Shell and tube heat exchanger. This exchanger is a vertical single-pass design. (Courtesy The Pfaudler Co.)

liquid flowing in series through the exchangers, giving a counter flow arrangement of cross flow elements.

TABLE 19-1

TYPICAL VALUES OF FILM COEFFICIENTS*

Fluid and Type of Process	h, Btu/sq ft hr °F
Heating or cooling, no state change:	
Water	300 to 2000
Gases	3 to 50
Organic solvents	60 to 500
Oils	10 to 120
Condensing vapor:	
Steam	1000 to 3000
Organic solvents	150 to 500
Light oils	200 to 400
Heavy oils	20 to 50
Ammonia	500 to 1000
Evaporating liquid (see note):	
Water	800 to 2000
Organic solvents	100 to 300
Light oils	150 to 300
Heavy oils	10 to 50
Ammonia	200 to 400

* Courtesy The Pfaudler Company

Note: In evaporation the maximum heat flux (Btu/sq ft hr) normally attainable is limited by the formation of a blanket of vapor on the heating surface when the temperature of the surface is heated too far above the saturation temperature of the liquid. This critical temperature difference is approximately 50°F in the case of water, and somewhat greater for some organic fluids. McAdams, Chap. 14, gives data.

Large exchangers for fluids under pressure are usually of the shell and tube type, Fig. 19-9, which may be built with any of a great variety of flow path arrangements. See also Fig. 17-7, p. 317.

Heat transfer coefficients for gases are generally much smaller than for liquids or condensing vapors. Therefore exchangers with gas on one side and liquid or condensing vapor on the other are often made with *extended surface*, or fins, on the gas side. Examples are the automobile radiator and the finned convector for room heating. Extended surface may decrease the gas side resistance to a small fraction of its value for bare tubes. Many data on extended surfaces are given in *Gas Turbine Plant Heat Exchangers*, by Kays, London, and Johnson, *ASME*, New York, 1951. Data on commercial extended surface exchangers are published by the manufacturers.

In multi-pass or cross flow exchangers the correct average temperature difference is not the logarithmic mean, but is a smaller value depending upon the pass arrangement and the heat capacities of the two fluid streams. Charts have been prepared by various investigators to give the correction factor to be applied to the logarithmic mean (or equivalent data) for all usual cases; examples are given in Figs. 19-10 and 19-11.

In general the overall heat transfer coefficient is not constant, but if the properties of the fluids do not change greatly in the exchanger the value of U may be based upon fluid properties at the arithmetic average temperature for each fluid. For fluids such as petroleum oils, which have large variation of viscosity with temperature, it is necessary to consider a series of steps of heat transfer in each of which the property variations are small; McAdams gives methods for taking account of variable U.

TABLE 19-2

RECOMMENDED MINIMUM FOULING RESISTANCES FOR DESIGN*

Fluid	R, hr °F sq ft/Btu
Water (over 3 fps velocity, under 120°F):	
Great Lakes or clear river water	0.0012
Soft surface water	0.0008
Silty river water	0.003
Brackish river and bay water	0.0025
Hard well water	0.0033
Treated cooling water and boiler feed water (212°F)	0.0015
Steam:	
Oil free steam	0.0005
Exhaust from reciprocating engine	0.001
Miscellaneous:	
Brine, untreated	0.003
Air, clean	0.0015
Air, reciprocating compressors	0.003
Refrigerant vapors, centrifugal compressors	0.0015
Refrigerant vapors, reciprocating compressors	0.0025
Clean organic solvents	0.001
Fuel oil	0.006

* Courtesy of The Pfaudler Company.

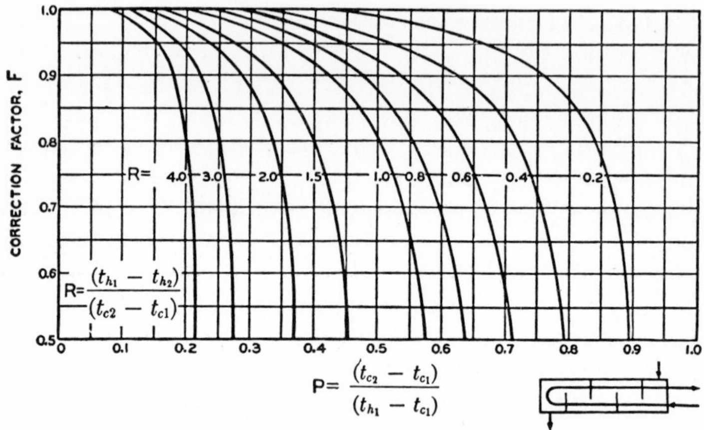

Fig. 19-10. Mean-temperature-difference correction factor for an exchanger with one shell pass and 2, 4, 6, etc. tube passes; multiply the logarithmic mean Δt by the correction factor. The plot is based on constant U and constant specific heats. The hot fluid may be in the shell or in the tubes. This is one of numerous plots given by Bowman, Mueller, and Nagle, "Mean Temperature Difference in Design," *Trans. ASME*, May 1940, p. 283.

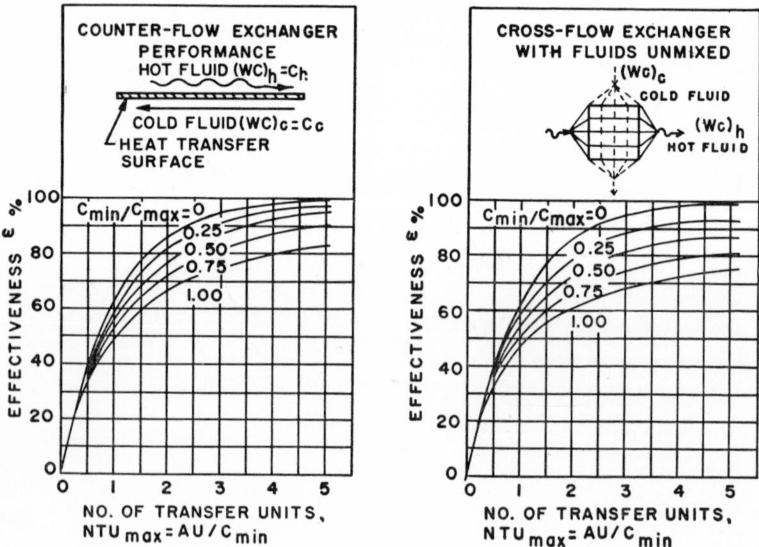

Fig. 19-11. (See caption on opposite page.)

The overall coefficient quoted by manufacturers usually includes allowances for dirt or scale on the tube surfaces and for failure of the fluid in the shell to follow the ideal path across the tubes. These factors are included on the basis of experience with the particular design and application involved. The allowances for dirt may be in the form of heat transfer coefficients for the dirt layers, or cleanliness factors which represent the fraction of the total exchanger surface actually required when clean; if the factor is 85 percent, 15 percent of the surface is excess when the exchanger is clean.

19-6 Mechanism of forced convection in a tube. It is assumed that the reader is familiar with the general properties of the flow pattern in tubes, as presented in undergraduate courses in fluid mechanics. In a circular tube the flow pattern far from the entrance becomes established in a shape that depends mainly on the Reynolds number N_{Re}. For values of N_{Re} less than 2100 the flow is laminar, with a parabolic velocity distribution, but for values of N_{Re} greater than about 4000 the flow is usually turbulent, with a blunt velocity distribution, Fig. 19-12. For intermediate values of N_{Re} the velocity distribution has some intermediate form.

Fig. 19-11. Heat exchanger performance characteristics (from *Gas Turbine Plant Heat Exchangers* by Kays, London, and Johnson, New York, *ASME*, 1951.)

Capacity rates, C_{max} and C_{min}:

C_{max} is the larger of $w_h c_{ph}$ and $w_c c_{pc}$

C_{min} is the smaller of $w_h c_{ph}$ and $w_c c_{pc}$

Effectiveness of the exchanger, ϵ, is the ratio of the actual heat transferred, to the maximum heat transfer permitted by the Second Law (i.e., when one fluid leaves at the entering temperature of the other fluid).

$$\epsilon = \frac{w_h c_{ph}\,(t_{h_1} - t_{h_2})}{C_{\mathrm{min}}\,(t_{h_1} - t_{c_1})} = \frac{w_c c_{pc}\,(t_{c_2} - t_{c_1})}{C_{\mathrm{min}}\,(t_{h_1} - t_{c_1})}$$

Number of transfer units, NTU, is the ratio of heat exchange capacity per degree temperature difference, to fluid stream heat capacity per degree temperature change. It is a measure of the relation between installed capacity and load.

$$\mathrm{NTU} = AU/C_{\mathrm{min}}$$

where A is the heat exchanger surface, and U is the corresponding overall coefficient.

The plots show that as NTU (or installed capacity/load) is increased the effectiveness increases, but at a decreasing rate. The plots can be used in the design of an exchanger, since with given fluid stream characteristics they show how much AU is required to obtain any specified effectiveness. They also reveal clearly the diminishing returns as AU is increased.

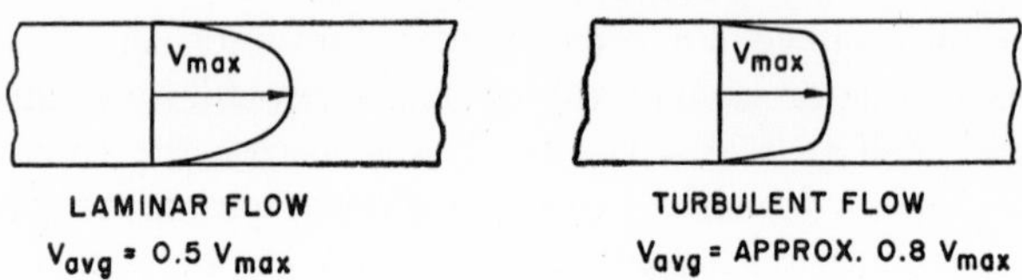

Fig. 19-12. Velocity distribution in pipe flow.

In laminar flow the stream has no eddies or cross currents, and heat transfer from the wall into the body of the stream is by conduction, which is dependent upon collisions between molecules in random motion. Momentum transfer (transfer of shear forces) across the stream also depends upon the collisions of the molecules, whereby molecules with greater momentum in the flow direction give up momentum to those with less. Thus it is not unreasonable that the rate of heat transfer should depend upon the same factors that influence the friction or pressure-drop of the stream.

In turbulent flow there are eddies and cross currents throughout the main body of the stream, but flow normal to the wall cannot persist immediately at the wall. Therefore the flow very close to the wall is laminar and parallel to the wall; between this *laminar sublayer* and the fully turbulent main stream there is a transition region called the *buffer layer* in which the turbulence increases with distance from the wall, Fig. 19-13. The transfer of fluid mass across the stream, in the eddies of the turbulent main stream and buffer layer, provides a vehicle for the transfer of both heat and momentum. If there is a temperature difference between the fluid near the wall and the fluid in the center of the stream, the eddies will carry hot fluid to the cold regions and cold fluid to the hot regions. This *eddy conductivity* is much more effective in heat transfer than the molecular conductivity which acts in the laminar sublayer close to the wall. Therefore when

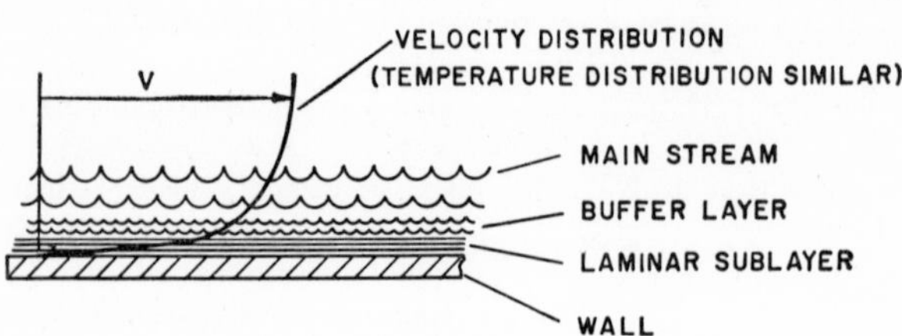

Fig. 19-13. Turbulent flow pattern.

heat is transferred between the wall and the stream most of the resistance, and most of the temperature drop, is in laminar layer, or film.

The momentum transfer in turbulent flow is also due to the eddies, being the result of the interchange of fluid between the slower and faster parts of the stream. As in the case of heat transfer, the momentum transfer requires a much greater driving force across the laminar layer than across the turbulent region. Thus the velocity variation is great in the laminar layer and small in the turbulent zone, as indicated in Figs. 19-12 and 19-13.

19-7 Formulas for forced convection in tubes. Convection processes are so complicated that the correlation of data would be impracticable without the use of dimensionless groups of variables. As in fluid mechanics, dimensional analysis is used to develop the appropriate groups, and experimental data are then plotted in terms of these quantities. The reader should see the references for details of the dimensional analysis, since we can give only a few examples of the results, taken, by permission, from McAdams.

Some of the variables which enter into convection equations are listed with their dimensions and units in Table 19-3.

TABLE 19-3

Variable	*Dimensions*	*Consistent Units*
Mass	M	lbm
Length or diameter	L	ft
Time	θ	hr
Temperature	T	°F
Energy (heat)	H	Btu
Velocity, $\overline{V}$	L/θ	ft/hr
Density, ρ	M/L^3	lbm/cu ft
Viscosity, μ	$M/L\theta$	lbm/hr ft
Conductivity, k	$H/L\theta T$	Btu/ft hr °F
Specific heat, c_p	H/MT	Btu/lbm °F
Heat transfer coefficient, h	$H/L^2\theta T$	Btu/sq ft hr °F
Mass velocity, G	$M/\theta L^2$	lbm/sq ft hr
Mass flow rate, w	M/θ	lbm/hr

The mass velocity G is defined as the mass flow rate per unit area and is equal to $\overline{V}\rho$, where $\overline{V}$ and ρ are average values over the stream cross section.

Some of the dimensionless groups used in forced convection equations are listed in Table 19-4.

TABLE 19-4

Name	Symbol	Formula
Reynolds number	N_{Re}	$\frac{D\overline{V}\rho}{\mu}$ or $\frac{DG}{\mu}$
Prandtl number	N_{Pr}	$\frac{c_p\mu}{k}$
Nusselt number	N_{Nu}	$\frac{hD}{k}$
Stanton number	N_{St}	$\frac{h}{c_p\overline{V}\rho}$ or $\frac{h}{c_pG}$
Colburn heat transfer factor	j	$N_{St}\,(N_{Pr})^{2/3}$

Forced convection occurs when heat is transferred to or from a moving fluid in which the motion is not dependent upon the local density gradients caused by the heat transfer.

For forced convection inside long tubes ($L/D > 60$), with turbulent flow ($N_{Re} > 10{,}000$) and moderate temperature difference between wall and fluid, a widely used equation is the following, which is applied for both heating and cooling the fluid:

$$\frac{hD}{k_b} = 0.023\left(\frac{DG}{\mu_b}\right)^{0.8}\left(\frac{c_p\mu}{k}\right)_b^{0.4} \tag{16}$$

The subscript b indicates that the properties are evaluated at the fluid *bulk* temperature t_b, which is the mass average temperature.

Example 5. Find the film coefficient for air flowing at 100 fps through a tube of 1 in. outside diameter, and 18 gage thickness, if the average bulk air temperature is 600°F, the pressure is 1 atmosphere, and the tube wall temperature is 200°F.

Solution: From Table A-17, the tube inside diameter is 0.902 in. and the internal cross section is 0.00443 sq ft. The properties of air at 600°F and 14.7 psia are: $c_p = 0.25$ Btu/lb °F; $\mu = 0.000020$ lbm/sec ft or 0.072 lbm/hr ft; $k = 0.027$ Btu/sq ft hr °F; $\rho = 0.0375$ lb/cu ft. Then

$$N_{Re} = \frac{D\overline{V}\rho}{\mu} = \frac{(0.902/12)(100)(0.0375)}{0.000020} = 1.41 \times 10^4$$

$$(N_{Re})^{0.8} = 2080, \qquad N_{Pr} = 0.66, \qquad (N_{Pr})^{0.4} = 0.847$$

$$h = \frac{k}{D}\,0.023N_{Re}^{0.8}N_{Pr}^{0.4} = \frac{0.027(0.023)(2080)(0.847)}{0.902/12}$$

$$= 14.5 \text{ Btu/sq ft hr °F}$$

Many modifications of Eq. (16) have been used because they correlated particular sets of data. Some investigators recommend slightly different values of the coefficient and exponents for different fluids or for heating and cooling processes. One modification, for large temperature differences between wall and fluid, is to evaluate the properties at the *film temperature* t_f, which is the arithmetic average between the bulk temperature and the tube wall temperature. Then

$$\frac{hD}{k_f} = 0.023 \left(\frac{DG}{\mu_f}\right)^{0.8} \left(\frac{c_p\mu}{k}\right)_f^{0.4} \tag{16a}$$

Another form of the same basic equation, using the Stanton number in place of the Nusselt number, has come into use because it is more convenient from the viewpoint of the person who is correlating experimental data, particularly in dealing with the analogies between heat transfer and fluid friction. This equation is

$$j = \frac{h}{c_{pb}G}\left(\frac{c_p\mu}{k}\right)_f^{2/3} = \frac{0.023}{(DG/\mu_f)^{0.2}} \tag{17}$$

The dimensionless factor j turns out to be the analog of the Fanning friction factor used in fluid mechanics. Equation (17) can be used like (16) for turbulent flow.

Example 6. Find the film coefficient for the conditions of Example 5, using Eq. (17).

Solution: $t_f = (600 + 200)/2 = 400°\text{F}$; $c_{pb} = 0.25$ Btu/lb °F; N_{Pr} at $t_f = 0.68$; $\mu_f = 0.000017$ lbm/sec ft.

$$N_{\text{Re}} = \frac{D\overline{V}\rho}{\mu_f} = \frac{(0.902/12)(100)(0.0375)}{0.000017} = 1.66 \times 10^4$$

$$N_{\text{Re}}^{0.2} = 6.98, \qquad N_{\text{Pr}}^{2/3} = (0.68)^{2/3} = 0.773$$

$$h = c_{pb}G\,\frac{0.023}{N_{\text{Re}}^{0.2}N_{\text{Pr}}^{2/3}} = \frac{(0.25)(100)(3600)(0.0375)(0.023)}{(6.98)(0.773)}$$

$$= 14.4 \text{ Btu/sq ft hr °F}$$

For Reynolds numbers less than 10,000 it is customary to present heat transfer data as fluid friction data are presented, in the form of a plot of j vs. N_{Re}, Fig. 19-14. Thus, variation of both the coefficient and the exponent on the right side of (17) can be accounted for. When fluid viscosity varies greatly with temperature, the flow pattern of the process may become a function of the rate of heat transfer; this

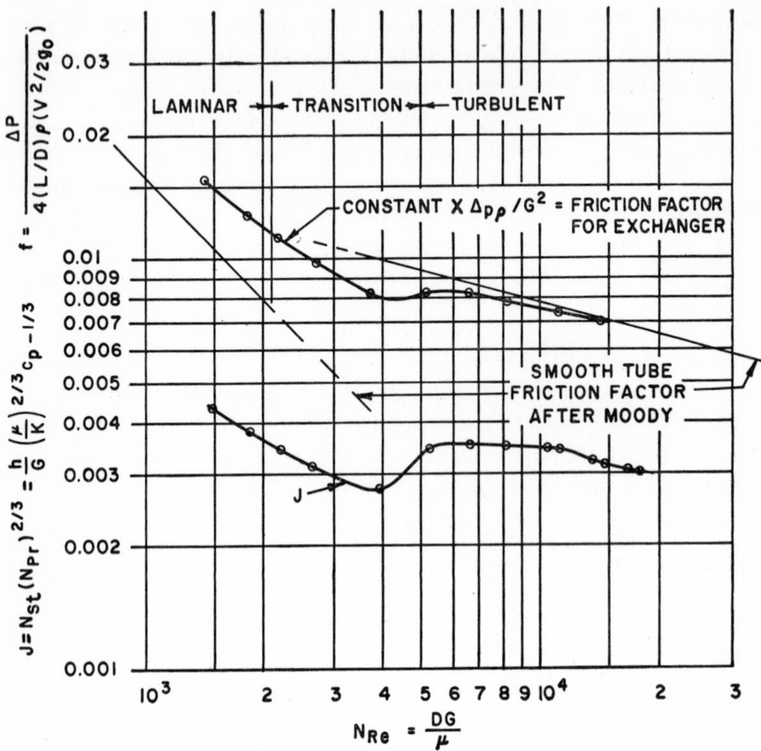

Fig. 19-14. Example of correlation of heat transfer data for flow in a tubular exchanger at low Reynolds numbers. Note the similarity of the plots for friction factor and j, indicating the analogous mechanisms for friction and heat transfer. Also note the dip in the j curve at the transition region; this makes it advisable to avoid operating exchangers with smooth tubes in the transition range. The dip is less pronounced, or absent, with irregular surfaces that promote turbulence. (Data from Green and King, *Trans. ASME*, February 1946, page 115.)

is often accounted for by introducing the dimensionless ratio (μ_w/μ_b), which is the ratio of viscosity at the wall temperature to viscosity at the bulk temperature. In working with short tubes $(L/D < 60)$, the ratio of tube length to diameter, L/D, is introduced to account for the difference in flow pattern near the entrance of a tube from that after the fluid has travelled some distance into the tube. These factors and others are considered in detail by McAdams.

19-8 Non-circular ducts. In fluid mechanics the friction factor and pressure drop for flow in non-circular ducts are computed by the same equations used for circular ducts, through the use of an equivalent diameter D_e defined as follows:

$$D_e = 4r_h = 4S/P \qquad (18)$$

where r_h is the *hydraulic radius* in feet, S is the cross section of the stream in square feet, and P is the perimeter (often called *wetted* perimeter because of its use in open-channel flow), in feet. It is seen that for a circular section, D_e reduces to D.

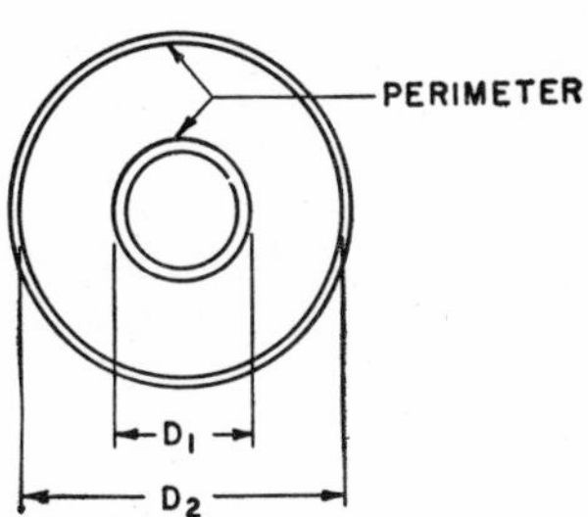

Fig. 19-15. Annulus.

For an annulus between two concentric circular tubes, Fig. 19-15, with inner and outer diameters D_1 and D_2 respectively,

$$D_e = 4\,\frac{\pi/4(D_2^2 - D_1^2)}{\pi(D_2 + D_1)} = D_2 - D_1 \qquad (19)$$

Heat transfer coefficients for forced convection in non-circular ducts may be obtained from the equations for circular tubes, using the effective diameter. Although the area for heat transfer may be only part of the perietmer, for example the inner wall of an annulus, the equivalent diameter, which is related to the character of the total flow pattern, is based upon the entire wetted perimeter.

19-9 Forced convection outside tubes: cross flow. In shell and tube or cross flow heat exchangers, the fluid outside the tubes flows across multiple rows of tubes which may be either in line or staggered, Fig. 19-16. For this case the form of equation generally used to correlate heat transfer data is the same as Eq. (16), but the coefficient and exponents are different:

$$\frac{hD_O}{k_f} = 0.33\left(\frac{D_O G_{\max}}{\mu_f}\right)^{0.6}\left(\frac{c_p\mu}{k}\right)_f^{0.33} \qquad (20)$$

where Do is the outside diameter of the tubes, in feet, and G_{max} is the mass velocity based upon the minimum opening between the tubes, whether, in the case of staggered tubes, the minimum opening is the diagonal or the transverse opening. Equation (20) is applicable for estimating the mean heat transfer coefficient for banks of staggered tubes under the following conditions: N_{Re} is between 2000 and

40,000, and there are at least ten rows of tubes in the bank. For banks of less than ten rows, a correction factor is applied from the following table:

Number of rows, N:	1	2	3	4	5	6	7	8	9
Ratio of h for N rows to h for 10 rows:	0.64	.73	.82	.88	.91	.94	.96	.98	.99

For banks of tubes in line the coefficient should be reduced about 25 percent. When Eq. (20) is used to estimate the shell side coefficient for a baffled shell and tube exchanger it is necessary to apply a correction factor of the order of 0.6 to account for the fact that flow around the ends of the baffles is parallel to the tubes, and there is also much baffle leakage and by-passing of the tube bundle in ordinary types of exchanger.

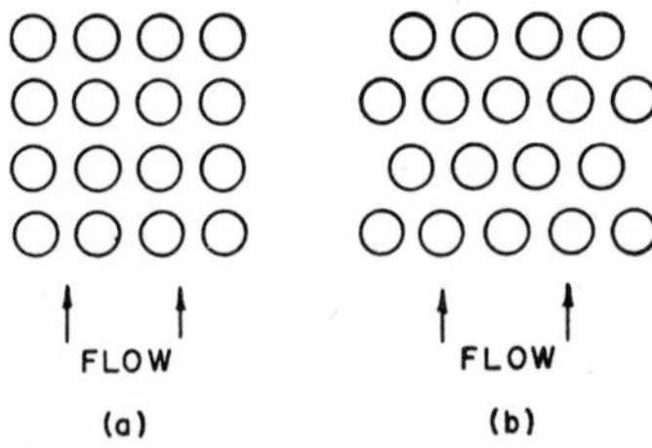

Fig. 19-16. Tube bank arrangement. (a) Tubes in line. (b) Staggered tubes.

For longitudinal flow outside tubes use Eq. (16) or Eq. (17) with an appropriate equivalent diameter.

19-10 Condensing vapors and boiling liquids. Local coefficients for condensing steam in power plant condensers and feed water heaters may be of the order of 1500 to 3000 Btu/sq ft hr °F, and coefficients for boiling water may be of similar magnitude. Therefore, in condensing and boiling processes the overall coefficient may often be controlled by other factors such as a gas or liquid convection coefficient or a dirt or scale coefficient.

In the case of condensers, experience has shown that the average vapor side coefficient may depend more upon the design with respect to distribution of the vapor to the tubes, and the removal of the liquid film from the tubes, than upon the local vapor side coefficients attainable. If air is allowed to accumulate in a condenser the condensation rate is eventually controlled by the rate of diffusion of vapor through the air film on the tubes.

Because coefficients for condensation and boiling are dependent upon numerous and varied factors any attempt to give detailed data in a brief space would be misleading. McAdams gives detailed explanations of the processes, and numerous data on individual coefficients. Overall coefficients attainable in commercial practice are given in handbooks and manufacturer's literature.

19-11 Natural convection. Heat transfer between a solid and a fluid results in a temperature gradient in the fluid near the solid wall. The temperature gradient is accompanied by a density gradient which, in most cases, will cause a certain amount of flow. If the fluid flow past the surface is caused solely by the local density gradient the heat flow is said to be by natural, or free, convection. Natural convection exists in the general circulation in a room heated by a radiator, and also in enclosed spaces such as the hollow spaces of a building wall. The outside surfaces of a house may transmit heat by both free and forced convection, the relative importance of the two modes depending upon the wind velocity. (The wind represents forced convection with respect to the house.)

For calculation of heat transfer in natural convection the usual equation, $q = hA\,\Delta t$, is used. The coefficient h is a function of the shape and orientation of the surface, of a characteristic dimension of the surface, of the properties of the fluid and of the temperature difference Δt. This may be written as follows:

$$\frac{hL}{k_f} = \phi\left[\left(\frac{c_p\mu}{k}\right)_f \left(\frac{L^3\rho^2\beta g(\Delta t)}{\mu^2}\right)_f\right] \tag{21}$$

where L is the characteristic dimension, in feet;

β is the coefficient of volume expansion of the fluid, cu ft/cu ft °F or 1/°F;

g is the local acceleration of gravity, ft/hr^2;

and the second dimensionless group on the right side is the Grashof number N_{Gr}. The functional relationship between N_{Nu}, N_{Pr} and N_{Gr} is usually presented in a plot as in McAdams. For the special case of natural convection in ambient air, the following dimensional equations are used for transfer *to* air.

For vertical plates more than 1 ft high:

$$h = 0.27\,(\Delta t)^{0.25} \tag{22a}$$

For vertical plates less than 1 ft high:

$$h = 0.28(\Delta t/L)^{0.25} \tag{22b}$$

For vertical or horizontal pipes:

$$h = 0.27\ (\Delta t/D_o)^{0.25} \tag{22c}$$

For horizontal plates facing upward:

$$h = 0.27(\Delta t/L)^{0.25} \tag{22d}$$

For horizontal plates facing downward:

$$h = 0.12\ (\Delta t/L)^{0.25} \tag{22e}$$

For heat transfer *from* the air the same equations are used, but the equations for horizontal plates should be interchanged. The coefficient for a vertical plate is not affected by the presence of another parallel plate more than about an inch away.

In enclosed vertical spaces with parallel walls more than about ¾ in. apart the rate of heat transfer is about half of the rate from a single vertical plate for which Δt is equal to the temperature difference between the two walls.

19-12 Radiation. Energy is transferred through space, without the aid of a material conductor, by the mechanism of radiation, which may be thought of as electromagnetic waves. Radiation is excited by moving electric charges; the motion may be caused by external electric potentials (radio waves), by other incident radiation (fluorescent light), or by the thermal vibrations of molecules (thermal radiation). Only thermal radiation, which depends entirely on the temperature of the radiating body, is of interest here.

Thermal radiation covers a broad band of wave lengths (approximately 10^{-2} to 10^{-5} cm) called *infra-red radiation*, and includes at the lower end the spectrum of visible light. Qualitatively, the behavior of thermal radiation is similar to that of light; it travels by line-of-sight, passing freely through a vacuum or some gases, and it may be reflected or transmitted or absorbed by the solid or liquid bodies in its path. However, the long waves of infra-red radiation do not pass as freely as the short waves of light through substances like glass or clear liquids.

19-13 Black body radiation. A body which absorbs all the radiation incident upon it, without reflection or transmission, is called a *black body*. This term has no connection with the color of a body as it appears to the eye. For example, the experimental physicist uses

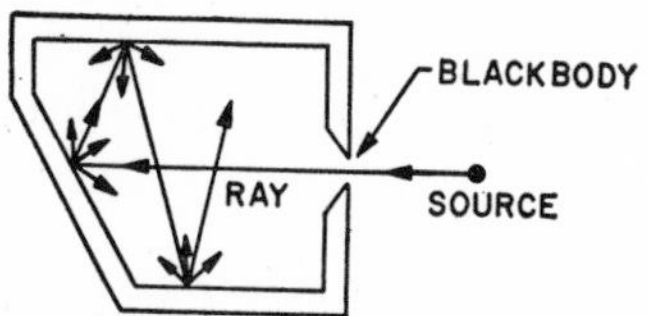

Fig. 19-17. A hole in the wall of an isothermal chamber acts as a black body. The ray that enters the hole cannot get out again until it has been reflected many times. If even a small fraction of the ray is absorbed at each point of incidence, the final effect must approach complete absorption as a limit.

a small hole in the wall of a large isothermal chamber, Fig. 19-17, to act as a black body, although the interior of the chamber, as seen through the hole, may be white hot.

The hole in the wall will not only absorb radiation from the outside, but will also emit radiation from the interior to the surroundings. Such *black body radiation* has some important properties which will be brought out in this section. Assume that the black body hole of Fig. 19-17 is covered with a plate of any material that is not transparent, and that the plate and black body chamber are then isolated from the surroundings. Eventually, the plate and the chamber will come to thermal equilibrium, with no net energy transfer between them. However, black body radiation will continually stream from the chamber, and must be either absorbed or reflected by the cover plate. Since there is no transmission, the sum of the energy absorbed and the energy reflected by the plate must be equal to the incident energy in the black body radiation. Now by the First Law, if the plate remains in a steady state, and is isolated except for radiation exchange with the chamber, the energy emitted from the plate must equal the energy absorbed by the plate. Therefore the emission from the plate will be less than the black body radiation incident upon it, if the reflected energy is greater than zero. If the reflected energy is zero, the plate will emit radiation equal to the incident black body radiation, being itself a black body in this case. Thus it is seen that at a given temperature the emission from a black body is greater than that from a nonblack body.

The rate of energy transfer per unit area by radiation from a body, taking account of all wavelengths and considering the complete hemisphere of space above each element of surface, is called the *total emissive power* of the body.

The total emissive power of a black body E_B is proportional to the fourth power of the absolute temperature;

$$E_B = \sigma T^4 \tag{23}$$

where σ is the *Stefan-Boltzmann constant*, which has the value 1.73×10^{-9} Btu/sq ft hr °R^4. Equation (23), which was deduced by Stefan from experimental data, and by Boltzmann from the Second Law, is called the *Stefan-Boltzmann law*. Because of its derivation from the Second Law, the Stefan-Boltzmann law provides a fundamental relation for the determination of temperatures on the absolute

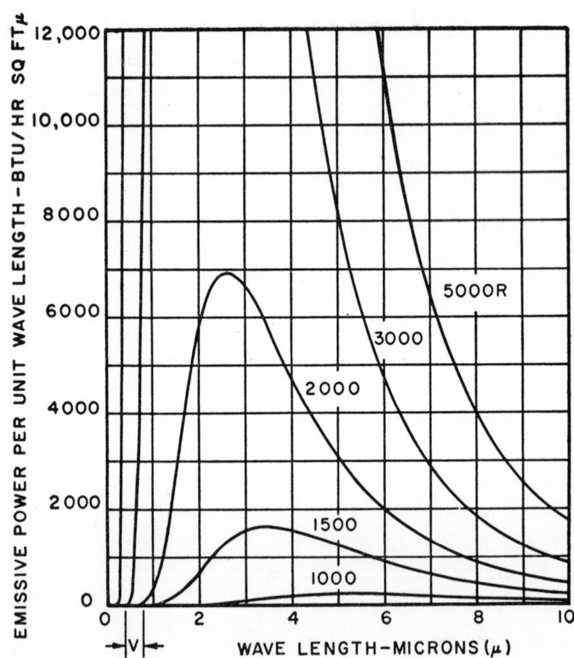

Fig. 19-18. Energy distribution of black body radiation. The curves follow Planck's law, $E_{B\lambda} = 2\pi hc^2\lambda^{-5}/(e^{ch/\kappa\lambda T} - 1)$, where h is Planck's constant, κ is Boltzmann's gas constant, c is the velocity of light, λ is the wave length, and $E_{B\lambda}$ is the monochromatic emissive power. The wave length at which the peak energy rate occurs is inversely proportional to the absolute temperature (Wien's displacement law). The range of wave lengths of visible light is indicated by the dimension V at the bottom of the plot. It is only at the temperature of the sun that the maximum energy rate occurs within the range of visible light.

thermodynamic scale, and it is so used at temperatures which are too high for the use of gas thermometers.

The *spectrum*, or distribution of energy among the wavelengths, for black body radiation is a function of the temperature, as shown in Fig. 19-18. It was through his search for a theory to fit this distribution (Planck's law) that Planck arrived at the quantum theory which underlies all modern physics.

19-14 Emissivity and absorptivity: Kirchhoff's law. The ratio of the total emissive power of any body to that of a black body at the same temperature is called the *emissivity* ϵ of the body at that temperature. The fraction of incident radiation which is absorbed by a body is called the *absorptivity* α of the body, at its particular temperature, for the particular type of radiation. By reasoning similar to that in Sec. 19-13, concerning the black body and the cover plate, it can be shown that the emissivity of a body *at thermal equilibrium with its surroundings* is equal to its absorptivity; this is *Kirchhoff's law.*

For bodies which transmit radiation, the fraction of the incident radiation which is transmitted is called the *transmissivity* τ. For bodies which reflect radiation, the fraction of the incident radiation reflected is called the *reflectivity* ρ. In general,

$$\alpha + \rho + \tau = 1$$

However, for solids and liquids τ is usually negligible, and for gases ρ is zero. For substances like glass or clear plastics it is necessary to consider all three fractors, and their dependence upon the wavelength of the incident radiation, and upon the thickness of the body.

The black body is a limiting case; for actual bodies the absorptivity is less than unity, and the *monochromatic absorptivity* (absorptivity at a particular wavelength) generally varies with wavelength. In ordinary heat transfer calculations, it is assumed that the monochromatic absorptivity is the same for all wavelengths; the body is then called a *gray body.*

19-15 Radiant heat transfer between black bodies. If the edge of the hole which acts as a black body is infinitely thin, the radiation from the hole will stream out into the entire hemisphere around the hole, Fig. 19-19. An observer at any point P on a hemispherical surface centered on the hole will see the same kind of radiation through the hole, wherever point P may be on the hemisphere; but the apparent area of the hole will be proportional to the cosine of the angle ϕ between the normal to the plane of the hole and the line of sight. Therefore the radiant energy reaching any element of area dA_p on the hemisphere of radius r, from an elementary black body of area dA_1 will be directly proportional to dA_1, to cosine ϕ, and to dA_p, and will be inversely proportional to the square of the radius r:

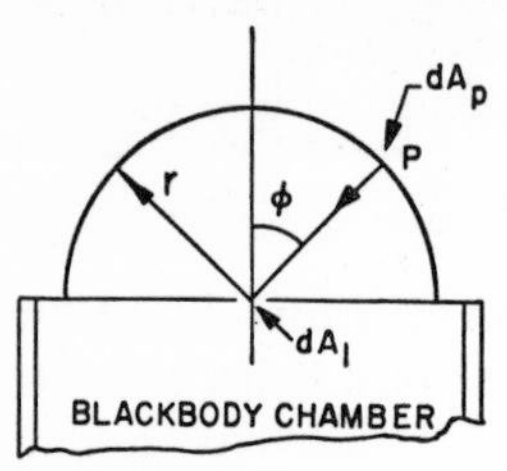

Fig. 19-19.

$$dq_{1\longrightarrow p} = I_1\, dA_1 \cos\phi\, dA_p/r^2 \tag{24}$$

where I_1 is the intensity, Btu/sq ft hr, of the radiation leaving the black body in the *normal* direction. By integrating Eq. (24) over the hemisphere, we obtain the total emissive power of the black body, E_B, which is thus found to equal πI_1.

The fraction of the total emissive power of dA_1 that is incident on any body A_2 which can be seen from dA_1 can be determined by finding the projection of A_2, as seen from dA_1, on the hemisphere, Fig. 19-19, and integrating Eq. (24) over the projected area. If the source is not an *element* of area dA_1, but is a finite area A_1, the process must be carried out for a number of elements dA_1 to obtain the average value for the entire area A_1. The result of this procedure is called, by Hottel, the *view factor* F_{12} for radiation from A_1 to A_2. It is a purely geometrical factor which can be evaluated once for all, for any given configuration of surfaces. A number of plots for common cases have been published, notably by Hottel; Fig. 19-20 is an example of such a plot.

When the view factor F_{12} has been determined, the heat transfer equation may be set up as follows. Since A_2 is a black body, all the incident radiation is absorbed, and the rate of radiant energy transfer *from* A_1 *to* A_2 is given by

$$q_{1\longrightarrow 2} = A_1 F_{12} \sigma T_1^4 \tag{25}$$

Similarly the transfer rate *from* A_2 *to* A_1 is given by

$$q_{2\longrightarrow 1} = A_2 F_{21} \sigma T_2^4 \tag{26}$$

But when $T_2 = T_1$ the net heat transfer must be zero, or

$$q_{1\longrightarrow 2} = q_{2\longrightarrow 1}$$

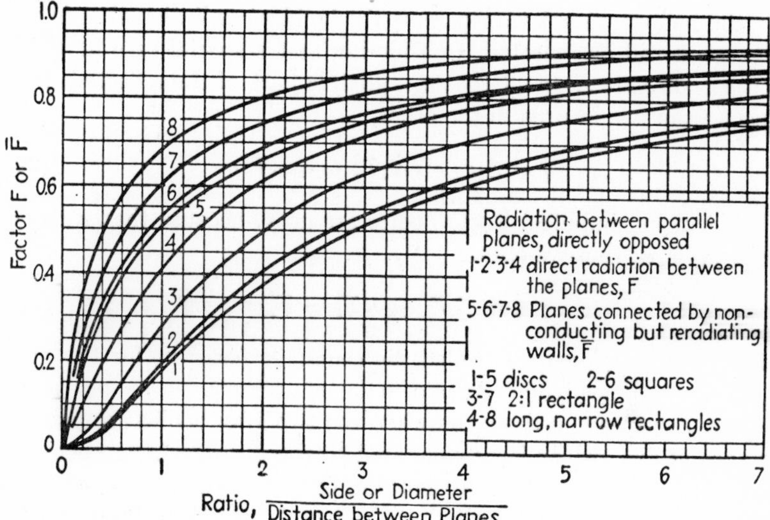

Fig. 19-20. Values of view factor F and interchange factor $\overline{F}$ for parallel planes. (By permission from *Mechanical Engineer's Handbook* by L. S. Marks, copyright 1951, McGraw-Hill Book Co.)

then it follows, independent of T, that

$$A_1F_{12} = A_2F_{21}$$

The equation for direct radiant-heat exchange between two black bodies separated by a nonabsorbing medium (vacuum or gas) is then

$$q = A_1F_{12}\sigma(T_1^4 - T_2^4) \tag{27}$$

This is often written as follows, for convenience in computation:

$$q = A_1F_{12} \times 0.173\left[\left(\frac{T_1}{100}\right)^4 - \left(\frac{T_2}{100}\right)^4\right] \tag{28}$$

19-16 Radiant heat transfer between nonblack bodies. In calculating the rate of heat transfer by radiation between bodies that are not black, the emissivities and absorptivities of the bodies must be taken into account. The radiation *emitted from A_1 and absorbed by A_2* is given by

$$q_{1\longrightarrow 2} = A_1F_{12}\epsilon_1\alpha_{21}\sigma T_1^4$$

where ϵ_1 is the emissivity of A_1 at T_1, and α_{21} is the absorptivity of A_2 at T_2 for the radiation from A_1 at T_1. Similarly,

$$q_{2\longrightarrow 1} = A_1F_{12}\epsilon_2\alpha_{12}\sigma T_2^4$$

If the bodies are not transparent, the incident radiation that is not absorbed will be reflected. Considering first a special case in which both bodies are so small that a negligible part of the reflected radiation from one returns to the other,

$$q = A_1 F_{12} \sigma (\epsilon_1 \alpha_{21} T_1^4 - \epsilon_2 \alpha_{12} T_2^4) \tag{29}$$

This can be simplified if the bodies are gray, because then $\alpha_{21} = \epsilon_2$ and $\alpha_{12} = \epsilon_1$; thus we obtain

$$q = A_1 F_{12} \epsilon_1 \epsilon_2 \sigma (T_1^4 - T_2^4) \tag{30}$$

The product $\epsilon_1 \epsilon_2$ may be called the *emissivity factor* F_E for this case (gray bodies, negligible return of reflected energy).

Another special case occurs when the two gray bodies are parallel planes of dimensions large compared to the distance between them. In this case each body sees nothing but the other so $F_{12} =$ unity; by taking account of the infinite series of reflections between the planes, F_E is found to be

$$\frac{1}{(1/\epsilon_1) + (1/\epsilon_2) - 1}$$

Then

$$q = A_1 \frac{1}{(1/\epsilon_1) + (1/\epsilon_2) - 1} \sigma (T_1^4 - T_2^4) \tag{31}$$

Equation (31) is also used for concentric spheres or infinitely long concentric cylinders which reflect specularly (like a mirror); for example, the silvered walls of a vacuum flask. For concentric spheres or cylinders which reflect diffusely (unpolished surfaces), the emissivity factor is modified because not all the reflected radiation from the outer wall returns to the inner wall; then, if A_1 is the inner surface,

$$q = \frac{A_1 \sigma}{\dfrac{1}{\epsilon_1} + \dfrac{A_1}{A_2}\left(\dfrac{1}{\epsilon_2} - 1\right)} (T_1^4 - T_2^4) \tag{31a}$$

Another special case is that of a small body A_1 entirely surrounded by a large enclosure. The view factor F_{12} is unity. To obtain the emissivity factor F_E note that only a small part of the reflected radiation from the walls of the large enclosure can be intercepted by A_1; most of it must strike the wall of the enclosure, which in effect is then a black body. As a result the small body controls the process, and the heat transfer equation is

$$q = A_1\epsilon_1\sigma(T_1^4 - T_2^4) \tag{32}$$

In the preceding discussion it has been assumed that each point on A_1 can see the entire projected area of A_2 lying within the hemisphere of space above the point on A_1. For example, if A_1 is a body inside an enclosure A_2, then A_1 must be everywhere convex or flat, so that no point on A_1 has its view of A_2 cut off by another part of A_1. When A_1 has a concave region, Fig. 19-21, the effective area of the body is found by substituting for the area of the depressed surface the area of the envelope which spans the depression, as indicated by the dash line in Fig. 19-21. This procedure is strictly valid only for a black body.

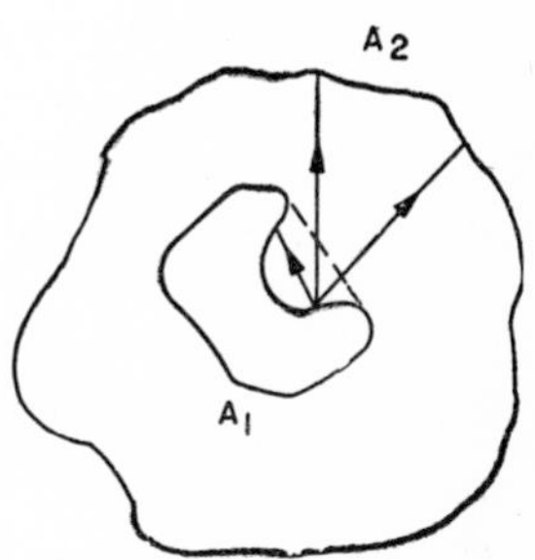

Fig. 19-21. A body with self-obstructed view; the effective area of the concave portion of A_1 is the area of the surface indicated by the dash line.

Example 7. An uninsulated 3 in. steam pipe passes through a room in which the air and all solid surfaces are at an average temperature of 70°F. If the surface temperature of the steam pipe is 200°F, estimate the heat loss per foot of pipe by radiation, and compare the relative magnitudes of the losses by radiation and by free convection.

Solution: A 3 in. pipe in an ordinary room may be considered a completely enclosed small body, so Eq. (32) will apply. Taking data from the appendix, the emissivity of the steel pipe is taken to be 0.8; the external area of a 3 in. pipe is 0.916 sq ft per ft. Then

$$\begin{aligned} q &= (0.916)(0.8)0.173\left[\left(\frac{660}{100}\right)^4 - \left(\frac{530}{100}\right)^4\right] \\ &= 0.127(1897 - 789) \\ &= 140 \text{ Btu/hr per foot of pipe} \end{aligned}$$

To compare this with the convection heat loss, we may first compute an equivalent coefficient h_r for the radiation:

$$h_r = \frac{q}{A\,\Delta t} = \frac{140}{0.196(130)} = 1.18 \text{ Btu/sq ft hr °F}$$

The free convection coefficient would be (Eq. 22c)

$$h_c = 0.27\left(\frac{\Delta t}{D_0}\right)^{0.25} = 0.27\left(\frac{130}{0.292}\right)^{0.25}$$
$$= 1.24 \text{ Btu/sq ft hr °F}$$

Thus, in this case, radiation would account for about half the heat loss from the pipe.

19-17 Adiabatic surfaces. In the preceding section we considered only *direct* radiation between bodies. When radiation between an energy source and a sink travels not only directly, but also by reflection from intermediate surfaces, the heat transfer rate is increased. In many practical cases, as in a furnace at steady operating conditions, the intermediate reflecting walls may be practically adiabatic, because the leakage of heat to the outside, and the transfer by convection on the inside, are very small compared to the incident radiation, or the combined emitted and reflected radiation. For some cases of this type, Hottel gives plots of an *interchange factor* $\overline{F}$ to take the place of the view factor and account for the effect of the reflecting walls; see Fig. 19-20 for an example.

Example 8. A peep hole in a furnace wall of 13.5 in. thickness is 8 in. square. The inside of the furnace is at a uniform temperature of 1500°F, and the external surroundings are at 120°F, on the average. Using Fig. 19-20, estimate the rate of heat loss by radiation through the open peep hole.

Solution: Since we are dealing with parallel planes, and all reflected radiation would eventually return to the source, the proper emissivity factor would be as given in Eq. (31); in this case, however, both planes are black bodies, so F_E is unity. From the plot, for the given configuration, $\overline{F} = 0.86$. Then

$$q = A_1\overline{F}F_E\sigma(T_1^4 - T_2^4)$$
$$= \frac{64}{144}\,0.86(1)(0.173)\left[\left(\frac{1960}{100}\right)^4 - \left(\frac{580}{100}\right)^4\right]$$
$$= 0.0661(14.74 \times 10^4 - 1130) = 9680 \text{ Btu/hr}$$

Note the small effect of the cold body temperature on the result. Since 3^4 is 81, the lower temperature could be taken as zero with little error when the ratio of the two temperatures is greater than three.

19-18 Other radiation problems. Many of the important practical problems in radiation involve gas radiation, flame radiation, and radiation from clouds of particles, as well as multi-dimensional problems involving several sources, sinks, and adiabatic reflecting surfaces. The interested reader should consult Hottel's chapter on radiation in McAdams for a discussion of these problems. However, a few general statements on gas radiation will be given here, to complete the discussion of radiation.

Some of the common gases which emit and absorb appreciable radiation are water vapor, carbon dioxide, carbon monoxide, sulfur dioxide, and ammonia. Oxygen, nitrogen, and hydrogen are transparent. Gases radiate in discrete bands of wavelengths rather than in a continuous spectrum; also their emissive power depends upon the number of molecules encountered along the path of a ray, which in turn depends upon both the partial pressure of the gas and the thickness of the gas body (beam length). Gas radiation becomes most important in large chambers, such as furnaces, where the beam length is great.

In calculating radiation between solid surfaces separated by a radiating gas, the absorption and emission of the gas must be taken into account. A somewhat similar problem exists in the case of radiation which passes through transparent substances like glass or plastics, since there is selective absorption, depending upon the wavelength.

19-19 Heat transfer—conclusion. The scope of this chapter is necessarily extremely limited, no information being given on many important aspects of heat transfer such as boiling, condensation, heat transfer in compressible flow, simultaneous heat and mass transfer as in cooling towers, time transient problems, and others. Information on these matters may be found in the references.

In heat exchangers there are many possible designs for any given heat transfer requirement, each design differing from the others in size, shape, weight, pressure losses, cost, and other factors such as convenience of cleaning. The art of heat exchanger design lies not in heat transfer calculations as such, but in balancing all the other factors to obtain the required heat transfer capacity in the simplest and most economical way compatible with the application. For example, in an aircraft heat exchanger, light weight and compactness are usually of first importance; in a stationary power plant, cost, low-pressure drop, and convenience of cleaning may be more important.

Because of the great variety of circumstances which affect practical heat transfer problems, detailed information on particular applications is best sought in the literature of the particular field of engineering involved, rather than in treatises on heat transfer in general.

PROBLEMS*

19-1. A brick wall 8 in. thick has room air on one side and outside air on the other. On a day when the outside air is at 0°F and the room air is at 70°F, the surface coefficients of heat transfer are 1.5 Btu/sq ft hr °F on the inside and 5.0 on the outside. Find the rate of heat transfer through the wall and the temperature of the inside surface.

19-2. For the wall of Problem 19-1 it is proposed to cement a layer of insulating board to the inner surface. If the conductivity of the board is 0.028 Btu/ft hr °F, find the rate of heat transfer and the inside surface temperature to be expected with a board of (a) 0.5 in., (b) 1.0 in. thickness.

19-3. A *heat meter* is a slab of material of known heat transfer coefficient, fitted with thermocouples to measure the temperature drop through the slab. It is sealed tightly to the surface of a wall through which heat is flowing and, after equilibrium is reached, the temperature difference is read and the heat flux is calculated. In a certain case, the coefficient of a heat meter is 1.0 Btu/sq ft hr °F, the temperature drop through the meter is 8.00°F, and the temperature difference between the surfaces of the wall in series with the meter is 20.0°F. Find the rate of heat flow per unit area and the heat transfer coefficient for the wall.

19-4. A hot-water tank in a room at 70°F is to contain water at 140°F. The liquid film coefficient inside the tank is estimated at 80 Btu/sq ft hr °F, and the total outside coefficient including, both radiation and convection, is estimated at 1.6 Btu/sq ft hr °F. The tank is 2 ft inside diameter and 6 ft long, and may be made of copper ⅛ in. thick, or of carbon steel ⅛ in. thick and coated on the inside with 0.015 in. of glass that has a thermal conductivity of 0.5 Btu/ft hr °F. Compare the expected heat loss from the copper tank with the loss from the glass-lined steel tank.

19-5. (a) How much heat loss would be saved in Problem 19-4 if the tank were covered by glass-wool insulation 1 in. thick? (b) If the heat transferred represented a net loss of 40 cents per million Btu, and the tank were in use 8700 hr per year, how much additional cost could be justified for the glass-wool insulation, on the basis that a dollar of investment balances 20 cents per year operating expense?

19-6. For the insulation of steam pipes it is customary to use 85 percent magnesia pipe covering at temperatures below 500°F, and diatomaceous earth at temperatures above 500°F. For a pipe of 6.625 in. outside diameter at 1000°F, a layer of diatomaceous earth insulation is applied first, with a

* Data may be found in the Appendix.

layer of magnesia outside. The thicknesses are selected so that the magnesia will not be heated above 500°F, and the outside surface temperature will not exceed 140°F. The surface coefficient may be taken as 2.0 Btu/sq ft hr °F. If the two materials are available in thicknesses of integral numbers of half inches, what thicknesses should be used? The room is at 100°F.

19-7. A small bee-hive oven of hemispherical shape is built of an inner layer of 2600-degree insulating firebrick $4\frac{1}{2}$ in. thick, an intermediate layer of $4\frac{1}{2}$ in. of 1600-degree insulating firebrick, and an outer covering of 85 percent magnesia 2 in. thick. The inside radius of the oven is 2 ft. If the inner surface of the brick is at 2200°F and the outside surface coefficient is 2.5 Btu/sq ft hr °F, find (a) the heat loss through the hemisphere, Btu/hr; (b) the maximum temperature of the 1600-degree brick; (c) the maximum temperature of the 85 percent magnesia; (d) the outside surface temperature if the room is at 80°F.

19-8. A 1-in. outside diameter tube carrying refrigerant passes through air at 40°F. In the course of time a concentric layer of frost builds up on the tube. If the tube wall remains at 0°F, the outside surface coefficient stays constant at 1.6 Btu/sq ft hr °F, and the thermal conductivity of the frost is 0.25 Btu/ft hr °F, plot the heat transfer rate, Btu/hr, for a linear foot of tube, vs. thickness of the frost layer. Frost thicknesses of 0, 1.0, 1.5, 2.0, and 2.5 in. are suggested for plotting. The radius corresponding to maximum heat transfer rate is called the *critical radius*; its value depends on the surface coefficient and the conductivity of the insulation.

19-9. Derive an expression for the critical radius of insulation on a cylinder in terms of the surface film coefficient and the thermal conductivity of the insulation, both assumed constant.

19-10. Two parts of a heat exchanger are connected by a conical tube of uniform wall thickness. The tube is 30 in. long, 16 in. outside diameter at the small end, 30 in. outside diameter at the large end, and 0.50 in. wall thickness. One end is at 1200°F and the other at 900°F, and there is negligible heat transfer from the surfaces between the two ends. If the material is type 321 stainless steel, how much heat will be transferred per hour from one end to the other?

19-11. A tubular heat exchanger in which water is evaporated at 1 atm by condensing steam is expected to operate so that the surface coefficients for the inside (condensing) and outside (boiling) sides of the tubes will be respectively 1500 Btu/sq ft hr °F and 1000 Btu/sq ft hr °F. The tubes will be 1.00 in. outside diameter by 18 Birmingham Wire Gage thickness. In the operation of the evaporator, scale will collect on the outside of the tubes; it will have an average heat transfer coefficient of 200 Btu/sq ft hr °F before the tubes are cleaned. For an overall temperature difference of 50°F, find the rate of heat transfer, Btu/sq ft hr, for both clean and scale-covered tubes, if the tubes are (a) copper; (b) admiralty metal; (c) 30 percent cupro-nickel; (d) type 304 stainless steel.

19-12. A counterflow heat exchanger for cooling air is to receive air at 350°F and cool it to 100°F, using water supplied at 75°F. If the water is to be heated to 150°F, and the overall heat transfer coefficient is predicted as 15 Btu/sq ft hr °F, how much heat transfer area will be required for an air flow rate of 10 lb/sec?

19-13. If it is desired to cool the air in Problem 19-12 to 85°F, all other conditions remaining unchanged, by what fraction will the required area increase? Calculate by the equations for heat transfer and mean temperature difference, then check by Fig. 19-11.

19-14. A heat exchanger with one shell pass and two tube passes is to be used for cooling the air in Problem 19-12, but the air is to be cooled to only 135°F, while the water rises from 75°F to 150°F. The overall coefficient is still the same. The water is in the tubes, and the air is in the shell. Using Fig. 19-10, find the area required, and compare with the result of Problem 19-12. From Fig. 19-10, does it seem possible to use a two-pass exchanger to cool the air to 100°F, as is done with the counter-flow exchanger? From consideration of the physical arrangement, can you estimate an approximate limiting temperature for the fluid leaving the shell side, in terms of the tube-side entering and leaving temperatures?

19-15. Lubricating oil, specific heat 0.5 Btu/lb °F, is to be cooled from 180°F to 160°F in the shell of an exchanger with water in the two tube passes, Fig. 19-10. The overall coefficient is expected to be 60 Btu/sq ft hr °F. The water is available at 140°F. For water exit temperatures of 150°F, 160°F, and 165°F, tabulate the water flow rate and the heat transfer area required to cool 20,000 lb/hr of oil.

19-16. Air is to be cooled by water in a cross-flow heat exchanger in which an overall coefficient of 10 Btu/sq ft hr °F is expected. The air enters at 80°F and is to be cooled to 50°F by water entering at 45°F. Both fluids are "unmixed" in passing through the exchanger, Fig. 19-11. The air flow rate is 10,000 lb/hr. (a) Plot heat transfer surface vs. water flow rate for various water flow rates that will satisfy the cooling requirement. (Assume various water outlet temperatures, and calculate the area and the water flow rate.) (b) If the heat transfer surface costs $1.00 per square foot and the capitalized cost of the water supply is $40.00 per gallon per minute how much surface should be used?

19-17. A regenerator for a gas turbine plant is to heat 300 lb/sec of air supplied at 550°F by cooling 310 lb/sec of combustion products, specific heat 0.27 Btu/lb °F, supplied at 840°F. An overall heat transfer coefficient of 9.0 Btu/sq ft hr °F is expected. Find the area of a counter-flow heat exchanger of 75 percent effectiveness for this application. Check your solution by Fig. 19-11.

19-18. Using the charts of Fig. 19-11, determine the relative areas required for counter-flow exchangers of 50, 60, 70, and 80 percent effectiveness, and for cross-flow exchangers of 50, 60, 70, and 75 percent effectiveness, for the application of Problem 19-17.

19-19. A condenser for a steam turbine is designed with 20 percent excess heat-transfer surface to allow for fouling. If the clean coefficient is 650 Btu/sq ft hr °F, what will be the average value of the *scale coefficient* or *dirt coefficient* which would exist when the condenser is just able to carry the design load?

19-20. A double-tube heat exchanger is to be constructed for heating air to 230°F for a laboratory test. The heating medium will be steam condensing at 30 psia in the annular space around a ⅝ in. outside diameter 18-gage copper tube; 3 lb/min of air will be supplied at 100 psia, 80°F, and will flow inside the tube. If the steam-side coefficient is taken as 2000 Btu/sq ft hr °F, and the exchanger is to have 25 percent excess area for safety factor, how long should the tube be?

19-21. In a steam condenser for a power plant the steam is condensing at 1.5 in. Hg (absolute pressure), and the cooling water is supplied at 60°F. If the condenser has ⅞ in. outside diameter 18 Birmingham Wire Gage tubes, and the water flows at 6.5 fps in the tubes, estimate the water-side film coefficient. For estimating purposes assume the *approach* or *terminal temperature difference* will be 10°F.

19-22. The tubes in the condenser of Problem 19-21 are 24 ft long, and the water makes a single pass through them. (a) If the tubes are of admiralty metal and the steam-side coefficient is assumed to be 1500 Btu/sq ft hr °F how much will the water temperature rise? (b) How many square feet of tube surface will be needed to transfer 200,000,000 Btu/hr? (c) How many gallons per minute of water will have to be supplied?

19-23. A tubular air preheater has 30 rows of 2 in. outside diameter tubes spaced on 3 in. centers on equilateral triangles. The air flows across the tubes with a velocity of 40 fps, based on the gross area of the tube bank. If the average air temperature is 300°F and the average tube wall temperature is 400°F, estimate the film coefficient.

19-24. The flue gases inside the tubes of Problem 19-23 have an average temperature of 500°F and flow at a velocity of 50 fps. The flue gas analysis may be taken as 0.12 carbon dioxide, 0.04 water, 0.06 oxygen, and the remainder nitrogen, on a volume basis. The tube walls are 0.109 in. thick, of carbon steel. Estimate the overall heat-transfer coefficient for clean tubes.

19-25. An electric transformer is in a tank 2 ft in diameter and 3 ft high, with flat top and bottom. If the tank transfers heat only by natural convection to the atmosphere, and the electrical losses are to be dissipated at the rate of 1 kw, how many degrees will the tank surface rise above the room temperature?

19-26. An insulated vertical wall has air spaces 3 in. thick, with surfaces of wood having emissivity 0.9. Estimate the rate of heat transfer per square foot of air space by radiation and natural convection, if the two surfaces are at 60°F and 35°F respectively.

19-27. Repeat Problem 19-26 for walls covered with aluminum foil, emissivity 0.09.

19-28. How much will the heat transfer rate in Problem 19-27 change if an intermediate sheet of aluminum foil is installed half way between the walls?

19-29. A thermometer well is at the center of a duct in which air is flowing at 1 atm, 250°F. The walls of the duct are at 150°F. The thermometer well is 0.5 in. in diameter, of oxidized steel having an emissivity of 0.9, and is placed with its axis across the stream. The well may be considered as equivalent to a single-row staggered-tube bank, insofar as convection is concerned. If there is negligible heat transfer to and from the well except by forced convection and radiation, estimate the error of the thermometer reading at gas velocities of 5, 10 and 50 fps.

19-30. Solve Problem 19-29 if the air and duct wall temperatures are 750°F and 200°F respectively.

19-31. A portion of the shell of a steam turbine has steam on one side at 1500 psi, 1050°F, and on the other side at 600 psi, 500°F. The convection heat-transfer coefficient on both sides is of the order of 500 Btu/sq ft hr °F because of high velocities and high density of the steam. The shell wall is 3 in. thick, of carbon steel. Estimate how much the maximum temperature of the wall will be reduced if a shield of type-304 stainless steel $\frac{1}{16}$ in. thick is mounted on the hot side of the wall, so that the hot steam heats the shield by convection, and the shield isolates the shell wall from the steam. Neglect conduction at the shield supports and convection in the $\frac{1}{2}$ in. space between the shield and the wall, but take account of conduction and radiation through the steam between the shield and the wall. Neglect radiation exchange with the steam.

19-32. An access door in the wall of a boiler furnace is to be protected from the heat by a water-cooled "pad", facing the inside of the furnace. The flame, at 2500°F, is so thick that it may be considered a black body. If the door is at 500°F, and can see only the flame, how many Btu/hr will be absorbed by the oxidized steel surface of 3 sq ft area? Assuming a water-side film coefficient of 400 Btu/sq ft hr °F and a $\frac{1}{4}$ in. thick carbon-steel wall, what must the average water temperature be to carry away the heat?

19-33. In Example 8, Sec. 19-17, what would have been the heat transfer rate if the hole in the furnace wall had tapered to a sharp edge instead of being $13\frac{1}{2}$ in. thick?

19-34. A vacuum bottle is made of two concentric cylinders with polished silver surfaces facing each other in the evacuated annular space. The two surfaces are of 4 and 5 in. diameter respectively. Considering the cylindrical surfaces only, compare the heat-transfer coefficient for radiation through the space, with the coefficient for conduction, using the best available insulation, silica aerogel, which has a thermal conductivity of 0.013 Btu/ft hr °F; the internal and external wall temperatures are 200°F and 80 F respectively. How thick would the aerogel insulation have to be to equal the vacuum insulation?

REFERENCES

General

McAdams, W. H., *Heat Transmission*, 3d ed. New York: McGraw-Hill, 1954.

Jakob, M., *Heat Transfer*. New York: Wiley, 1949.

Eckert, E. R. G., *Introduction to the Transfer of Heat and Mass*. New York: McGraw-Hill, 1950.

Boelter, L. M. K., Cherry, Johnson, and Martinelli, *Heat Transfer*. University of California Press, 1946.

Conduction

Dusinberre, G. M., *Numerical Analysis of Heat Flow*. New York: McGraw-Hill, 1949.

Carslaw, H. S., and Jaeger, *Conduction of Heat in Solids*. New York: Oxford, 1947.

Heat Transfer Data

Society of Automotive Engineers, *Airplane Air Conditioning Engineering Data—Heat Transfer*. New York, 1952. (Much more comprehensive than the title indicates.)

American Society of Heating and Ventilating Engineers, *Heating and Ventilating Guide*, published annually

American Society of Refrigerating Engineers, *Refrigerating Data Book*, published periodically.

Perry, J. H. *Chemical Engineers Handbook*. New York: McGraw-Hill, 1950.

Tubular Exchanger Manufacturers Association, *TEMA Standards*. New York, 1952.

Heat Insulation

Wilkes, G. B., *Heat Insulation*. New York: Wiley, 1950.

APPENDIX

LIST OF TABLES AND GRAPHS

APPENDIX

DEFINITIONS OF SYMBOLS

Symbols of general interest are defined below, while symbols of limited interest have been defined where they first occur in the text. The units given are as customarily used, but their exclusive use is not implied.

a or A	area, sq ft
c	heat capacity, Btu/lb °F
c_p	specific heat at constant pressure, Btu/lb °F
c_v	specific heat at constant volume, Btu/lb °F
$C_{\bar{V}}$	coefficient of velocity, dimensionless
C_u	coefficient of discharge, dimensionless
C	centigrade temperature, degrees
E	internal energy of a *system*, Btu or ft lbf
F	Fahrenheit temperature, degrees
F	force, lbf
g	acceleration of gravity, ft/sec^2
g_0	constant in Newton's second law; 32.17 lbm ft/lbf sec^2
h	specific enthalpy, Btu/lbm or ft lbf/lbm
H	enthalpy, Btu or ft lbf
J	ratio of work unit to heat unit; 778.16 ft lb/Btu
k	ratio of specific heats, c_p/c_v
K	Kelvin temperature (absolute centigrade)
l	length or distance, ft
ln	natural logarithm
m	mass, lbm
M	molecular weight, lbm/lb mol
n	number of mols
n	exponent in pv^n = constant
p	pressure, lbf/sq ft
Q	heat transferred, Btu
R	Rankine temperature (Fahrenheit absolute)
R	gas constant, ft lbf/lbm °R
$\bar{R}$	universal gas constant, 1545.3 ft lbf/lb mol °R
s	specific entropy, Btu/lbm °R

S	entropy, Btu/°R
t	temperature, F or C
T	absolute temperature, R or K
u	specific internal energy of a *substance*, Btu/lbm
U	internal energy of a *substance*, Btu
v	specific volume, cu ft/lbm
V	volume, cu ft
$\overline{V}$	velocity, ft/sec
w	mass flow rate, lbm/sec
w	power plant fluid rate, lbm/hphr or lbm/kwhr
W	work, ft lb or Btu
W_x	shaft work, ft lb or Btu
x	vapor fraction of liquid-vapor mixture, fraction by mass
x	mol fraction of a gas component of a mixture
z	elevation above a datum level, ft

GREEK LETTERS

α	alpha	nozzle angle, degrees
Δ	delta	mathematical symbol for "change of"
γ	gamma	specific humidity
η	eta	efficiency
ϕ	phi	relative humidity

SUBSCRIPTS

A, B, etc., or 1, 2, etc., identify a quantity with a certain point or path in a process

f	saturated liquid state
fg	difference between saturated liquid and saturated vapor
g	saturated vapor state
h	constant enthalpy
0	stagnation state
	see also g_0
p	constant pressure
s	constant entropy
t or T	constant temperature
u	constant internal energy
x	mixture state of quality x
	see also W_x

Table A1: Properties of Gases*

Gas	Formula	Molecular weight	Specific heats, Btu/lbm °F at 1 atm, ordinary room temperatures c_p	c_v	c_p/c_v	Gas constant, R, ft lbf/lbm °R† $\overline{R}/M = R$ at 0 p	$pv/T = R$ 1 atm 32°F
Air	‡	28.97	0.240	0.171	1.40	53.35	53.34
Monatomic gases:							
Argon	A	39.94	0.123	0.074	1.67	38.68	38.65
Helium	He	4.003	1.25	0.75	1.66	386.2	386.3
Diatomic gases:							
Carbon monoxide	CO	28.01	0.249	0.178	1.40	55.18	55.13
Hydrogen	H_2	2.016	3.42	2.43	1.41	766.6	767.0
Nitrogen	N_2	28.02	0.248	0.177	1.40	55.16	55.13
Oxygen	O_2	32.00	0.219	0.156	1.40	48.29	48.24
Triatomic gases:							
Carbon dioxide	CO_2	44.01	0.202	0.156	1.30	35.12	34.88
Sulfur dioxide	SO_2	64.07	0.154	0.122	1.26	24.12	23.55
Water vapor	H_2O	18.016	0.446§	0.336§	1.33	85.78	85.58§
Hydrocarbons:							
Acetylene	C_2H_2	26.04	0.383	0.303	1.26	59.35	58.77
Methane	CH_4	16.04	0.532	0.403	1.32	96.35	96.07
Ethane	C_2H_6	30.07	0.419	0.342	1.22	51.40	50.82
Iso-butane	C_4H_{10}	58.12	0.398	0.358	1.11	26.59	25.79

* Data mainly from U.S. Department of Commerce, Bureau of Standards Circular No. C 461; and from Eshbach, *Handbook of Engineering Fundamentals.* New York: Wiley, 1952.

† The universal gas constant is taken as $\overline{R}$ = 1.986 Btu/lb mol deg R = 1545.3 ft lbf/lb mol deg R.

‡ The composition of air, percent by volume, is taken as N_2, 78.03; O_2, 20.99; A, 0.98; following Keenan and Kaye.

§ pv/T for water vapor at 1 psia, 300°F, data from Keenan and Keyes. c_p and c_v for water vapor at pressures below 1 psia.

Table A1 (Concluded)

Gas	Mol volume cu ft/lb mol, at 1 atm, 32°F	Critical constants		Van de Waals constants	
		Temp. °F	Pressure atm	(a) $\frac{\text{atm ft}^3}{(\text{lb mol})^2}$	(b) $\frac{\text{ft}^3}{\text{lb mol}}$
Air‡	359.0	−221.3	37.2	343.5	0.585
Monatomic gases:					
Argon	359.6	−187.7	48.0	346	0.517
Helium	359.2	−450.2	2.26	8.57	0.372
Diatomic gases:					
Carbon monoxide	358.8	−218.2	35.0	381	0.639
Hydrogen	359.3	−399.8	12.8	62.8	0.426
Nitrogen	358.9	−232.8	33.5	346	0.618
Oxygen	358.7	−181.8	49.7	349.5	0.510
Triatomic gases:					
Carbon dioxide	356.6	88.0	73	925	0.686
Sulfur dioxide	350.6	315.0	77.7	1737	0.910
Water vapor	....	705.5	218.5	1400	0.488
Hydrocarbons:					
Acetylene	355.6	103.5	62.0	1129	0.8232
Methane	358.0	−116.5	45.8	581.2	0.6855
Ethane	355.1	90.1	48.2	1391	1.028
Iso-butane	348.2	273.2	36.9	3265	1.807

Table A2: Combustion Data*

Substance	Formula	Molecular weight	Combustion reaction
Hydrogen	H_2 (gas)	2.016	$2H_2 + O_2 \rightarrow 2H_2O$
Sulfur	S (solid)	32.07	$S + O_2 \rightarrow SO_2$
Carbon	C (solid)	12.01	$C + O_2 \rightarrow CO_2$
			$C + \frac{1}{2}O_2 \rightarrow CO$
Carbon monoxide	CO (gas)	28.01	$CO + \frac{1}{2}O_2 \rightarrow CO_2$
Methane	CH_4 (gas)	16.04	$CH_4 + 2O_2 \rightarrow CO_2 + 2H_2O$
Ethane	C_2H_6 (gas)	30.07	$C_2H_6 + 3\frac{1}{2}O_2 \rightarrow 2CO_2 + 3H_2O$
Propane	C_3H_8 (gas)	44.09	$C_3H_8 + 5O_2 \rightarrow 3CO_2 + 4H_2O$
Butane	C_4H_{10} (gas)	58.12	$C_4H_{10} + 6\frac{1}{2}O_2 \rightarrow 4CO_2 + 5H_2O$
Octane	C_8H_{18} (liq)	114.2	$C_8H_{18} + 12\frac{1}{2}O_2 \rightarrow 8CO_2 + 9H_2O$
Benzene	C_6H_6 (liq)	78.11	$C_6H_6 + 7\frac{1}{2}O_2 \rightarrow 6CO_2 + 3H_2O$
Acetylene	C_2H_2 (gas)	26.04	$C_2H_2 + 2\frac{1}{2}O_2 \rightarrow 2CO_2 + H_2O$

Substance	Material balance (lb/lb of fuel)				Heat of combustion ($-\Delta h$) (Btu/lb of fuel, at 77°F)	
	Required		Produced		Liquid H_2O in products	All gaseous products
	O_2	Air	CO_2	H_2O		
Hydrogen	7.94	34.2	...	8.94	60,958	51,571
Sulfur	1.00	4.31	2.00(SO_2)	...	...	3,895
Carbon	2.66	11.5	3.66	...	...	14,087
	1.33	5.75	2.33(CO)	...	...	3,952
Carbon monoxide	0.571	2.46	1.57	...	...	4,344
Methane	3.99	17.2	2.74	2.25	23,861	21,502
Ethane	3.73	16.1	2.93	1.80	22,304	20,416
Propane	3.63	15.7	3.00	1.63	21,646	19,929
Butane	3.58	15.5	3.03	1.55	21,293	19,665
Octane	3.50	15.1	3.08	1.42	20,591	19,100
Benzene	3.07	13.3	3.38	0.69	17,986	17,259
Acetylene	3.07	13.3	3.38	0.69	21,460	20,734

* Small differences between the material quantities in this table and those computed in the examples in the text are due to rounding off the values of the molecular weights in the examples.

Data mainly from U.S. Department of Commerce, Bureau of Standards Circular No. C 461.

Table A2 (Continued)

Latent Heats of Vaporization, h_{fg}, Btu/lb, at 77°F *and Boiling Point at* 1 atm (NBP) *for Some Hydrocarbons*

Compound	h_{fg}, Btu/lb	NBP, °F
C_3H_8	147.07	−44
C_4H_{10}	155.83	31
C_8H_{18}	156.14	258
C_6H_6	186.31	176

Specific Heats of Certain Fuel Substances, Btu/lb °F

Substance	c_p at 77°F	c_p avg 77°F to 250°F
Carbon	0.172	0.205
Sulfur	0.168	0.178
Average for petroleum liquid fuels	0.45	0.50

Table A2 (Concluded): Analyses of Typical Fuels*

Fuel	Ultimate Analysis (percent by mass) C	H	O	S	N	*A*	Higher Heating Value ($-\Delta H$, Btu/lb, dry basis)
Coals:							
Anthracite	84.2	2.8	2.2	0.6	0.8	9.4	13,810
Bituminous	86.4	4.9	3.6	0.6	1.6	2.9	15,178
Bituminous	73.1	4.8	8.9	2.6	1.5	9.1	13,469
Lignite	64.1	4.6	18.3	0.8	1.2	11.0	11,084
Oils:							
Heavy Fuel Oil	84.0	11.3	3.0	0.9	2.1	...	18,370
Light Fuel Oil	85.4	13.1	1.1	0.2	...	...	19,230
Gasoline	85.5	14.4	...	0.1	...	...	20,160
Gas:							
Natural Gas	CH_4, 96; N_2, 3.2; CO_2, 0.8 percent by volume						
Natural Gas	CH_4, 75.5; C_6H_6, 22; N_2, 1.2; CO_2, 1.3 percent by volume						
Coal Gas	H_2, 46.5; CH_4, 32.1; CO, 6.3; N_2, 8.1; CO_2, 2.2; other gases, 4.8 percent by volume.						

* Actual analyses vary widely. (From O. de Lorenzi, ed., *Combustion Engineering.* New York: Combustion Engineering-Superheater, Inc., 1948; and John Griswold, *Fuels, Furnaces and Combustion.* New York: McGraw-Hill, 1946.)

Table A3: Conversion of Units*

Class of Units	To obtain	multiply	by
Length	cm	feet	30.480
Volume	cu in.	cu ft	1,728
	cu cm	cu ft	28,320
	cu in.	gallons	231
Mass	grams	lbm	453.59
	lbm	slugs	32.17
Specific volume	cu cm/gm	cu ft/lbm	62.428
Pressure	psf	psi	144
	in. Hg	psi	2.036
	ft H_2O	psi	2.309
	psi	atm	14.696
	psi	kg/sq cm	14.223
Energy	IT calorie†	Btu	251.996
	ft lbf	Btu	778.16
	hphr	Btu	3.93010×10^{-4}
	kwhr	Btu	2.93018×10^{-4}
	Btu	hphr	2544.46
	Btu	kwhr	3412.76
Specific energy	IT calorie/gm	Btu/lbm	0.555556
Specific energy/deg	IT cal/gm °C	Btu/lbm °F	1
Specific entropy	IT cal/gm °K	Btu/lbm °R	1
Thermal conductivity	Btu/ft hr °F	cal/cm sec °C	241.9
	Btu/sq ft hr (°F/in.)	Btu/ft hr °F	12
Coefficient of heat transfer	Btu/sq ft hr °F	cal/sq cm sec °C	7,373
Viscosity	lbm/hr ft	centipoise	2.42
	centipoise	gm/sec cm	100
	lbm/hr ft	lbf sec/sq ft	116,000

* From U.S. Department of Commerce, Bureau of Standards Circular No. C 461, and other sources.

† IT = International steam tables.

Temperature Conversions

$$°F = 1.8°C + 32$$
$$°R = °F + 459.69$$
$$°K = °C + 273.16$$
$$°R = 1.8°K$$

Table A4: Specific Heat at Constant Pressure, and Specific Heat Ratio, for Gases at Low Pressure* (as functions of temperature)

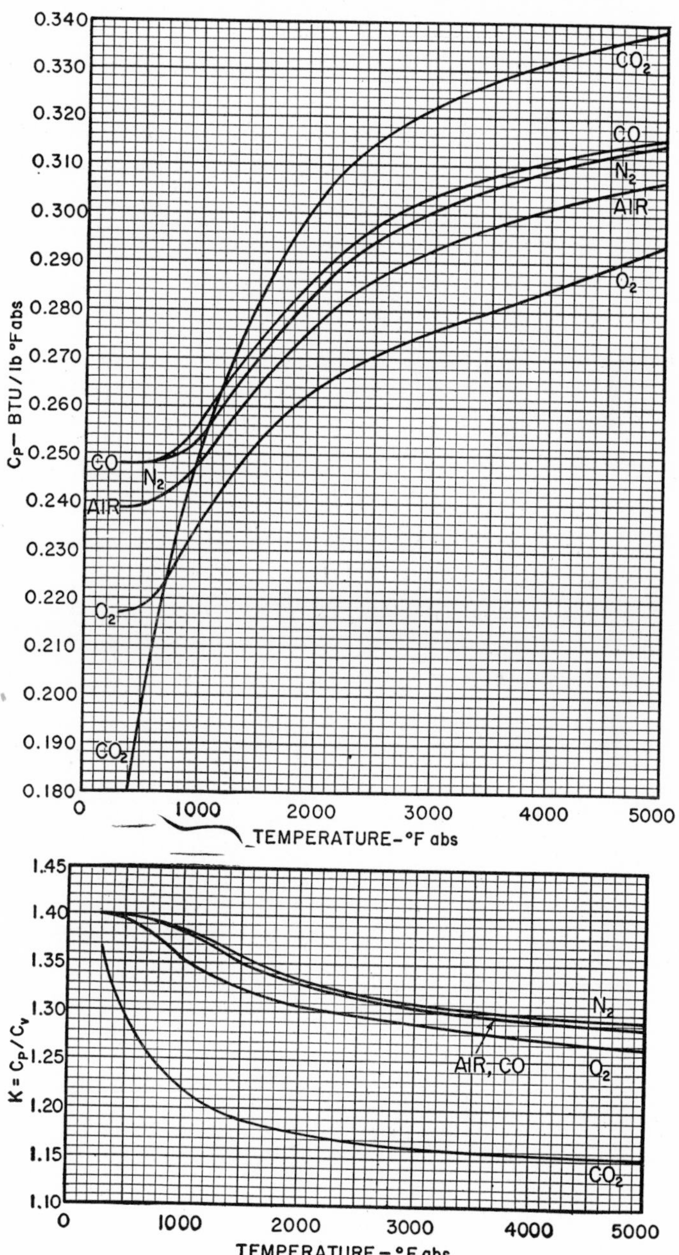

* Based on data from the Gas Tables of Keenan and Kaye.

Table A5: h/T for Gases at Low Pressure*

(as a function of temperature)

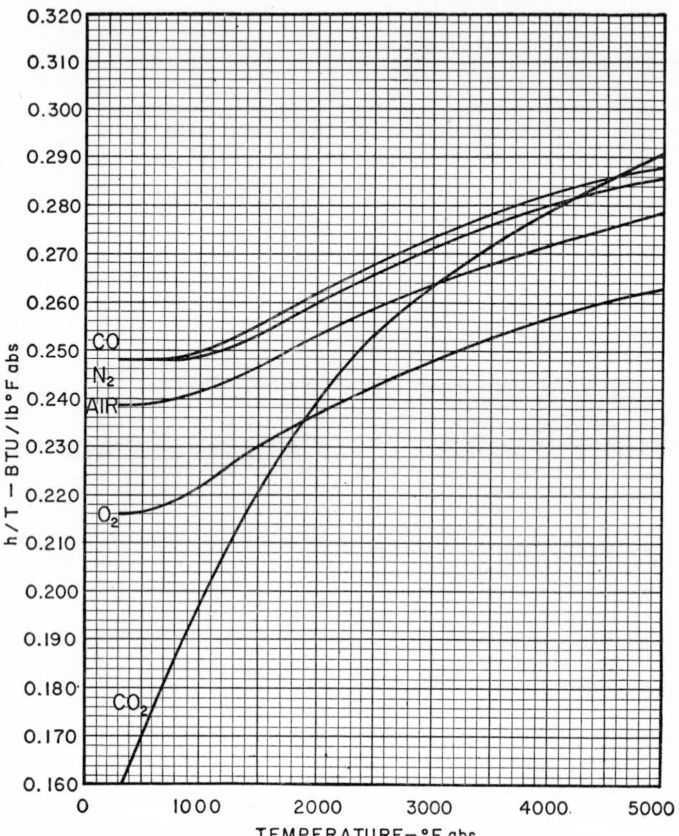

* Based on data from the Gas Tables of Keenan and Kaye.

Table A6: Specific Heat at Constant Pressure, h/T, and Specific Heat Ratio* (for hydrogen and water vapor at low pressure)

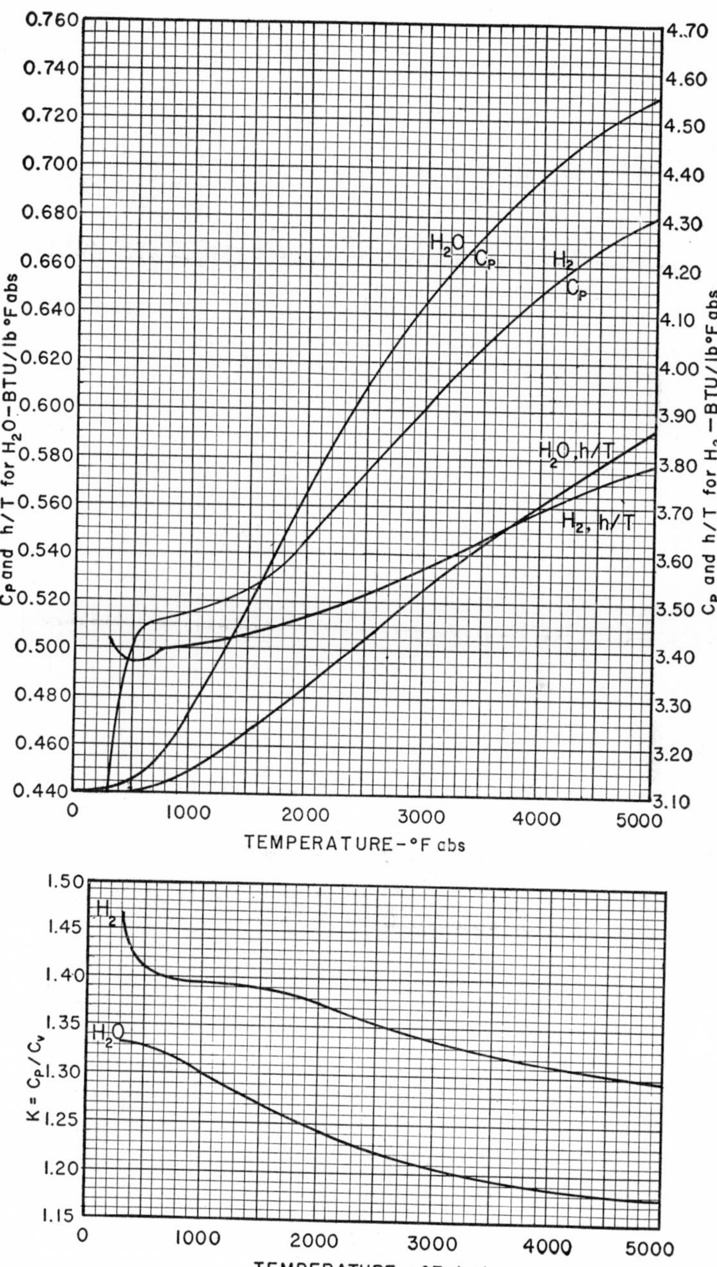

* Based on data from the Gas Tables of Keenan and Kaye.

Table A7: Air Tables (for one pound)

(The properties given here are condensed by permission of authors and publisher from *Gas Tables*, by J. H. Keenan and J. Kaye, published by John Wiley and Sons, 1948.)

T, °F abs	t, °F	h, Btu/lb	p_r	u, Btu/lb	v_r	ϕ, Btu/lb °F	T, °F abs	t, °F	h, Btu/lb	p_r	u, Btu/lb	v_r	ϕ, Btu/lb °F
100	−360	23.7	.00384	16.9	9640	.1971	1200	740	291.3	24.0	209.0	18.51	.7963
120	−340	28.5	.00726	20.3	6120	.2408	1220	760	296.4	25.5	212.8	17.70	.8005
140	−320	33.3	.01244	23.7	4170	.2777	1240	780	301.5	27.1	216.5	16.93	.8047
160	−300	38.1	.01982	27.1	2990	.3096	1260	800	306.6	28.8	220.3	16.20	.8088
180	−280	42.9	.0299	30.6	2230	.3378	1280	820	311.8	30.6	224.0	15.52	.8128
200	−260	47.7	.0432	34.0	1715	.3630	1300	840	316.9	32.4	227.8	14.87	.8168
220	−240	52.5	.0603	37.4	1352	.3858	1320	860	322.1	34.3	231.6	14.25	.8208
240	−220	57.2	.0816	40.8	1089	.4067	1340	880	327.3	36.3	235.4	13.67	.8246
260	−200	62.0	.1080	44.2	892	.4258	1360	900	332.5	38.4	239.2	13.12	.8285
280	−180	66.8	.1399	47.6	742	.4436	1380	920	337.7	40.6	243.1	12.59	.8323
300	−160	71.6	.1780	51.0	624	.4601	1400	940	342.9	42.9	246.9	12.10	.8360
320	−140	76.4	.2229	54.5	532	.4755	1420	960	348.1	45.3	250.8	11.62	.8398
340	−120	81.2	.2754	57.9	457	.4900	1440	980	353.4	47.8	254.7	11.17	.8434
360	−100	86.0	.336	61.3	397	.5037	1460	1000	358.6	50.3	258.5	10.74	.8470
380	−80	90.8	.406	64.7	347	.5166	1480	1020	363.9	53.0	262.4	10.34	.8506
400	−60	95.5	.486	68.1	305	.5289	1500	1040	369.2	55.9	266.3	9.95	.8542
420	−40	100.3	.576	71.5	270	.5406	1520	1060	374.5	58.8	270.3	9.58	.8568
440	−20	105.1	.678	74.9	241	.5517	1540	1080	379.8	61.8	274.2	9.23	.8611
460	0	109.9	.791	78.4	215.3	.5624	1560	1100	385.1	65.0	278.1	8.89	.8646
480	20	114.7	.918	81.8	193.6	.5726	1580	1120	390.4	68.3	282.1	8.57	.8679
500	40	119.5	1.059	85.2	174.9	.5823	1600	1140	395.7	71.7	286.1	8.26	.8713
520	60	124.3	1.215	88.6	158.6	.5917	1620	1160	401.1	75.3	290.0	7.97	.8746
540	80	129.1	1.386	92.0	144.3	.6008	1640	1180	406.4	79.0	294.0	7.69	.8779
560	100	133.9	1.574	95.5	131.8	.6095	1660	1200	411.8	82.8	298.0	7.42	.8812
580	120	138.7	1.780	98.9	120.7	.6179	1680	1220	417.2	86.8	302.0	7.17	.8844
600	140	143.5	2.00	102.3	110.9	.6261	1700	1240	422.6	91.0	306.1	6.92	.8876
620	160	148.3	2.25	105.8	102.1	.6340	1720	1260	428.0	95.2	310.1	6.69	.8907
640	180	153.1	2.51	109.2	94.3	.6416	1740	1280	433.4	99.7	314.1	6.46	.8939
660	200	157.9	2.80	112.7	87.3	.6490	1760	1300	438.8	104.3	318.2	6.25	.8970
680	220	162.7	3.11	116.1	81.0	.6562	1780	1320	444.3	109.1	322.2	6.04	.9000
700	240	167.6	3.45	119.6	75.2	.6632	1800	1340	449.7	114.0	326.3	5.85	.9031
720	260	172.4	3.81	123.0	70.1	.6700	1820	1360	455.2	119.2	330.4	5.66	.9061
740	280	177.2	4.19	126.5	65.4	.6766	1840	1380	460.6	124.5	334.5	5.48	.9091
760	300	182.1	4.61	130.0	61.1	.6831	1860	1400	466.1	130.0	338.6	5.30	.9120
780	320	186.9	5.05	133.5	57.2	.6894	1880	1420	471.6	135.6	342.7	5.13	.9150
800	340	191.8	5.53	137.0	53.6	.6956	1900	1440	477.1	141.5	346.8	4.97	.9179
820	360	196.7	6.03	140.5	50.4	.7016	1920	1460	482.6	147.6	351.0	4.82	.9208
840	380	201.6	6.57	144.0	47.3	.7075	1940	1480	488.1	153.9	355.1	4.67	.9236
860	400	206.5	7.15	147.5	44.6	.7132	1960	1500	493.6	160.4	359.3	4.53	.9264
880	420	211.4	7.76	151.0	42.0	.7189	1980	1520	499.1	167.1	363.4	4.39	.9293
900	440	216.3	8.41	154.6	39.6	.7244	2000	1540	504.7	174.0	367.6	4.26	.9320
920	460	221.2	9.10	158.1	37.4	.7298	2020	1560	510.3	181.2	371.8	4.13	.9348
940	480	226.1	9.83	161.7	35.4	.7351	2040	1580	515.8	188.5	376.0	4.01	.9376
960	500	231.1	10.61	165.3	33.5	.7403	2060	1600	521.4	196.2	380.2	3.89	.9403
980	520	236.0	11.43	168.8	31.8	.7454	2080	1620	527.0	204.0	384.4	3.78	.9430
1000	540	241.0	12.30	172.4	30.1	.7504	2100	1640	532.6	212	388.6	3.67	.9456
1020	560	246.0	13.22	176.0	28.6	.7554	2120	1660	538.2	220	392.8	3.56	.9483
1040	580	251.0	14.18	179.7	27.2	.7602	2140	1680	543.7	229	397.0	3.46	.9509
1060	600	256.0	15.20	183.3	25.8	.7650	2160	1700	549.4	238	401.3	3.36	.9535
1080	620	261.0	16.28	186.9	24.6	.7696	2180	1720	555.0	247	405.5	3.27	.9561
1100	640	266.0	17.41	190.6	23.4	.7743	2200	1740	560.6	257	409.8	3.18	.9587
1120	660	271.0	18.60	194.2	22.3	.7788	2220	1760	566.2	266	414.0	3.09	.9612
1140	680	276.1	19.86	197.9	21.3	.7833	2240	1780	571.9	276	418.3	3.00	.9638
1160	700	281.1	21.2	201.6	20.29	.7877	2260	1800	577.5	287	422.6	2.92	.9663
1180	720	286.2	22.6	205.3	19.38	.7920	2280	1820	583.2	297	426.9	2.84	.9688

Table A7 (Concluded): Air Tables

(For One Pound)

(The properties given here are condensed by permission of authors and publisher from *Gas Tables*, by J. H. Keenan and J. Kaye, published by John Wiley and Sons, 1948.)

T, °F abs	t, °F	h, Btu/lb	p_r	u, Btu/lb	v_r	φ, Btu/lb °F	T, °F abs	t, °F	h, Btu/lb	p_r	u, Btu/lb	v_r	φ, Btu/lb °F
2300	1840	588.8	308	431.2	2.76	.9712	3000	2540	790.7	941	585.0	1.180	1.0478
2320	1860	594.5	319	435.5	2.69	.9737	3020	2560	796.5	969	589.5	1.155	1.0497
2340	1880	600.2	331	439.8	2.62	.9761	3040	2580	802.4	996	594.0	1.130	1.0517
2360	1900	605.8	343	444.1	2.55	.9785	3060	2600	808.3	1025	598.5	1.106	1.0536
2380	1920	611.5	355	448.4	2.48	.9809	3080	2620	814.2	1054	603.0	1.083	1.0555
2400	1940	617.2	368	452.7	2.42	.9833	3100	2640	820.0	1083	607.5	1.060	1.0574
2420	1960	622.9	380	457.0	2.36	.9857	3120	2660	825.9	1114	612.0	1.038	1.0593
2440	1980	628.6	394	461.4	2.30	.9880	3140	2680	831.8	1145	616.6	1.016	1.0612
2460	2000	634.3	407	465.7	2.24	.9904	3160	2700	837.7	1176	621.1	.995	1.0630
2480	2020	640.0	421	470.0	2.18	.9927	3180	2720	843.6	1209	625.6	.975	1.0649
2500	2040	645.8	436	474.4	2.12	.9950	3200	2740	849.5	1242	630.1	.955	1.0668
2520	2060	651.5	450	478.8	2.07	.9972	3220	2760	855.4	1276	634.6	.935	1.0686
2540	2080	657.2	466	483.1	2.02	.9995	3240	2780	861.3	1310	639.2	.916	1.0704
2560	2100	663.0	481	487.5	1.971	1.0018	3260	2800	867.2	1345	643.7	.898	1.0722
2580	2120	668.7	497	491.9	1.922	1.0040	3280	2820	873.1	1381	648.3	.880	1.0740
2600	2140	674.5	514	496.3	1.876	1.0062	3300	2840	879.0	1418	652.8	.862	1.0758
2620	2160	680.2	530	500.6	1.830	1.0084	3320	2860	884.9	1455	657.4	.845	1.0776
2640	2180	686.0	548	505.0	1.786	1.0106	3440	2880	890.9	1494	661.9	.828	1.0794
2660	2200	691.8	565	509.4	1.743	1.0128	3360	2900	896.8	1533	666.5	.812	1.0812
2680	2220	697.6	583	513.8	1.702	1.0150	3380	2920	902.7	1573	671.0	.796	1.0830
2700	2240	703.4	602	518.3	1.662	1.0171	3400	2940	908.7	1613	675.6	.781	1.0847
2720	2260	709.1	621	522.7	1.623	1.0193	3420	2960	914.6	1655	680.2	.766	1.0864
2740	2280	714.9	640	527.1	1.585	1.0214	3440	2980	920.6	1697	684.8	.751	1.0882
2760	2300	720.7	660	531.5	1.548	1.0235	3460	3000	926.5	1740	689.3	.736	1.0899
2780	2320	726.5	681	536.0	1.512	1.0256	3480	3020	932.4	1784	693.9	.722	1.0916
2800	2340	732.3	702	540.4	1.478	1.0277	3500	3040	938.4	1829	698.5	.709	1.0933
2820	2360	738.2	724	544.8	1.444	1.0297	3520	3060	944.4	1875	703.1	.695	1.0950
2840	2380	744.0	746	549.3	1.411	1.0318	3540	3080	950.3	1922	707.6	.682	1.0967
2860	2400	749.8	768	553.7	1.379	1.0338	3560	3100	956.3	1970	712.2	.670	1.0984
2880	2420	755.6	791	558.2	1.348	1.0359	3580	3120	962.2	2018	716.8	.657	1.1000
2900	2440	761.4	815	562.7	1.318	1.0379	3600	3140	968.2	2068	721.4	.645	1.1017
2920	2460	767.3	839	567.1	1.289	1.0399	3620	3160	974.2	2118	726.0	.633	1.1034
2940	2480	773.1	864	571.6	1.261	1.0419	3640	3180	980.2	2170	730.6	.621	1.1050
2960	2500	779.0	889	576.1	1.233	1.0439	3660	3200	986.1	2222	735.3	.610	1.1066
2980	2520	784.8	915	580.6	1.206	1.0458	3680	3220	992.1	2276	739.9	.599	1.1083

EXAMPLE 1. COMPRESSION OF AIR IN STEADY FLOW. Air at a pressure of 1 atm abs and a temperature of 520 F abs is compressed in steady flow to a pressure of 6 atm abs. Find the work of compression and the temperature after compression for (1) 100% efficiency of compression and (2) 60% efficiency of compression. The efficiency of compression is here defined as the ratio of the isentropic work of compression to the actual work of compression.

Solution. (1) From Table 2 we get for $T_1 = 520$ F abs,

$$p_{r1} = 1.215, \qquad h_1 = 124.3 \text{ Btu/lb}$$

where subscript 1 refers to the state at the compressor inlet. To determine the properties at the compressor outlet for isentropic compression we compute the relative pressure there

$$p_{r2} = 6/1 \times 1.215 = 7.29$$

Interpolating in Table 2 with this value of p_r, we find, for h_{2s} and T_{2s}, the enthalpy and temperature at the compressor outlet for isentropic compression

$$h_{2s} = 207.6 \text{ Btu/lb}, \qquad T_{2s} = 864.6 \text{ F abs}$$

The work of compression for 100% efficiency is then

$$h_{2s} - h_1 = 83.3 \text{ Btu/lb}$$

(2) Since the efficiency of compression η is defined by the equation $\eta = (h_{2s} - h_1)/\text{work per pound}$, we have for 60% efficiency

$$\text{Work per pound} = \frac{83.3}{0.60} = 138.8 \text{ Btu/lb}$$

Table A8: Steam Table I

Dry Saturated Steam: Temperature Table *

Temp., F	Abs Press., Lb/Sq In.	Specific Volume			Enthalpy			Entropy			Temp., F
		Sat. Liquid	Evap.	Sat. Vapor	Sat. Liquid	Evap.	Sat. Vapor	Sat. Liquid	Evap	Sat. Vapor	
t	p	v_f	v_{fg}	v_g	h_f	h_{fg}	h_g	s_f	s_{fg}	s_g	t
32	0.08854	0.01602	3306	3306	0.00	1075.8	1075.8	0.0000	2.1877	2.1877	**32**
35	0.09995	0.01602	2947	2947	3.02	1074.1	1077.1	0.0061	2.1709	2.1770	**35**
40	0.12170	0.01602	2444	2444	8.05	1071.3	1079.3	0.0162	2.1435	2.1597	**40**
45	0.14752	0.01602	2036.4	2036.4	13.06	1068.4	1081.5	0.0262	2.1167	2.1429	**45**
50	0.17811	0.01603	1703.2	1703.2	18.07	1065.6	1083.7	0.0361	2.0903	2.1264	**50**
60	0.2563	0.01604	1206.6	1206.7	28.06	1059.9	1088.0	0.0555	2.0393	2.0948	**60**
70	0.3631	0.01606	867.8	867.9	38.04	1054.3	1092.3	0.0745	1.9902	2.0647	**70**
80	0.5069	0.01608	633.1	633.1	48.02	1048.6	1096.6	0.0932	1.9428	2.0360	**80**
90	0.6982	0.01610	468.0	468.0	57.99	1042.9	1100.9	0.1115	1.8972	2.0087	**90**
100	0.9492	0.01613	350.3	350.4	67.97	1037.2	1105.2	0.1295	1.8531	1.9826	**100**
110	1.2748	0.01617	265.3	265.4	77.94	1031.6	1109.5	0.1471	1.8106	1.9577	**110**
120	1.6924	0.01620	203.25	203.27	87.92	1025.8	1113.7	0.1645	1.7694	1.9339	**120**
130	2.2225	0.01625	157.32	157.34	97.90	1020.0	1117.9	0.1816	1.7296	1.9112	**130**
140	2.8886	0.01629	122.99	123.01	107.89	1014.1	1122.0	0.1984	1.6910	1.8894	**140**
150	3.718	0.01634	97.06	97.07	117.89	1008.2	1126.1	0.2149	1.6537	1.8685	**150**
160	4.741	0.01639	77.27	77.29	127.89	1002.3	1130.2	0.2311	1.6174	1.8485	**160**
170	5.992	0.01645	62.04	62.06	137.90	996.3	1134.2	0.2472	1.5822	1.8293	**170**
180	7.510	0.01651	50.21	50.23	147.92	990.2	1138.1	0.2630	1.5480	1.8109	**180**
190	9.339	0.01657	40.94	40.96	157.95	984.1	1142.0	0.2785	1.5147	1.7932	**190**
200	11.526	0.01663	33.62	33.64	167.99	977.9	1145.9	0.2938	1.4824	1.7762	**200**
210	14.123	0.01670	27.80	27.82	178.05	971.6	1149.7	0.3090	1.4508	1.7598	**210**
212	14.696	0.01672	26.78	26.80	180.07	970.3	1150.4	0.3120	1.4446	1.7566	**212**
220	17.186	0.01677	23.13	23.15	188.13	965.2	1153.4	0.3239	1.4201	1.7440	**220**
230	20.780	0.01684	19.365	19.382	198.23	958.8	1157.0	0.3387	1.3901	1.7288	**230**
240	24.969	0.01692	16.306	16.323	208.34	952.2	1160.5	0.3531	1.3609	1.7140	**240**
250	29.825	0.01700	13.804	13.821	216.48	945.5	1164.0	0.3675	1.3323	1.6998	**250**
260	35.429	0.01709	11.746	11.763	228.64	938.7	1167.3	0.3817	1.3043	1.6860	**260**
270	41.858	0.01717	10.044	10.061	238.84	931.8	1170.6	0.3958	1.2769	1.6727	**270**
280	49.203	0.01726	8.628	8.645	249.06	924.7	1173.8	0.4096	1.2501	1.6597	**280**
290	57.556	0.01735	7.444	7.461	259.31	917.5	1176.8	0.4234	1.2238	1.6472	**290**

300	67.013	0.01745	6.449	6.466	269.59	910.1	1179.7	0.4369	1.1980	1.6350	**300**
310	77.68	0.01755	5.609	5.626	279.92	902.6	1182.5	0.4504	1.1727	1.6231	**310**
320	89.66	0.01765	4.896	4.914	290.28	894.9	1185.2	0.4637	1.1478	1.6115	**320**
330	103.06	0.01776	4.289	4.307	300.68	887.0	1187.7	0.4769	1.1233	1.6002	**330**
340	118.01	0.01787	3.770	3.788	311.13	879.0	1190.1	0.4900	1.0992	1.5891	**340**
350	134.63	0.01799	3.324	3.342	321.63	870.7	1192.3	0.5029	1.0754	1.5783	**350**
360	153.04	0.01811	2.939	2.957	332.18	862.2	1194.4	0.5158	1.0519	1.5677	**360**
370	173.37	0.01823	2.606	2.625	342.79	853.5	1196.3	0.5286	1.0287	1.5573	**370**
380	195.77	0.01836	2.317	2.335	353.45	844.6	1198.1	0.5413	1.0059	1.5471	**380**
390	220.37	0.01850	2.0651	2.0836	364.17	835.4	1199.6	0.5539	0.9832	1.5371	**390**
400	247.31	0.01864	1.8447	1.8633	374.97	826.0	1201.0	0.5664	0.9608	1.5272	**400**
410	276.75	0.01878	1.6512	1.6700	385.83	816.3	1202.1	0.5788	0.9386	1.5174	**410**
420	308.83	0.01894	1.4811	1.5000	396.77	806.3	1203.1	0.5912	0.9166	1.5078	**420**
430	343.72	0.01910	1.3308	1.3499	407.79	796.0	1203.8	0.6035	0.8947	1.4982	**430**
440	381.59	0.01926	1.1979	1.2171	418.90	785.4	1204.3	0.6158	0.8730	1.4887	**440**
450	422.6	0.0194	1.0799	1.0993	430.1	774.5	1204.6	0.6280	0.8513	1.4793	**450**
460	466.9	0.0196	0.9748	0.9944	441.4	763.2	1204.6	0.6402	0.8298	1.4700	**460**
470	514.7	0.0198	0.8811	0.9009	452.8	751.5	1204.3	0.6523	0.8083	1.4606	**470**
480	566.1	0.0200	0.7972	0.8172	464.4	739.4	1203.7	0.6645	0.7868	1.4513	**480**
490	621.4	0.0202	0.7221	0.7423	476.0	726.8	1202.8	0.6766	0.7653	1.4419	**490**
500	680.8	0.0204	0.6545	0.6749	487.8	713.9	1201.7	0.6887	0.7438	1.4325	**500**
520	812.4	0.0209	0.5385	0.5594	511.9	686.4	1198.2	0.7130	0.7006	1.4136	**520**
540	962.5	0.0215	0.4434	0.4649	536.6	656.6	1193.2	0.7374	0.6568	1.3942	**540**
560	1133.1	0.0221	0.3647	0.3868	562.2	624.2	1186.4	0.7621	0.6121	1.3742	**560**
580	1325.8	0.0228	0.2989	0.3217	588.9	588.4	1177.3	0.7872	0.5659	1.3532	**580**
600	1542.9	0.0236	0.2432	0.2668	617.0	548.5	1165.5	0.8131	0.5176	1.3307	**600**
620	1786.6	0.0247	0.1955	0.2201	646.7	503.6	1150.3	0.8398	0.4664	1.3062	**620**
640	2059.7	0.0260	0.1538	0.1798	678.6	452.0	1130.5	0.8679	0.4110	1.2789	**640**
660	2365.4	0.0278	0.1165	0.1442	714.2	390.2	1104.4	0.8987	0.3485	1.2472	**660**
680	2708.1	0.0305	0.0810	0.1115	757.3	309.9	1067.2	0.9351	0.2719	1.2071	**680**
700	3093.7	0.0369	0.0392	0.0761	823.3	172.1	995.4	0.9905	0.1484	1.1389	**700**
705.4	3206.2	0.0503	0	0.0503	902.7	0	902.7	1.0580	0	1.0580	**705.4**

* Abridged from "Thermodynamic Properties of Steam" by Joseph H. Keenan and Frederick G. Keyes.
Published by John Wiley & Sons, Inc., New York.

Table A8: Steam Table II

Dry Saturated Steam: Pressure Table*

Abs Press., Lb Sq In. p	Temp., F t	Specific Volume Sat. Liquid v_f	Specific Volume Sat. Vapor v_g	Enthalpy Sat. Liquid h_f	Enthalpy Evap h_{fg}	Enthalpy Sat. Vapor h_g	Entropy Sat. Liquid s_f	Entropy Evap s_{fg}	Entropy Sat. Vapor s_g	Internal Energy Sat. Liquid u_f	Internal Energy Sat. Vapor u_g	Abs Press., Lb Sq In. p
1.0	101.74	0.01614	333.6	69.70	1036.3	1106.0	0.1326	1.8456	1.9782	69.70	1044.3	1.0
2.0	126.08	0.01623	173.73	93.99	1022.2	1116.2	0.1749	1.7451	1.9200	93.98	1051.9	2.0
3.0	141.48	0.01630	118.71	109.37	1013.2	1122.6	0.2008	1.6855	1.8863	109.36	1056.7	3.0
4.0	152.97	0.01636	90.63	120.86	1006.4	1127.3	0.2198	1.6427	1.8625	120.85	1060.2	4.0
5.0	162.24	0.01640	73.52	130.13	1001.0	1131.1	0.2347	1.6094	1.8441	130.12	1063.1	5.0
6.0	170.06	0.01645	61.98	137.96	996.2	1134.2	0.2472	1.5820	1.8292	137.94	1065.4	6.0
7.0	176.85	0.01649	53.64	144.76	992.1	1136.9	0.2581	1.5586	1.8167	144.74	1067.4	7.0
8.0	182.86	0.01653	47.34	150.79	988.5	1139.3	0.2674	1.5383	1.8057	150.77	1069.2	8.0
9.0	188.28	0.01656	42.40	156.22	985.2	1141.4	0.2759	1.5203	1.7962	156.19	1070.8	9.0
10	193.21	0.01659	38.42	161.17	982.1	1143.3	0.2835	1.5041	1.7876	161.14	1072.2	10
14.696	212.00	0.01672	26.80	180.07	970.3	1150.4	0.3120	1.4446	1.7566	180.02	1077.5	14.696
15	213.03	0.01672	26.29	181.11	969.7	1150.8	0.3135	1.4415	1.7549	181.06	1077.8	15
20	227.96	0.01683	20.089	196.16	960.1	1156.3	0.3356	1.3962	1.7319	196.10	1081.9	20
25	240.07	0.01692	16.303	208.42	952.1	1160.6	0.3533	1.3606	1.7139	208.34	1085.1	25
30	250.33	0.01701	13.746	218.82	945.3	1164.1	0.3680	1.3313	1.6993	218.73	1087.8	30
35	259.28	0.01708	11.898	227.91	939.2	1167.1	0.3807	1.3063	1.6870	227.80	1090.1	35
40	267.25	0.01715	10.498	236.03	933.7	1169.7	0.3919	1.2844	1.6763	235.90	1092.0	40
45	274.44	0.01721	9.401	243.36	928.6	1172.0	0.4019	1.2650	1.6669	243.22	1093.7	45
50	281.01	0.01727	8.515	250.09	924.0	1174.1	0.4110	1.2474	1.6585	249.93	1095.3	50
55	287.07	0.01732	7.787	256.30	919.6	1175.9	0.4193	1.2316	1.6509	256.12	1096.7	55
60	292.71	0.01738	7.175	262.09	915.5	1177.6	0.4270	1.2168	1.6438	261.90	1097.9	60
65	297.97	0.01743	6.655	267.50	911.6	1179.1	0.4342	1.2032	1.6374	267.29	1099.1	65
70	302.92	0.01748	6.206	272.61	907.9	1180.6	0.4409	1.1906	1.6315	272.38	1100.2	70
75	307.60	0.01753	5.816	277.43	904.5	1181.9	0.4472	1.1787	1.6259	277.19	1101.2	75
80	312.03	0.01757	5.472	282.02	901.1	1183.1	0.4531	1.1676	1.6207	281.76	1102.1	80
85	316.25	0.01761	5.168	286.39	897.8	1184.2	0.4587	1.1571	1.6158	286.11	1102.9	85
90	320.27	0.01766	4.896	290.56	894.7	1185.3	0.4641	1.1471	1.6112	290.27	1103.7	90
95	324.12	0.01770	4.652	294.56	891.7	1186.2	0.4692	1.1376	1.6068	294.25	1104.5	95
100	327.81	0.01774	4.432	298.40	888.8	1187.2	0.4740	1.1286	1.6026	298.08	1105.2	100
110	334.77	0.01782	4.049	305.66	883.2	1188.9	0.4832	1.1117	1.5948	305.30	1106.5	110

120	341.25	0.01789	3.728	312.44	877.9	1190.4	0.4916	1.0962	1.5878	312.05	1107.6	**120**
130	347.32	0.01796	3.455	318.81	872.9	1191.7	0.4995	1.0817	1.5812	318.38	1108.6	**130**
140	353.02	0.01802	3.220	324.82	868.2	1193.0	0.5069	1.0682	1.5751	324.35	1109.6	**140**
150	358.42	0.01809	3.015	330.51	863.6	1194.1	0.5138	1.0556	1.5694	330.01	1110.5	**150**
160	363.53	0.01815	2.834	335.93	859.2	1195.1	0.5204	1.0436	1.5640	335.39	1111.2	**160**
170	368.41	0.01822	2.675	341.09	854.9	1196.0	0.5266	1.0324	1.5590	340.52	1111.9	**170**
180	373.06	0.01827	2.532	346.03	850.8	1196.9	0.5325	1.0217	1.5542	345.42	1112.5	**180**
190	377.51	0.01833	2.404	350.79	846.8	1197.6	0.5381	1.0116	1.5497	350.15	1113.1	**190**
200	381.79	0.01839	2.288	355.36	843.0	1198.4	0.5435	1.0018	1.5453	354.68	1113.7	**200**
250	400.95	0.01865	1.8438	376.00	825.1	1201.1	0.5675	0.9588	1.5263	375.14	1115.8	**250**
300	417.33	0.01890	1.5433	393.84	809.0	1202.8	0.5879	0.9225	1.5104	392.79	1117.1	**300**
350	431.72	0.01913	1.3260	409.69	794.2	1203.9	0.6056	0.8910	1.4966	408.45	1118.0	**350**
400	444.59	0.0193	1.1613	424.0	780.5	1204.5	0.6214	0.8630	1.4844	422.6	1118.5	**400**
450	456.28	0.0195	1.0320	437.2	767.4	1204.6	0.6356	0.8378	1.4734	435.5	1118.7	**450**
500	467.01	0.0197	0.9278	449.4	755.0	1204.4	0.6487	0.8147	1.4634	447.6	1118.6	**500**
550	476.94	0.0199	0.8424	460.8	743.1	1203.9	0.6608	0.7934	1.4542	458.8	1118.2	**550**
600	486.21	0.0201	0.7698	471.6	731.6	1203.2	0.6720	0.7734	1.4454	469.4	1117.7	**600**
650	494.90	0.0203	0.7083	481.8	720.5	1202.3	0.6826	0.7548	1.4374	479.4	1117.1	**650**
700	503.10	0.0205	0.6554	491.5	709.7	1201.2	0.6925	0.7371	1.4296	488.8	1116.3	**700**
750	510.86	0.0207	0.6092	500.8	699.2	1200.0	0.7019	0.7204	1.4223	598.0	1115.4	**750**
800	518.23	0.0209	0.5687	509.7	688.9	1198.6	0.7108	0.7045	1.4153	506.6	1114.4	**800**
850	525.26	0.0210	0.5327	518.3	678.8	1197.1	0.7194	0.6891	1.4085	515.0	1113.3	**850**
900	531.98	0.0212	0.5006	526.6	668.8	1195.4	0.7275	0.6744	1.4020	523.1	1112.1	**900**
950	538.43	0.0214	0.4717	534.6	659.1	1193.7	0.7355	0.6602	1.3957	530.9	1110.8	**950**
1000	544.61	0.0216	0.4456	542.4	649.4	1191.8	0.7430	0.6467	1.3897	538.4	1109.4	**1000**
1100	556.31	0.0220	0.4001	557.4	630.4	1187.8	0.7575	0.6205	1.3780	552.9	1106.4	**1100**
1200	567.22	0.0223	0.3619	571.7	611.7	1183.4	0.7711	0.5956	1.3667	566.7	1103.0	**1200**
1300	577.46	0.0227	0.3293	585.4	593.2	1178.6	0.7840	0.5719	1.3559	580.0	1099.4	**1300**
1400	587.10	0.0231	0.3012	598.7	574.7	1173.4	0.7963	0.5491	1.3454	592.7	1095.4	**1400**
1500	596.23	0.0235	0.2765	611.6	556.3	1167.9	0.8082	0.5269	1.3351	605.1	1091.2	**1500**
2000	635.82	0.0257	0.1878	671.7	463.4	1135.1	0.8619	0.4230	1.2849	662.2	1065.6	**2000**
2500	668.13	0.0287	0.1307	730.6	360.5	1091.1	0.9126	0.3197	1.2322	717.3	1030.6	**2500**
3000	695.36	0.0346	0.0858	802.5	217.8	1020.3	0.9731	0.1885	1.1615	783.4	972.7	**3000**
3206.2	705.40	0.0503	0.0503	902.7	0	902.7	1.0580	0	1.0580	872.9	872.9	**3206.2**

***** Abridged from "Thermodynamic Properties of Steam" by Joseph H. Keenan and Frederick G. Keyes. **Copyright, 1937, by Joseph H. Keenan and Frederick G. Keyes.**
Published by John Wiley & Sons, Inc., New York.

Table A8: Steam Table III

Properties of Superheated Steam *

Abs Press., Lb Sq In. (Sat. Temp.)		Temperature—Degrees Fahrenheit												
		200	300	400	500	600	700	800	900	1000	1100	1200	1400	1600
1	v	392.6	452.3	512.0	571.6	631.2	690.8	750.4	809.9	869.5	929.1	988.7	1107.8	1227.0
	h	1150.4	1195.8	1241.7	1288.3	1335.7	1383.8	1432.8	1482.7	1533.5	1585.2	1637.7	1745.7	1857.5
(101.74)	s	2.0512	2.1153	2.1720	2.2233	2.2702	2.3137	2.3542	2.3923	2.4283	2.4625	2.4952	2.5566	2.6137
5	v	78.16	90.25	102.26	114.22	126.16	138.10	150.03	161.95	173.87	185.79	197.71	221.6	245.4
	h	1148.8	1195.0	1241.2	1288.0	1335.4	1383.6	1432.7	1482.6	1533.4	1585.1	1637.7	1745.7	1857.4
(162.24)	s	1.8718	1.9370	1.9942	2.0456	2.0927	2.1361	2.1767	2.2148	2.2509	2.2851	2.3178	2.3792	2.4363
10	v	38.85	45.00	51.04	57.05	63.03	69.01	74.98	80.95	86.92	92.88	98.84	110.77	122.69
	h	1146.6	1193.9	1240.6	1287.5	1335.1	1383.4	1432.5	1482.4	1533.2	1585.0	1637.6	1745.6	1857.3
(193.21)	s	1.7927	1.8595	1.9172	1.9689	2.0160	2.0596	2.1002	2.1383	2.1744	2.2086	2.2413	2.3028	2.3598
14.696	v	......	30.53	34.68	38.78	42.86	46.94	51.00	55.07	59.13	63.19	67.25	75.37	83.48
	h	......	1192.8	1239.9	1287.1	1334.8	1383.2	1432.3	1482.3	1533.1	1584.8	1637.5	1745.5	1857.3
(212.00)	s	......	1.8160	1.8743	1.9261	1.9734	2.0170	2.0576	2.0958	2.1319	2.1662	2.1989	2.2603	2.3174
20	v	......	22.36	25.43	28.46	31.47	34.47	37.46	40.45	43.44	46.42	49.41	55.37	61.34
	h	......	1191.6	1239.2	1286.6	1334.4	1382.9	1432.1	1482.1	1533.0	1584.7	1637.4	1745.4	1857.2
(227.96)	s	......	1.7808	1.8396	1.8918	1.9392	1.9829	2.0235	2.0618	2.0978	2.1321	2.1648	2.2263	2.2834
40	v	......	11.040	12.628	14.168	15.688	17.198	18.702	20.20	21.70	23.20	24.69	27.68	30.66
	h	......	1186.8	1236.5	1284.8	1333.1	1381.9	1431.3	1481.4	1532.4	1584.3	1637.0	1745.1	1857.0
(267.25)	s	......	1.6994	1.7608	1.8140	1.8619	1.9058	1.9467	1.9850	2.0212	2.0555	2.0883	2.1498	2.2069
60	v	......	7.259	8.357	9.403	10.427	11.441	12.449	13.452	14.454	15.453	16.451	18.446	20.44
	h	......	1181.6	1233.6	1283.0	1331.8	1380.9	1430.5	1480.8	1531.9	1583.8	1636.6	1744.8	1856.7
(292.71)	s	......	1.6492	1.7135	1.7678	1.8162	1.8605	1.9015	1.9400	1.9762	2.0106	2.0434	2.1049	2.1621
80	v	......	......	6.220	7.020	7.797	8.562	9.322	10.077	10.830	11.582	12.332	13.830	15.325
	h	......	......	1230.7	1281.1	1330.5	1379.9	1429.7	1480.1	1531.3	1583.4	1636.2	1744.5	1856.5
(312.03)	s	......	......	1.6791	1.7346	1.7836	1.8281	1.8694	1.9079	1.9442	1.9787	2.0115	2.0731	2.1303
100	v	......	......	4.937	5.589	6.218	6.835	7.446	8.052	8.656	9.259	9.860	11.060	12.258
	h	......	......	1227.6	1279.1	1329.1	1378.9	1428.9	1479.5	1530.8	1582.9	1635.7	1744.2	1856.2
(327.81)	s	......	......	1.6518	1.7085	1.7581	1.8029	1.8443	1.8829	1.9193	1.9538	1.9867	2.0484	2.1056
120	v	......	......	4.081	4.636	5.165	5.683	6.195	6.702	7.207	7.710	8.212	9.214	10.213
	h	......	......	1224.4	1277.2	1327.7	1377.8	1428.1	1478.8	1530.2	1582.4	1635.3	1743.9	1856.0
(341.25)	s	......	......	1.6287	1.6869	1.7370	1.7822	1.8237	1.8625	1.8990	1.9335	1.9664	2.0281	2.0854

Abs. press. (sat. temp.)														
140	v	......	......	3.468	3.954	4.413	4.861	5.301	5.738	6.172	6.604	7.035	7.895	8.752
(353.02)	h	......	......	1221.1	1275.2	1326.4	1376.8	1427.3	1478.2	1529.7	1581.9	1634.9	1743.5	1855.7
	s	......	......	1.6087	1.6683	1.7190	1.7645	1.8063	1.8451	1.8817	1.9163	1.9493	2.0110	2.0683
160	v	......	......	3.008	3.443	3.849	4.244	4.631	5.015	5.396	5.775	6.152	6.906	7.656
(363.53)	h	......	......	1217.6	1273.1	1325.0	1375.7	1426.4	1477.5	1529.1	1581.4	1634.5	1743.2	1855.5
	s	......	......	1.5908	1.6519	1.7033	1.7491	1.7911	1.8301	1.8667	1.9014	1.9344	1.9962	2.0535
180	v	......	......	2.649	3.044	3.411	3.764	4.110	4.452	4.792	5.129	5.466	6.136	6.804
(373.06)	h	......	......	1214.0	1271.0	1323.5	1374.7	1425.6	1476.8	1528.6	1581.0	1634.1	1742.9	1855.2
	s	......	......	1.5745	1.6373	1.6894	1.7355	1.7776	1.8167	1.8534	1.8882	1.9212	1.9831	2.0404
200	v	......	......	2.361	2.726	3.060	3.380	3.693	4.002	4.309	4.613	4.917	5.521	6.123
(381.79)	h	......	......	1210.3	1268.9	1322.1	1373.6	1424.8	1476.2	1528.0	1580.5	1633.7	1742.6	1855.0
	s	......	......	1.5594	1.6240	1.6767	1.7232	1.7655	1.8048	1.8415	1.8763	1.9094	1.9713	2.0287
220	v	......	......	2.125	2.465	2.772	3.066	3.352	3.634	3.913	4.191	4.467	5.017	5.565
(389.86)	h	......	......	1206.5	1266.7	1320.7	1372.6	1424.0	1475.5	1527.5	1580.0	1633.3	1742.3	1854.7
	s	......	......	1.5453	1.6117	1.6652	1.7120	1.7545	1.7939	1.8308	1.8656	1.8987	1.9607	2.0181
240	v	......	......	1.9276	2.247	2.533	2.804	3.068	3.327	3.584	3.839	4.093	4.597	5.100
(397.37)	h	......	......	1202.5	1264.5	1319.2	1371.5	1423.2	1474.8	1526.9	1579.6	1632.9	1742.0	1854.5
	s	......	......	1.5319	1.6003	1.6546	1.7017	1.7444	1.7839	1.8209	1.8558	1.8889	1.9510	2.0084
260	v	......	......	......	2.063	2.330	2.582	2.827	3.067	3.305	3.541	3.776	4.242	4.707
(404.42)	h	......	......	......	1262.3	1317.7	1370.4	1422.3	1474.2	1526.3	1579.1	1632.5	1741.7	1854.2
	s	......	......	......	1.5897	1.6447	1.6922	1.7352	1.7748	1.8118	1.8467	1.8799	1.9420	1.9995
280	v	......	......	......	1.9047	2.156	2.392	2.621	2.845	3.066	3.286	3.504	3.938	4.370
(411.05)	h	......	......	......	1260.0	1316.2	1369.4	1421.5	1473.5	1525.8	1578.6	1632.1	1741.4	1854.0
	s	......	......	......	1.5796	1.6354	1.6834	1.7265	1.7662	1.8033	1.8383	1.8716	1.9337	1.9912
300	v	......	......	......	1.7675	2.005	2.227	2.442	2.652	2.859	3.065	3.269	3.674	4.078
(417.33)	h	......	......	......	1257.6	1314.7	1368.3	1420.6	1472.8	1525.2	1578.1	1631.7	1741.0	1853.7
	s	......	......	......	1.5701	1.6268	1.6751	1.7184	1.7582	1.7954	1.8305	1.8638	1.9260	1.9835
350	v	......	......	......	1.4923	1.7036	1.8980	2.084	2.266	2.445	2.622	2.798	3.147	3.493
(431.72)	h	......	......	......	1251.5	1310.9	1365.5	1418.5	1471.1	1523.8	1577.0	1630.7	1740.3	1853.1
	s	......	......	......	1.5481	1.6070	1.6563	1.7002	1.7403	1.7777	1.8130	1.8463	1.9086	1.9663
400	v	......	......	......	1.2851	1.4770	1.6508	1.8161	1.9767	2.134	2.290	2.445	2.751	3.055
(444.59)	h	......	......	......	1245.1	1306.9	1362.7	1416.4	1469.4	1522.4	1575.8	1629.6	1739.5	1852.5
	s	......	......	......	1.5281	1.5894	1.6398	1.6842	1.7247	1.7623	1.7977	1.8311	1.8936	1.9513

* Abridged from "Thermodynamic Properties of Steam," by Joseph H. Keenan and Frederick G. Keyes.
Published by John Wiley & Sons, Inc., New York.

Table A8: Steam Table III (Concluded)

Properties of Superheated Steam *

Abs Press. Lb Sq In. (Sat. Temp.)		Temperature—Degrees Fahrenheit 500	550	600	620	640	660	680	700	800	900	1000	1200	1400	1600
450	v	1.1231	1.2155	1.3005	1.3332	1.3652	1.3967	1.4278	1.4584	1.6074	1.7516	1.8928	2.170	2.443	2.714
	h	1238.4	1272.0	1302.8	1314.6	1326.2	1337.5	1348.8	1359.9	1414.3	1467.7	1521.0	1628.6	1738.7	1851.9
(456.28)	s	1.5095	1.5437	1.5735	1.5845	1.5951	1.6054	1.6153	1.6250	1.6699	1.7108	1.7486	1.8177	1.8803	1.9381
500	v	0.9927	1.0800	1.1591	1.1893	1.2188	1.2478	1.2763	1.3044	1.4405	1.5715	1.6996	1.9504	2.197	2.442
	h	1231.3	1266.8	1298.6	1310.7	1322.6	1334.2	1345.7	1357.0	1412.1	1466.0	1519.6	1627.6	1737.9	1851.3
(467.01)	s	1.4919	1.5280	1.5588	1.5701	1.5810	1.5915	1.6016	1.6115	1.6571	1.6982	1.7363	1.8056	1.8683	1.9262
550	v	0.8852	0.9686	1.0431	1.0714	1.0989	1.1259	1.1523	1.1783	1.3038	1.4241	1.5414	1.7706	1.9957	2.219
	h	1223.7	1261.2	1294.3	1306.8	1318.9	1330.8	1342.5	1354.0	1409.9	1464.3	1518.2	1626.6	1737.1	1850.6
(476.94)	s	1.4751	1.5131	1.5451	1.5568	1.5680	1.5787	1.5890	1.5991	1.6452	1.6868	1.7250	1.7946	1.8575	1.9155
600	v	0.7947	0.8753	0.9463	0.9729	0.9988	1.0241	1.0489	1.0732	1.1899	1.3013	1.4096	1.6208	1.8279	2.033
	h	1215.7	1255.5	1289.9	1302.7	1315.2	1327.4	1339.3	1351.1	1407.7	1462.5	1516.7	1625.5	1736.3	1850.0
(486.21)	s	1.4586	1.4990	1.5323	1.5443	1.5558	1.5667	1.5773	1.5875	1.6343	1.6762	1.7147	1.7846	1.8476	1.9056
700	v	......	0.7277	0.7934	0.8177	0.8411	0.8639	0.8860	0.9077	1.0108	1.1082	1.2024	1.3853	1.5641	1.7405
	h	......	1243.2	1280.6	1294.3	1307.5	1320.3	1332.8	1345.0	1403.2	1459.0	1513.9	1623.5	1734.8	1848.8
(503.10)	s	......	1.4722	1.5084	1.5212	1.5333	1.5449	1.5559	1.5665	1.6147	1.6573	1.6963	1.7666	1.8299	1.8881
800	v	......	0.6154	0.6779	0.7006	0.7223	0.7433	0.7635	0.7833	0.8763	0.9633	1.0470	1.2088	1.3662	1.5214
	h	......	1229.8	1270.7	1285.4	1299.4	1312.9	1325.9	1338.6	1398.6	1455.4	1511.0	1621.4	1733.2	1847.5
(518.23)	s	......	1.4467	1.4863	1.5000	1.5129	1.5250	1.5366	1.5476	1.5972	1.6407	1.6801	1.7510	1.8146	1.8729
900	v	......	0.5264	0.5873	0.6089	0.6294	0.6491	0.6680	0.6863	0.7716	0.8506	0.9262	1.0714	1.2124	1.3509
	h	......	1215.0	1260.1	1275.9	1290.9	1305.1	1318.8	1332.1	1393.9	1451.8	1508.1	1619.3	1731.6	1846.3
(531.98)	s	......	1.4216	1.4653	1.4800	1.4938	1.5066	1.5187	1.5303	1.5814	1.6257	1.6656	1.7371	1.8009	1.8595
1000	v	......	0.4533	0.5140	0.5350	0.5546	0.5733	0.5912	0.6084	0.6878	0.7604	0.8294	0.9615	1.0893	1.2146
	h	......	1198.3	1248.8	1265.9	1281.9	1297.0	1311.4	1325.3	1389.2	1448.2	1505.1	1617.3	1730.0	1845.0
(544.61)	s	......	1.3961	1.4450	1.4610	1.4757	1.4893	1.5021	1.5141	1.5670	1.6121	1.6525	1.7245	1.7886	1.8474
1100	v	......	......	0.4532	0.4738	0.4929	0.5110	0.5281	0.5445	0.6191	0.6866	0.7503	0.8716	0.9885	1.1031
	h	......	......	1236.7	1255.3	1272.4	1288.5	1303.7	1318.3	1384.3	1444.5	1502.2	1615.2	1728.4	1843.8
(556.31)	s	......	......	1.4251	1.4425	1.4583	1.4728	1.4862	1.4989	1.5535	1.5995	1.6405	1.7130	1.7775	1.8363
1200	v	......	......	0.4016	0.4222	0.4410	0.4586	0.4752	0.4909	0.5617	0.6250	0.6843	0.7967	0.9046	1.0101
	h	......	......	1223.5	1243.9	1262.4	1279.6	1295.7	1311.0	1379.3	1440.7	1499.2	1613.1	1726.9	1842.5
(567.22)	s	......	......	1.4052	1.4243	1.4413	1.4568	1.4710	1.4843	1.5409	1.5879	1.6293	1.7025	1.7672	1.8263

1400	*v*	......	......	0.3174	0.3390	0.3580	0.3753	0.3912	0.4062	0.4714	0.5281	0.5805	0.6789	0.7727	0.8640
	h	......	......	1193.0	1218.4	1240.4	1260.3	1278.5	1295.5	1369.1	1433.1	1493.2	1608.9	1723.7	1840.0
(587.10)	*s*	......	......	1.3639	1.3877	1.4079	1.4258	1.4419	1.4567	1.5177	1.5666	1.6093	1.6836	1.7489	1.8083
1600	*v*	......	......	......	0.2733	0.2936	0.3112	0.3271	0.3417	0.4034	0.4553	0.5027	0.5906	0.6738	0.7545
	h	......	......	......	1187.8	1215.2	1238.7	1259.6	1278.7	1358.4	1425.3	1487.0	1604.6	1720.5	1837.5
(604.90)	*s*	......	......	......	1.3489	1.3741	1.3952	1.4137	1.4303	1.4964	1.5476	1.5914	1.6669	1.7328	1.7926
1800	*v*	......	......	......	......	0.2407	0.2597	0.2760	0.2907	0.3502	0.3986	0.4421	0.5218	0.5968	0.6693
	h	......	......	......	......	1185.1	1214.0	1238.5	1260.3	1347.2	1417.4	1480.8	1600.4	1717.3	1835.0
(621.03)	*s*	......	......	......	......	1.3377	1.3638	1.3855	1.4044	1.4765	1.5301	1.5752	1.6520	1.7185	1.7786
2000	*v*	......	......	......	......	0.1936	0.2161	0.2337	0.2489	0.3074	0.3532	0.3935	0.4668	0.5352	0.6011
	h	......	......	......	......	1145.6	1184.9	1214.8	1240.0	1335.5	1409.2	1474.5	1596.1	1714.1	1832.5
(635.82)	*s*	......	......	......	......	1.2945	1.3300	1.3564	1.3783	1.4576	1.5139	1.5603	1.6384	1.7055	1.7660
2500	*v*	......	......	......	......	......	......	0.1484	0.1686	0.2294	0.2710	0.3061	0.3678	0.4244	0.4784
	h	......	......	......	......	......	......	1132.3	1176.8	1303.6	1387.8	1458.4	1585.3	1706.1	1826.2
(668.13)	*s*	......	......	......	......	......	......	1.2687	1.3073	1.4127	1.4772	1.5273	1.6088	1.6775	1.7389
3000	*v*	......	......	......	......	......	......	......	0.0984	0.1760	0.2159	0.2476	0.3018	0.3505	0.3966
	h	......	......	......	......	......	......	......	1060.7	1267.2	1365.0	1441.8	1574.3	1698.0	1819.9
(695.36)	*s*	......	......	......	......	......	......	......	1.1966	1.3690	1.4439	1.4984	1.5837	1.6540	1.7163
3206.2	*v*	......	......	......	......	......	......	......	......	0.1583	0.1981	0.2288	0.2806	0.3267	0.3703
	h	......	......	......	......	......	......	......	......	1250.5	1355.2	1434.7	1569.8	1694.6	1817.2
(705.40)	*s*	......	......	......	......	......	......	......	......	1.3508	1.4309	1.4874	1.5742	1.6452	1.7080
3500	*v*	......	......	......	......	......	......	......	0.0306	0.1364	0.1762	0.2058	0.2546	0.2977	0.3381
	h	......	......	......	......	......	......	......	780.5	1224.9	1340.7	1424.5	1563.3	1689.8	1813.6
	s	......	......	......	......	......	......	......	0.9515	1.3241	1.4127	1.4723	1.5615	1.6336	1.6968
4000	*v*	......	......	......	......	......	......	......	0.0287	0.1052	0.1462	0.1743	0.2192	0.2581	0.2943
	h	......	......	......	......	......	......	......	763.8	1174.8	1314.4	1406.8	1552.1	1681.7	1807.2
	s	......	......	......	......	......	......	......	0.9347	1.2757	1.3827	1.4482	1.5417	1.6154	1.6795
4500	*v*	......	......	......	......	......	......	......	0.0276	0.0798	0.1226	0.1500	0.1917	0.2273	0.2602
	h	......	......	......	......	......	......	......	753.5	1113.9	1286.5	1388.4	1540.8	1673.5	1800.9
	s	......	......	......	......	......	......	......	0.9235	1.2204	1.3529	1.4253	1.5235	1.5990	1.6640
5000	*v*	......	......	......	......	......	......	......	0.0268	0.0593	0.1036	0.1303	0.1696	0.2027	0.2329
	h	......	......	......	......	......	......	......	746.4	1047.1	1256.5	1369.5	1529.5	1665.3	1794.5
	s	......	......	......	......	......	......	......	0.9152	1.1622	1.3231	1.4034	1.5066	1.5839	1.6499
5500	*v*	......	......	......	......	......	......	......	0.0262	0.0463	0.0880	0.1143	0.1516	0.1825	0.2106
	h	......	......	......	......	......	......	......	741.3	985.0	1224.1	1349.3	1518.2	1657.0	1788.1
	s	......	......	......	......	......	......	......	0.9090	1.1093	1.2930	1.3821	1.4908	1.5699	1.6369

Table A9: Ammonia Table—Saturated States†

Temp.	Pressure		Volume	Density	Enthalpy from −40 F			Entropy from −40 F	
F	Abs. lb/in.²	Gage lb/in.²	Vapor ft³/lb	Vapor lb/ft³	Liquid Btu/lb	Vapor Btu/lb	Latent Btu/lb	Liquid Btu/lb F	Vapor Btu/lb F
t	p	p_d	v_g	$1/v_g$	h_f	h_g	h_{fg}	s_f	s_g
−60	5.55	*18.6	44.73	0.02235	−21.2	589.6	610.8	−0.0517	1.4769
−55	6.54	*16.6	38.38	0.02605	−15.9	591.6	607.5	−0.0386	1.4631
−50	7.67	*14.3	33.08	0.03023	−10.6	593.7	604.3	−0.0256	1.4497
−45	8.95	*11.7	28.62	0.03494	− 5.3	595.6	600.9	−0.0127	1.4368
−40	10.41	*8.7	24.86	0.04022	0.0	597.6	597.6	0.0000	1.4242
−38	11.04	*7.4	23.53	.04251	2.1	598.3	596.2	.0051	.4193
−36	11.71	*6.1	22.27	.04489	4.3	599.1	594.8	.0101	.4144
−34	12.41	*4.7	21.10	.04739	6.4	599.9	593.5	.0151	.4096
−32	13.14	*3.2	20.00	.04999	8.5	600.6	592.1	.0201	.4048
−30	13.90	*1.6	18.97	0.05271	10.7	601.4	590.7	0.0250	1.4001
−28	14.71	0.0	18.00	.05555	12.8	602.1	589.3	.0300	.3955
−26	15.55	0.8	17.09	.05850	14.9	602.8	587.9	.0350	.3909
−24	16.42	1.7	16.24	.06158	17.1	603.6	586.5	.0399	.3863
−22	17.34	2.6	15.43	.06479	19.2	604.3	585.1	.0448	.3818
−20	18.30	3.6	14.68	0.06813	21.4	605.0	583.6	0.0497	1.3774
−18	19.30	4.6	13.97	.07161	23.5	605.7	582.2	.0545	.3729
−16	20.34	5.6	13.29	.07522	25.6	606.4	580.8	.0594	.3686
−14	21.43	6.7	12.66	.07898	27.8	607.1	579.3	.0642	.3643
−12	22.56	7.9	12.06	.08289	30.0	607.8	577.8	.0690	.3600
−10	23.74	9.0	11.50	0.08695	32.1	608.5	576.4	0.0738	1.3558
− 8	24.97	10.3	10.97	.09117	34.3	609.2	574.9	.0786	.3516
− 6	26.26	11.6	10.47	.09555	36.4	609.8	573.4	.0833	.3474
− 4	27.59	12.9	9.991	.1001	38.6	610.5	571.9	.0880	.3433
− 2	28.98	14.3	9.541	.1048	40.7	611.1	570.4	.0928	.3393
0	30.42	15.7	9.116	0.1097	42.9	611.8	568.9	0.0975	1.3352
2	31.92	17.2	8.714	.1148	45.1	612.4	567.3	.1022	.3312
4	33.47	18.8	8.333	.1200	47.2	613.0	565.8	.1069	.3273
6	35.09	20.4	7.971	.1254	49.4	613.6	564.2	.1115	.3234
8	36.77	22.1	7.629	.1311	51.6	614.3	562.7	.1162	.3195
10	38.51	23.8	7.304	0.1369	53.8	614.9	561.1	0.1208	1.3157
12	40.31	25.6	6.996	.1429	56.0	615.5	559.5	.1254	.3118
14	42.18	27.5	6.703	.1492	58.2	616.1	557.9	.1300	.3081
16	44.12	29.4	6.425	.1556	60.3	616.6	556.3	.1346	.3043
18	46.13	31.4	6.161	.1623	62.5	617.2	554.7	.1392	.3006
20	48.21	33.5	5.910	0.1692	64.7	617.8	553.1	0.1437	1.2969
22	50.36	35.7	5.671	.1763	66.9	618.3	551.4	.1483	.2933
24	52.59	37.9	5.443	.1837	69.1	618.9	549.8	.1528	.2897
26	54.90	40.2	5.227	.1913	71.3	619.4	548.1	.1573	.2861
28	57.28	42.6	5.021	.1992	73.5	619.9	546.4	.1618	.2825
30	59.74	45.0	4.825	0.2073	75.7	620.5	544.8	0.1663	1.2790
32	62.29	47.6	4.637	.2156	77.9	621.0	543.1	.1708	.2755
34	64.91	50.2	4.459	.2243	80.1	621.5	541.4	.1753	.2721
36	67.63	52.9	4.289	.2332	82.3	622.0	539.7	.1797	.2686
38	70.43	55.7	4.126	.2423	84.6	622.5	537.9	.1841	.2652

* Inches of mercury below one atmosphere.

† Abstracted, by permission, from "Tables of Thermodynamic Properties of Ammonia," U. S. Department of Commerce, Bureau of Standards Circular No. 142, 1945.

Temp.	Pressure		Volume	Density	Enthalpy from −40 F			Entropy from −40 F	
F	Abs. lb/in²	Gage lb/in²	Vapor ft³/lb.	Vapor lb/ft³	Liquid Btu/lb	Vapor Btu/lb	Latent Btu/lb	Liquid Btu/lb F	Vapor Btu/lb F
t	p	p_d	v_g	$1/v_g$	h_f	h_g	h_{fg}	s_f	s_g
40	73.32	58.6	3.971	0.2518	86.8	623.0	536.2	0.1885	1.2618
42	76.31	61.6	3.823	.2616	89.0	623.4	534.4	.1930	.2585
44	79.38	64.7	3.682	.2716	91.2	623.9	532.7	.1974	.2552
46	82.55	67.9	3.547	.2819	93.5	624.4	530.9	.2018	.2519
48	85.82	71.1	3.418	.2926	95.7	624.8	529.1	.2062	.2486
50	89.19	74.5	3.294	0.3036	97.9	625.2	527.3	0.2105	1.2453
52	92.66	78.0	3.176	.3149	100.2	625.7	525.5	.2149	.2421
54	96.23	81.5	3.063	.3265	102.4	626.1	523.7	.2192	.2389
56	99.91	85.2	2.954	.3385	104.7	626.5	521.8	.2236	.2357
58	103.7	89.0	2.851	.3508	106.9	626.9	520.0	.2279	.2325
60	107.6	92.9	2.751	0.3635	109.2	627.3	518.1	0.2322	1.2294
62	111.6	96.9	2.656	.3765	111.5	627.7	516.2	.2365	.2262
64	115.7	101.0	2.565	.3899	113.7	628.0	514.3	.2408	.2231
66	120.0	105.3	2.477	.4037	116.0	628.4	512.4	.2451	.2201
68	124.3	109.6	2.393	.4179	118.3	628.8	510.5	.2494	.2170
70	128.8	114.1	2.312	0.4325	120.5	629.1	508.6	0.2537	1.2140
72	133.4	118.7	2.235	.4474	122.8	629.4	506.6	.2579	.2110
74	138.1	123.4	2.161	.4628	125.1	629.8	504.7	.2622	.2080
76	143.0	128.3	2.089	.4786	127.4	630.1	502.7	.2664	.2050
78	147.9	133.2	2.021	.4949	129.7	630.4	500.7	.2706	.2020
80	153.0	138.3	1.955	0.5115	132.0	630.7	498.7	0.2749	1.1991
82	158.3	143.6	1.892	.5287	134.3	631.0	496.7	.2791	.1962
84	163.7	149.0	1.831	.5462	136.6	631.3	494.7	.2833	.1933
86	169.2	154.5	1.772	.5643	138.9	631.5	492.6	.2875	.1904
88	174.8	160.1	1.716	.5828	141.2	631.8	490.6	.2917	.1875
90	180.6	165.9	1.661	0.6019	143.5	632.0	488.5	0.2958	1.1846
92	186.6	171.9	1.609	.6214	145.8	632.2	486.4	.3000	.1818
94	192.7	178.0	1.559	.6415	148.2	632.5	484.3	.3041	.1789
96	198.9	184.2	1.510	.6620	150.5	632.6	482.1	.3083	.1761
98	205.3	190.6	1.464	.6832	152.9	632.9	480.0	.3125	.1733
100	211.9	197.2	1.419	0.7048	155.2	633.0	477.8	0.3166	1.1705
102	218.6	203.9	1.375	.7270	157.6	633.2	475.6	.3207	.1677
104	225.4	210.7	1.334	.7498	159.9	633.4	473.5	.3248	.1649
106	232.5	217.8	1.293	.7732	162.3	633.5	471.2	.3289	.1621
108	239.7	225.0	1.254	.7972	164.6	633.6	469.0	.3330	.1593
110	247.0	232.3	1.217	0.8219	167.0	633.7	466.7	0.3372	1.1566
112	254.5	239.8	1.180	.8471	169.4	633.8	464.4	.3413	.1538
114	262.2	247.5	1.145	.8730	171.8	633.9	462.1	.3453	.1510
116	270.1	255.4	1.112	.8996	174.2	634.0	459.8	.3495	.1483
118	278.2	263.5	1.079	.9269	176.6	634.0	457.4	.3535	.1455
120	286.4	271.7	1.047	0.9549	179.0	634.0	455.0	0.3576	1.1427
122	294.8	280.1	1.017	.9837	181.4	634.0	452.6	.3618	.1400
124	303.4	288.7	0.987	1.0132	183.9	634.0	450.1	.3659	.1372

Table A9: Ammonia Table—Superheated Ammonia

Absolute Pressure in lb/in.² (Saturation Temperature in italics)

Temp. F	50 *21.67*			60 *30.21*			70 *37.70*			80 *44.40*		
t	*v*	*h*	*s*	*v*	*h*	*s*	*v*	*h*	*s*	*v*	*h*	*s*
Sat.	*5.710*	*618.2*	*1.2939*	*4.805*	*620.5*	*1.2787*	*4.151*	*622.4*	*1.2658*	*3.655*	*624.0*	*1.2545*
30	5.838	623.4	1.3046									
40	5.988	629.5	.3169	4.933	626.8	1.2913	4.177	623.9	1.2688			
50	6.135	635.4	1.3286	5.060	632.9	1.3035	4.290	630.4	1.2816	3.712	627.7	1.2619
60	6.280	641.2	.3399	5.184	639.0	.3152	4.401	636.6	.2937	3.812	634.3	.2745
70	6.423	646.9	.3508	5.307	644.9	.3265	4.509	642.7	.3054	3.909	640.6	.2866
80	6.564	652.6	.3613	5.428	650.7	.3373	4.615	648.7	.3166	4.005	646.7	.2981
90	6.704	658.2	.3716	5.547	656.4	.3479	4.719	654.6	.3274	4.098	652.8	.3092
100	6.843	663.7	1.3816	5.665	662.1	1.3581	4.822	660.4	1.3378	4.190	658.7	1.3199
110	6.980	669.2	.3914	5.781	667.7	.3681	4.924	666.1	.3480	4.281	664.6	.3303
120	7.117	674.7	.4009	5.897	673.3	.3778	5.025	671.8	.3579	4.371	670.4	.3404
130	7.252	680.2	.4103	6.012	678.9	.3873	5.125	677.5	.3676	4.460	676.1	.3502
140	7.387	685.7	.4195	6.126	684.4	.3966	5.224	683.1	.3770	4.548	681.8	.3598
150	7.521	691.1	1.4286	6.239	689.9	1.4058	5.323	688.7	1.3863	4.635	687.5	1.3692
160	7.655	696.6	.4374	6.352	695.5	.4148	5.420	694.3	.3954	4.722	693.2	.3784
170	7.788	702.1	.4462	6.464	701.0	.4236	5.518	699.9	.4043	4.808	698.8	.3874
180	7.921	707.5	.4548	6.576	706.5	.4323	5.615	705.5	.4131	4.893	704.4	.3963
190	8.053	713.0	.4633	6.687	712.0	.4409	5.711	711.0	.4217	4.978	710.0	.4050
200	8.185	718.5	1.4716	6.798	717.5	1.4493	5.807	716.6	1.4302	5.063	715.6	1.4136
210	8.317	724.0	.4799	6.909	723.1	.4576	5.902	722.2	.4386	5.147	721.3	.4220
220	8.448	729.4	.4880	7.019	728.6	.4658	5.998	727.7	.4469	5.231	726.9	.4304
240	8.710	740.5	.5040	7.238	739.7	.4819	6.187	738.9	.4631	5.398	738.1	.4467
260	8.970	751.6	.5197	7.457	750.9	.4976	6.376	750.1	.4789	5.565	749.4	.4626
280	9.230	762.7	1.5350	7.675	762.1	1.5130	6.563	761.4	1.4943	5.730	760.7	1.4781
300	9.489	774.0	.5500	7.892	773.3	.5281	6.750	772.7	.5095	5.894	772.1	.4933

Temp. F	90 *50.47*			100 *56.05*			120 *66.02*			140 *74.79*		
Sat.	*3.266*	*625.3*	*1.2445*	*2.952*	*626.5*	*1.2356*	*2.476*	*628.4*	*1.2201*	*2.132*	*629.9*	*1.2068*
50												
60	3.353	631.8	1.2571	2.985	629.3	1.2409						
70	3.442	638.3	.2695	3.068	636.0	.2539	2.505	631.3	1.2255			
80	3.529	644.7	.2814	3.149	642.6	.2661	2.576	638.3	.2386	2.166	633.8	1.2140
90	3.614	650.9	.2928	3.227	649.0	.2778	2.645	645.0	.2510	2.228	640.9	.2272
100	3.698	657.0	1.3038	3.304	655.2	1.2891	2.712	651.6	1.2628	2.288	647.8	1.2396
110	3.780	663.0	.3144	3.380	661.3	.2999	2.778	658.0	.2741	2.347	654.5	.2515
120	3.862	668.9	.3247	3.454	667.3	.3104	2.842	664.2	.2850	2.404	661.1	.2628
130	3.942	674.7	.3347	3.527	673.3	.3206	2.905	670.4	.2956	2.460	667.4	.2738
140	4.021	680.5	.3444	3.600	679.2	.3305	2.967	676.5	.3058	2.515	673.7	.2843
150	4.100	686.3	1.3539	3.672	685.0	1.3401	3.029	682.5	1.3157	2.569	679.9	1.2945
160	4.178	692.0	.3633	3.743	690.8	.3495	3.089	688.4	.3254	2.622	686.0	.3045
170	4.255	697.7	.3724	3.813	696.6	.3588	3.149	694.3	.3348	2.675	692.0	.3141
180	4.332	703.4	.3813	3.883	702.3	.3678	3.209	700.2	.3441	2.727	698.0	.3236
190	4.408	709.0	.3901	3.952	708.0	.3767	3.268	706.0	.3531	2.779	704.0	.3328
200	4.484	714.7	1.3988	4.021	713.7	1.3854	3.326	711.8	1.3620	2.830	709.9	1.3418
210	4.560	720.4	.4073	4.090	719.4	.3940	3.385	717.6	.3707	2.880	715.8	.3507
220	4.635	726.0	.4157	4.158	725.1	.4024	3.442	723.4	.3793	2.931	721.6	.3594
230	4.710	731.7	.4239	4.226	730.8	.4108	3.500	729.2	.3877	2.981	727.5	.3679
240	4.785	737.3	.4321	4.294	736.5	.4190	3.557	734.9	.3960	3.030	733.3	.3763
250	4.859	743.0	1.4401	4.361	742.2	1.4271	3.614	740.7	1.4042	3.080	739.2	1.3846
260	4.933	748.7	.4481	4.428	747.9	.4350	3.671	746.5	.4123	3.129	745.0	.3928
280	5.081	760.0	.4637	4.562	759.4	.4507	3.783	758.0	.4281	3.227	756.7	.4088
300	5.228	771.5	.4789	4.695	770.8	.4660	3.895	769.6	.4435	3.323	768.3	.4243

Absolute Pressure in lb/in.² (Saturation Temperatures in italics)

Temp. F	160 *82.64*			180 *89.78*			200 *96.34*			220 *102.42*		
t	*v*	*h*	*s*	*v*	*h*	*s*	*v*	*h*	*s*	*v*	*h*	*s*
Sat.	*1.872*	*631.1*	*1.1952*	*1.667*	*632.0*	*1.1850*	*1.502*	*632.7*	*1.1756*	*1.367*	*633.2*	*1.1671*
90	1.914	636.6	1.2055	1.668	632.2	1.1853						
100	1.969	643.9	1.2186	1.720	639.9	1.1992						
110	2.023	651.0	.2311	1.770	647.3	.2123	1.567	643.4	1.1947	1.400	639.4	1.1781
120	2.075	657.8	.2429	1.818	654.4	.2247	1.612	650.9	.2077	1.443	647.3	.1917
130	2.125	664.4	.2542	1.865	661.3	.2364	1.656	658.1	.2200	1.485	654.8	.2045
140	2.175	670.9	.2652	1.910	668.0	.2477	1.698	665.0	.2317	1.525	662.0	.2167
150	2.224	677.2	1.2757	1.955	674.6	1.2586	1.740	671.8	1.2429	1.564	669.0	1.2281
160	2.272	683.5	.2859	1.999	681.0	.2691	1.780	678.4	.2537	1.601	675.8	.2394
170	2.319	689.7	.2958	2.042	687.3	.2792	1.820	684.9	.2641	1.638	682.5	.2501
180	2.365	695.8	.3054	2.084	693.6	.2891	1.859	691.3	.2742	1.675	689.1	.2604
190	2.411	701.9	.3148	2.126	699.8	.2987	1.897	697.7	.2840	1.710	695.5	.2704
200	2.457	707.9	1.3240	2.167	705.9	1.3081	1.935	703.9	1.2935	1.745	701.9	1.2801
210	2.502	713.9	.3331	2.208	712.0	.3172	1.972	710.1	.3029	1.780	708.2	.2896
220	2.547	719.9	.3419	2.248	718.1	.3262	2.009	716.3	.3120	1.814	714.4	.2989
230	2.591	725.8	.3506	2.288	724.1	.3350	2.046	722.4	.3209	1.848	720.6	.3079
240	2.635	731.7	.3591	2.328	730.1	.3436	2.082	728.4	.3296	1.881	726.8	.3168
250	2.679	737.6	1.3675	2.367	736.1	1.3521	2.118	734.5	1.3382	1.914	732.9	1.3255
260	2.723	743.5	.3757	2.407	742.0	.3605	2.154	740.5	.3467	1.947	739.0	.3340
270	2.766	749.4	.3838	2.446	748.0	.3687	2.189	746.5	.3550	1.980	745.1	.3424
280	2.809	755.3	.3919	2.484	753.9	.3768	2.225	752.5	.3631	2.012	751.1	.3507
290	2.852	761.2	.3998	2.523	759.9	.3847	2.260	758.5	.3712	2.044	757.2	.3588
300	2.895	767.1	1.4076	2.561	765.8	1.3926	2.295	764.5	1.3791	2.076	763.2	1.3668
320	2.980	778.9	.4229	2.637	777.7	.4081	2.364	776.5	.3947	2.140	775.3	.3825
340	3.064	790.7	.4379	2.713	789.6	.4231	2.432	788.5	.4099	2.203	787.4	.3978
360	......	.....	......	......	.....	......	2.500	800.5	.4247	2.265	799.5	.4127
380	......	.....	......	......	.....	......	2.568	812.5	.4392	2.327	811.6	.4273

Temp. F	240 *108.09*			260 *113.42*			280 *118.45*			300 *123.21*		
Sat.	*1.253*	*633.6*	*1.1592*	*1.155*	*633.9*	*1.1518*	*1.072*	*634.0*	*1.1449*	*0.999*	*634.0*	*1.1383*
110	1.261	635.3	1.1621									
120	1.302	643.5	.1764	1.182	639.5	1.1617	1.078	635.4	1.1473			
130	1.342	651.3	.1898	1.220	647.8	.1757	1.115	644.0	.1621	1.023	640.1	1.1487
140	1.380	658.8	.2025	1.257	655.6	.1889	1.151	652.2	.1759	1.058	648.7	.1632
150	1.416	666.1	1.2145	1.292	663.1	1.2014	1.184	660.1	1.1888	1.091	656.9	1.1767
160	1.452	673.1	.2259	1.326	670.4	.2132	1.217	667.6	.2011	1.123	664.7	.1894
170	1.487	680.0	.2369	1.359	677.5	.2245	1.249	674.9	.2127	1.153	672.2	.2014
180	1.521	686.7	.2475	1.391	684.4	.2354	1.279	681.9	.2239	1.183	679.5	.2129
190	1.554	693.3	.2577	1.422	691.1	.2458	1.309	688.9	.2346	1.211	686.5	.2239
200	1.587	699.8	1.2677	1.453	697.7	1.2560	1.339	695.6	1.2449	1.239	693.5	1.2344
210	1.619	706.2	.2773	1.484	704.3	.2658	1.367	702.3	.2550	1.267	700.3	.2447
220	1.651	712.6	.2867	1.514	710.7	.2754	1.396	708.8	.2647	1.294	706.9	.2546
230	1.683	718.9	.2959	1.543	717.1	.2847	1.424	715.3	.2742	1.320	713.5	.2642
240	1.714	725.1	.3049	1.572	723.4	.2938	1.451	721.8	.2834	1.346	720.0	.2736
250	1.745	731.3	1.3137	1.601	729.7	1.3027	1.478	728.1	1.2924	1.372	726.5	1.2827
260	1.775	737.5	.3224	1.630	736.0	.3115	1.505	734.4	.3013	1.397	732.9	.2917
270	1.805	743.6	.3308	1.658	742.2	.3200	1.532	740.7	.3099	1.422	739.2	.3004
280	1.835	749.8	.3392	1.686	748.4	.3285	1.558	747.0	.3184	1.447	745.5	.3090
290	1.865	755.9	.3474	1.714	754.5	.3367	1.584	753.2	.3268	1.472	751.8	.3175
300	1.895	762.0	1.3554	1.741	760.7	1.3449	1.610	759.4	1.3350	1.496	758.1	1.3257
320	1.954	774.1	.3712	1.796	772.9	.3608	1.661	771.7	.3511	1.544	770.5	.3419
340	2.012	786.3	.3866	1.850	785.2	.3763	1.712	784.0	.3667	1.592	782.9	.3576
360	2.069	798.4	.4016	1.904	797.4	.3914	1.762	796.3	.3819	1.639	795.3	.3729
380	2.126	810.6	.4163	1.957	809.6	.4062	1.811	808.7	.3967	1.686	807.7	.3878
400	......	.....	......	2.009	821.9	1.4206	1.861	821.0	1.4112	1.732	820.1	1.4024

Table A10: Mean Specific Heats of Various Solids and Liquids between 32 and 212 F

Substance	Specific heat
Solids	
Alloys:	
Bismuth-tin	0.040–0.045
Bell metal	0.086
Brass, yellow	0.0883
Brass, red	0.090
Bronze	0.104
Constantan	0.098
D'Arcet's metal	0.050
German silver	0.095
Lipowitz's metal	0.040
Nickel steel	0.109
Rose's metal	0.050
Solders (Pb and Sn)	0.040–0.045
Type metal	0.0388
Wood's metal	0.040
40 Pb + 60 Bi	0.0317
25 Pb + 75 Bi	0.030
Asbestos	0.20
Ashes	0.20
Bakelite	0.3–0.4
Basalt (lava)	0.20
Borax	0.229
Brick	0.22
Carbon–coke	0.203
Chalk	0.215
Charcoal	0.20
Cinders	0.18
Coal	0.3
Concrete	0.156
Cork	0.485
Corundum	0.198
Dolomite	0.222
Ebonite	0.33
Glass:	
Normal	0.199
Crown	0.16
Flint	0.12
Gneiss	0.18
Granite	0.195
Graphite	0.201
Gypsum	0.259
Hornblende	0.195
Humus (soil)	0.44
Ice:	
−4 F	0.465
32 F	0.487
India rubber (Para)	0.27–0.48
Kaolin	0.224
Limestone	0.217
Marble	0.210
Oxides:	
Alumina (Al_2O_3)	0.183
Cu_2O	0.111
Lead oxide (PbO)	0.055
Lodestone	0.156
Magnesia	0.222
Magnetite (Fe_3O_4)	0.168
Silica	0.191
Soda	0.231
Zinc oxide (ZnO)	0.125
Paraffin wax	0.69
Porcelain	0.22
Quartz	0.17–0.28
Quicklime	0.217
Salt, rock	0.21
Sand	0.195
Sandstone	0.22
Serpentine	0.25
Sulphur	0.180
Talc	0.209
Tufa	0.33
Vulcanite	0.331
Wood:	
Fir	0.65
Oak	0.57
Pine	0.67
Liquids	
Acetic acid	0.51
Acetone	0.544
Alcohol (absolute)	0.58
Aniline	0.49
Benzol	0.40
Chloroform	0.23
Ether	0.54
Ethyl acetate	0.478
Ethylene glycol	0.602
Fusel oil	0.56
Gasoline	0.50
Glycerin	0.58
Hydrochloric acid	0.60
Kerosine	0.50
Naphthalene	0.31
Machine oil	0.40
Mercury	0.033
Olive oil	0.40
Paraffin oil	0.52
Petroleum	0.50
Sulphuric acid	0.336
Sea water	0.94
Toluene	0.40
Turpentine	0.42
Molten metals:	
Bismuth (535–725 F)	0.036
Lead (590–680 F)	0.041
Sulphur (246–297 F)	0.235
Tin (460–660 F)	0.058

Table A11: Heat Transfer Properties of Gases

Thermal conductivity, k, in Btu/ft hr °F
Viscosity, μ, in lb/hr ft
Prandtl number, N_{Pr} dimensionless
} at low pressure

	Air			*Nitrogen*			*Carbon Dioxide*		
T, °R	k	μ	N_{Pr}	k	μ	N_{Pr}	k	μ	N_{Pr}
180	0.0053	0.0168	0.770	0.0055	0.0166	0.786			
360	0.0105	0.0321	0.739	0.0105	0.0313	0.747	0.0055	0.0244	0.830
540	0.0152	0.0447	0.708	0.0151	0.0432	0.713	0.0096	0.0362	0.770
720	0.0195	0.0553	0.689	0.0193	0.0532	0.691	0.0142	0.0467	0.738
900	0.0234	0.0646	0.680	0.0230	0.0622	0.684	0.0194	0.0562	0.702
1080	0.0269	0.0730	0.680	0.0265	0.0705	0.686	0.0243	0.0649	0.685
1260	0.0303	0.0806	0.684	0.0296	0.0781	0.695	0.0285	0.0729	0.687
1440	0.0334	0.0877	0.689	0.0325	0.0854	0.708	0.0324	0.0804	0.692
1620	0.0363	0.0943	0.696	0.0351	0.0923	0.723	0.0359	0.0874	0.698
1800	0.0390	0.1005	0.702	0.0374	0.0990	0.740	0.0392	0.0941	0.705
1980		0.1063		0.0396	0.1053	0.757	0.0423	0.1004	0.712
2160		0.1119		0.0416	0.1115	0.774	0.0451	0.1066	0.721
2340		0.1172			0.1174		0.0477	0.1125	0.729

	Oxygen			*Hydrogen*			*Carbon Monoxide*		
T, °R	k	μ	N_{Pr}	k	μ	N_{Pr}	k	μ	N_{Pr}
180	0.0052	0.0188	0.815	0.0384	0.0102	0.712	0.0050	0.0159	
360	0.0105	0.0360	0.745	0.0741	0.0165	0.719	0.0101	0.0308	0.764
540	0.0150	0.0499	0.729	0.1051	0.0217	0.706	0.0146	0.0432	0.737
720	0.0200	0.0618	0.696	0.1319	0.0263	0.690	0.0186	0.0537	0.722
900	0.0241	0.0724	0.697	0.1572	0.0306	0.675	0.0223	0.0631	0.718
1080	0.0280	0.0821	0.703	0.1820	0.0346	0.664	0.0257	0.0716	0.724
1260		0.0911		0.2070	0.0383	0.661		0.0796	

	Helium			*Argon*			*Steam*		
T, °R	k	μ	N_{Pr}	k	μ	N_{Pr}	k	μ	N_{Pr}
180	0.0422	0.0230	0.677	0.0037	0.0201	0.692			
360	0.0672	0.0366	0.674	0.0072	0.0393	0.676			
540	0.0866	0.0480	0.686	0.0102	0.0555	0.670		0.0237	
720	0.1029	0.0582	0.700	0.0129	0.0693	0.664	0.0150	0.0325	1.037
900	0.1171	0.0676	0.715	0.0153	0.0816	0.658	0.0196	0.0412	0.995
1080	0.1299	0.0764	0.729	0.0174	0.0928	0.655	0.0244	0.0499	0.992
1260		0.0848			0.1032	0.654	0.0293	0.0587	1.000
1440		0.0928			0.1130	0.653	0.0342	0.0674	1.015

Data by permission from Hilsenrath and Touloukian, *Trans. ASME*, August, 1954, p. 967.

Table A12: Thermal Conductivity of Metals

(k in Btu/hr ft °F)

Metal	t, °F	k
Aluminum-14S alloy 4.4 Cu, 0.8 Si, 0.75 Mn, 0.35 Mg (Ref. b)	200	113
	400	111
	600	107
Aluminum-355 alloy 1.2 Cu, 5.0 Si, 0.5 Mg (Ref. b)	200	95
	400	97
	600	97
Aluminum (Ref. a)	65	116
	212	118
Note: Data on aluminum and its alloys vary appreciably, probably due to differences in composition and treatment.		
Brass—61.5 Cu, 35.5 Zn, 3.0 Pb. (Ref. b)	200	67
	400	73.5
	600	80.5
Constantan—60 Cu, 40 Ni. (Ref. a)	65	13
	212	15.5
Copper (decreases with rising temperature, about 2.3 per 100°F)	68	228
Gold (Ref. a)	65	169
	212	170
Inconel X (Ref. b) (Typical of several high-temperature alloys used in gas turbine construction)	200	9.1
	400	10.4
	600	11.7
	800	13.0
	1000	14.3
Lead (Ref. a)	65	20
	212	19.8
Nickel (Ref. a)	65	34
	212	34
	572	30
Silver, 99.9% (Ref. b)	200	241
	400	231
	600	220
	800	209
	1000	198
Tin (Ref. a)	65	37
	212	35
Zinc (Ref. a)	65	64
	212	63

Data by permission from:
Ref. a, *Handbook of Chemistry and Physics*, Chemical Rubber Publishing Co.
Ref. b, J. E. Evans, Jr., "Thermal Conductivity of Several Metals and Alloys," *Product Engineering Handbook for 1955*, copyright 1954, McGraw-Hill Publishing Co., Inc.

TYPICAL STEELS

(k varies linearly with temperature)

Description of steel	k at 100°F	k at 1100°F
0.2 Carbon	28.7	22.7
0.5 Mo	25.5	21.
1.0 Cr, 0.5 Mo	19.1	17.9
5 Cr, 0.5 Mo	15.6	16.4
2.5 Cr, 0.5 Mo, 0.75 Si	14.7	15.8
12 Cr, 0.15 C (Type 410)	16.0	14.3
27 Cr (Type 446)	11.0	17.6
18 Cr, 8 Ni (Types 304, 316, 321, 347)	9.7	14.2
25 Cr, 20 Ni (Type 310)	7.8	13.5

Data on steels mainly courtesy of National Tube Division, U.S. Steel Corp.

CONDENSER TUBE ALLOYS

(Commercial standards for k at 68°F)

Alloy	k
Phosphorized copper, 99.9+ Cu	227
Arsenical copper, 0.35 As, rem. Cu	102
Red brass, 85 Cu, 15 Zn	92
Muntz metal, 60 Cu, 40 Zn	71
Admiralty metal, 71 Cu, 28 Zn, 1 Sn	64
Naval brass, 60 Cu, 39.25 Zn, 0.75 Sn	67
Copper-nickel, 80 Cu, 20 Ni	21.6
Copper-nickel, 70 Cu, 30 Ni	16.8
Aluminum brass, 77 Cu, 2 Al, 0.04 As, rem. Zn	58

Table A13: Thermal Conductivities of Miscellaneous Solid Substances

(Values of k are to be regarded as rough average values for the temperature range indicated)

Material	Apparent density, lb per cu ft	Temp, deg F	k
Air spaces (¾ in.) faced with aluminum foil		100	0.025
Asbestos	36.0	500	0.122
Asbestos mill board	60.5	86	0.07
Ashes, wood		32–212	0.041
Brick masonry		70	0.33–0.42
Carborundum brick	129	1112	10.7
Cardboard, corrugated			0.37
Celluloid	87.3	86	0.12
Cement			0.17
Chalk			0.48
Charcoal flakes	11.9	176	0.043
Charcoal flakes	15.0	172	0.051
Clinker, granular		32–1300	0.27
Coke, petroleum		212	3.4
Coke, powdered		32–212	0.106
Concrete, cinder			0.167–0.42
Concrete, stone			0.5–0.75
Cotton wool	5.06	32	0.024
Ebonite			0.10
Felt, wool			0.03
Fiber, red	80.5	68	0.27
(With binder, baked)		70–208	0.097
Enamel, silicate		100	0.5–0.75
Firebrick		2000	0.7–1.0
Gas carbon		32–212	2.0
Glass			0.333–0.5
Granite			1.0–2.32
Graphite, powdered (through 100 mesh)	30.0	104	0.104
Graphite, solid	93.5	122	87
Gypsum, moulded and dry	78.0	68	0.25
Hair felt	17	86	0.021
Ice	57.5		1.26
Infusorial earth	20	32–1200	0.06
Kapok	0.88	86	0.020
Lampblack	10.0	104	0.038
Lava			0.49
Leather, sole	62.4		0.092
Limestone			0.3–0.75
Linen			0.05
Magnesia, light carbonate	13	100	0.034
Magnesite brick	158	400	2.2
Mica			0.34
Marble			1.2–1.7
Mill shavings			0.033–0.05
Mineral wool	9.4	86	0.023
Paper			0.075
Paraffin wax	55.6	86	0.145
Plaster			0.25–0.5
Porcelain		392	0.88
Pumice stone		70–150	0.135
Pyrex	139	127	0.63
Quartz, parallel to axis			7.25
Quartz, perpendicular to axis			3.90
Rubber, hard	74.3	100	0.092
Rubber, soft, vulcanized	68.6	86	0.08
Sand, dry	94.8	68	0.188
Sandstone	140	104	1.1
Sawdust, dry	13.4	68	0.042
Silica brick		2000	0.85–1.0
Silk	6.3		0.026
Slag, blast-furnace		75–260	0.064
Slate		201	0.86
Woods: (across grain except as otherwise noted)			
Balsa	8.8	86	0.03
Balsa	20.0	86	0.048
Cypress	28.7	86	0.056
Fir, white	28.1	140	0.062
Fir (along grain)	34.4	77	0.215
Maple	44.3	86	0.11
Oak	51.5		0.12
White pine	34	59	0.087
Yellow pine			0.085
Wool, pure	6.9	86	0.021

The thermal conductivity of different materials varies greatly. For metals and alloys k is high, while for certain insulating materials, as asbestos, cork, and kapok, it is very low. In general, k varies with the temperature. In the case of metals, k usually decreases with rising temperature, while for most other substances the reverse is true. By permission from *Mechanical Engineer's Handbook* by L. S. Marks, copyright 1951, McGraw-Hill Book Co.

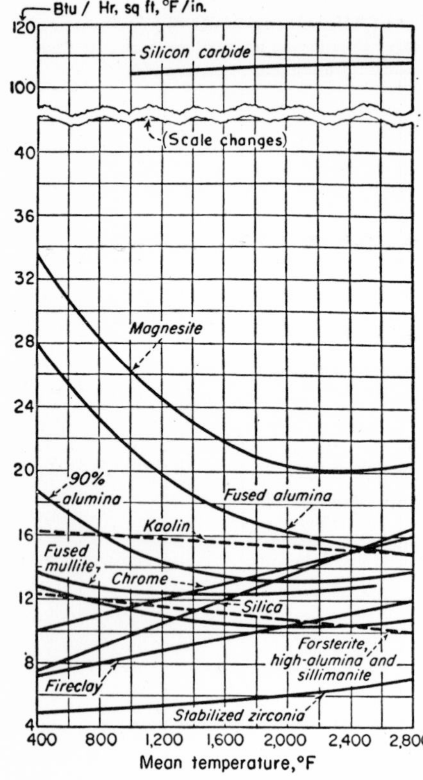

A-14. Thermal Conductivities of Refractory Brick.* Note that k is given in Btu/sq ft hr (°F/in.).

Bulk density of brick:	lb/cu ft
Silica,	100–105
Fireclay and kaolin,	125–150
Magnesite, mullite, forsterite, and silicon carbide,	150–160
Fused alumina, 90% alumina, and chrome,	170–195

Specific heats of common refractories vary from about 0.20 at room temperature to about 0.28 at 2500°F.

A-14a. Thermal Conductivities of Insulating Firebrick.* Note that k is given in Btu/sq ft hr (°F/in.).

Insulating firebrick is made porous to decrease conductivity and heat capacity.

Bulk density of brick:	lb/cu ft
1600° brick	21–37
2000° brick	26–45
2300° brick	27–47
2600° brick	43–64
2800° brick	45–65
2900° brick	52

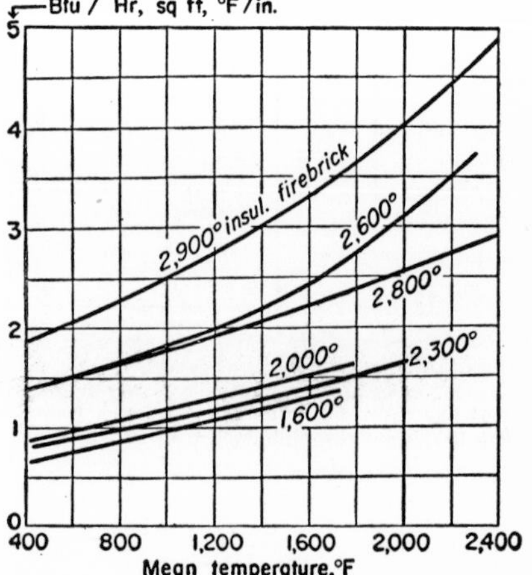

* Reproduced by permission from article by C. L. Norton, Jr. in *Chemical Engineering*, June 1953, p. 216. Copyright 1953, McGraw-Hill Publishing Co.

Table A15: Thermal Conductivities of Liquids and Gases

Substance	Temp, deg F	k	Substance	Temp, deg F	k
LIQUIDS			GASES		
Acetone	68	0.103	Air	32	0.0140
Ammonia	45	0.29	Ammonia, vapor	32	0.0128
Aniline	32	0.104	Ammonia	212	0.0175
Benzol	86	0.089	Argon	32	0.00912
Carbon bisulphide	68	0.0931	Carbon dioxide	32	0.0085
Ethyl alcohol	68	0.105		212	0.0133
Ether	68	0.0798	Carbon monoxide	32	0.0135
Glycerin, USP, 95%	68	0.165	Chlorine	32	0.0043
Kerosene	68	0.086	Ethane	32	0.0106
Methyl alcohol	68	0.124	Ethylene	32	0.0101
n-Pentane	68	0.0787	Helium	32	0.0802
Petroleum ether	68	0.0758	n-Hexane	32	0.0061
Toluene	86	0.088	Hydrogen	32	0.0917
Water	32	0.335		212	0.115
	140	0.377	Methane	32	0.0175
Oil, castor	39	0.104	Neon	32	0.00256
Oil, olive	39	0.101	Nitrogen	32	0.0140
Oil, turpentine	54	0.0734	Nitrous oxide	32	0.0080
Vaseline	59	0.106		212	0.0090
			Nitric oxide	32	0.0138
			Oxygen	32	0.0142
			n-Pentane	32	0.0074
			Sulphur dioxide	32	0.005

By permission from *Mechanical Engineer's Handbook* by L. S. Marks, copyright 1951, McGraw-Hill Book Co.

Note: An average value of k for petroleum oils at ordinary temperatures is 0.08 Btu/ft hr °F.

Table A15a: Thermal Conductivities of Miscellaneous Insulating Materials (k is in Btu/ft hr °F)

Material	Mean Temperature, °F				Density lb/cu ft
	32	70	212	500	
Fiber-glass blanket	0.018	0.023	0.027	0.050	6
Mineral-wool, molded pipe covering		0.033	0.035	0.035	18
85% Magnesia pipe covering		0.034	0.038	0.048	11
Diatomaceous-earth pipe covering		0.040	0.045	0.053	23
Cork pipe covering	0.024	0.025			11
Air-cell pipe covering		0.041	0.046		9
Silica aerogel	0.011	0.012	0.016	0.025	8

Data mainly from an article by Ray Thomas in *Chemical Engineering*, June, 1953, p. 221.

A16: Viscosity of Liquids

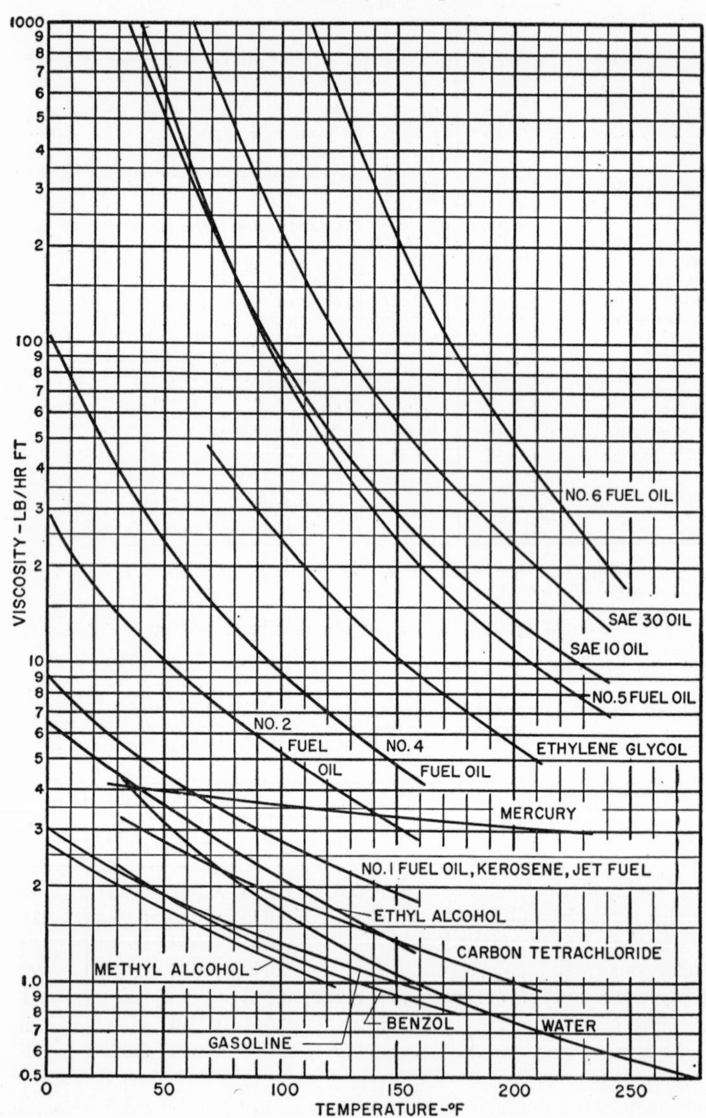

Viscosity of liquids. Viscosities of petroleum products are for typical commercial samples; considerable variation occurs. Sources: petroleum products, compiled from commercial data; other liquids, by permission from *Handbook of Chemistry and Physics.*

Table A17: Tube and Pipe Data

HEAT EXCHANGER TUBES

OD (in.)	Thickness BWG	Thickness In.	ID (in.)	Internal cross-section (sq ft)	External surface (sq ft/ft)	Velocity, fps for 1 gpm
0.375	20	0.035	0.305	0.000506	0.0982	4.40
	18	0.049	0.277	0.000416	0.0982	5.35
0.500	20	0.035	0.430	0.00101	0.1309	2.209
	18	0.049	0.402	0.00088	0.1309	2.528
0.625	20	0.035	0.555	0.00168	0.1636	1.33
	18	0.049	0.527	0.00151	0.1636	1.47
	16	0.065	0.495	0.00133	0.1636	1.67
0.750	18	0.049	0.652	0.00232	0.1963	0.96
	16	0.065	0.620	0.00210	0.1963	1.06
	14	0.083	0.584	0.00186	0.1963	1.20
0.875	18	0.049	0.777	0.00329	0.2291	0.677
	16	0.065	0.745	0.00303	0.2291	0.736
	14	0.083	0.709	0.00274	0.2291	0.813
1.000	18	0.049	0.902	0.00443	0.2618	0.502
	16	0.065	0.870	0.00413	0.2618	0.540
	14	0.083	0.834	0.00379	0.2618	0.587
	12	0.109	0.782	0.00334	0.2618	0.668
1.250	16	0.065	1.120	0.00684	0.3272	0.3255
	14	0.083	1.084	0.00640	0.3272	0.3476
	12	0.109	1.032	0.00580	0.3272	0.3835
1.500	16	0.065	1.370	0.01020	0.3927	0.2176
	14	0.083	1.334	0.00970	0.3927	0.2295
	12	0.109	1.282	0.00895	0.3927	0.2485
	10	0.134	1.232	0.00830	0.3927	0.2691
2.000	16	0.065	1.870	0.0191	0.5236	0.1168
	14	0.083	1.834	0.0183	0.5236	0.1214
	12	0.109	1.782	0.0173	0.5236	0.1286
	10	0.134	1.732	0.0163	0.5236	0.1362

IRON-PIPE-SIZE PIPE AND TUBES

Nominal size (in.)	OD (in.)	Thickness (in.)	ID (in.)	Internal cross-section (sq ft)	External surface (sq ft/ft)	Velocity, fps for 1 gpm
⅛	0.405	0.068	0.269	0.000396	0.106	5.63
¼	0.540	0.088	0.364	0.000723	0.141	3.09
⅜	0.675	0.091	0.493	0.00133	0.177	1.68
½	0.840	0.109	0.622	0.00211	0.220	1.055
¾	1.050	0.113	0.824	0.00370	0.275	0.602
1	1.315	0.133	1.049	0.00600	0.334	0.372
1¼	1.660	0.140	1.380	0.0104	0.434	0.214
1½	1.900	0.145	1.610	0.0141	0.497	0.158
2	2.375	0.154	2.067	0.0233	0.622	0.0956
2½	2.875	0.203	2.469	0.0332	0.753	0.0670
3	3.500	0.216	3.068	0.0513	0.916	0.0435
4	4.500	0.237	4.026	0.0885	1.178	0.0254
5	5.563	0.258	5.047	0.139	1.456	0.0160
6	6.625	0.280	6.065	0.201	1.734	0.0111
8	8.625	0.322	7.981	0.348	2.258	0.00640

Table A18: Emissivity of Surfaces

Surface	Temp,[a] deg F	Emissivity[a]
METALS AND THEIR OXIDES		
Aluminum:		
Highly polished	440–1070	0.039–0.057
Polished	73	0.040
Rough plate	78	0.055
Oxidized at 1110 F	390–1110	0.11–0.19
Brass:		
Highly polished	494–710	0.033–0.037
Rolled plate, natural	72	0.06
Rubbed with coarse emery	72	0.20
Oxidized at 1110 F	390–1110	0.61–0.59
Chromium	100–1000	0.08–0.26
Copper:		
Carefully polished	176	0.018
Commercial polish	66	0.030
Heated at 1110 F	390–1110	0.57–0.57
Thick oxide coating	77	0.78
Cuprous oxide	1470–2010	0.66–0.54
Molten copper	1970–2330	0.16–0.13
Gold, highly polished	440–1160	0.018–0.035
Iron and Steel:		
Highly polished pure Fe	300–1800	0.05–0.37
Freshly emeried Fe	68	0.24
Ground sheet steel	1400–2000	0.52–0.60
Turned cast iron	70–1800	0.43–0.70
Rolled sheet steel	70	0.65 to 0.82
Well oxidized, smooth	70–2000	0.80–0.90
Molten cast iron	2370–2550	0.29–0.29
Molten mild steel	2910–3270	0.28–0.28
Lead:		
Pure unoxidized	260–440	0.057–0.075
Gray oxidized	75	0.28
Oxidized at 390 F	390	0.63
Mercury, pure clean	32–212	0.09–0.12
Molybdenum filament	1340–4700	0.096–0.292
Monel metal, oxidized at 1110 F	390–1110	0.41–0.46
Nickel:		
Pure polished	70–700	0.045–0.087
Electroplated, not polished	68	0.11
Wire	368–1844	0.096–0.186
Plate, oxidized at 1110 F	390–1110	0.37–0.48
Nickel oxide	1200–2290	0.59–0.86
Nickel alloys		
"Chromnickel"	125–1894	0.64–0.76
Nickelin, gray oxide	70	0.26
KA-2S alloy steel (8Ni, 18Cr; light silvery rough; brown after heating)	420–914	0.44–0.36
Same, after 24 hr at 980 F.	420–980	0.62–0.73
NCT-3(20Ni, 25Cr); brown, splotched, oxidized from service	420–980	0.90–0.97
NCT-6(60Ni, 12Cr); smooth black firm adhesive oxide coat from service	520–1045	0.89–0.82
Platinum, polished plate	440–2960	0.054–0.17
Platinum filament	80–2240	0.036–0.192
Silver, pure polished	440–1160	0.020–0.032
Tantalum filament	2420–5430	0.194–0.33
Tin, bright	76	0.043–0.064
Tungsten, aged filament	80–6000	{0.032–0.35 0.018
Zinc:		
Comm'l polished	440–620	0.045–0.053
Oxid. at 750 F	750	0.11
Galv. iron, fairly bright	82	0.23
Galv. iron, gray oxidized	75	0.28
REFRACTORIES, BUILDING MATERIALS, PAINTS, MISC.		
Aluminum paints (vary with am't of lacquer body and age)	212	0.27–0.67
Asbestos	100–700	0.93–0.95
Candle soot; lampblack-water glass	70–700	0.95 ± 0.01
Lubricating oil, layer 0.01 in. thick	68	0.82
Linseed oil, 1 and 2 coats on Al foil	68	0.56 and 0.57
Rubber, soft gray reclaimed	76	0.86
Misc. I: shiny black lacquer, planed oak, white enamel, serpentine, gypsum, white enamel paint, roofing paper, lime plaster, black matte shellac	70	0.87–0.91
Misc. II: glazed porcelain, white paper, fused quartz, polished marble, rough red brick, smooth glass, hard glossy rubber, flat black lacquer, water, electrographite	70	0.92–0.96

[a] When two temperatures and two emissivities are given they correspond, first to first and second to second, and linear interpolation is permissible. By permission from *Mechanical Engineer's Handbook* by L. S. Marks, copyright 1951, McGraw-Hill Book Co.

Table A-18 (Concluded): Approximate Absorptivity for Solar Radiation*

Surface	*Emissivity*
Magnesium carbonate	0.02
Silver, polished	0.07
White paper	0.25
Whitewash	0.25
Aluminum paint	0.35
Nickel, polished	0.40
Steel, polished	0.45
Copper, polished	0.50
Galvanized iron, new	0.65
Red brick or tile	0.65
Slate, gray	0.90
Lampblack	0.97

* From King, W. J., *The Basic Laws and Data of Heat Transmission*. New York; *ASME*, 1932.

Notes: Emissivity is a surface phenomenon, so even thin layers of foreign matter such as dirt or oil may affect the emissivity appreciably.

At ordinary temperatures the emissivity is not a function of color because, as shown in Fig. 19-18, a negligible portion of the radiant energy is in the range of visible light. Solar radiation, however, is at its peak in the visible range, so color is important with respect to absorptivity for solar radiation. White objects reflect solar radiation, while dark objects absorb it. At temperatures of 100 to 200°F, on the other hand, Heilman (*Trans. ASME*, 1929, p. 287) found identical emissivities for both white and black lacquer.

The intensity of solar radiation at normal incidence is about 425 Btu/sq ft hr; however, when the sky is not clear, or the sun is far from the zenith, much of the energy is absorbed before it reaches the earth's surface. See *Airplane Air Conditioning Data—Heat Transfer*, New York, Society of Automotive Engineers, for data on solar and nocturnal radiation.

INDEX